AF830935

Particle Physics Phenomenology

Proceedings of the Fourth International Workshop

Particle Physics Phenomenology

Held at National Sun Yat-Sen University
June 18 – 21, 1998

editors

Hai-Yang Cheng
Academia Sinica, Taiwan, R.O.C.

Wei-Shu Hou
National Taiwan University, Taiwan, R.O.C.

Hsiang-Nan Li
National Cheng-Kung University, Taiwan, R.O.C.

Guey-Lin Lin
National Chiao-Tung University, Taiwan, R.O.C.

World Scientific
Singapore • New Jersey • London • Hong Kong

Published by

World Scientific Publishing Co. Pte. Ltd.

P O Box 128, Farrer Road, Singapore 912805

USA office: Suite 1B, 1060 Main Street, River Edge, NJ 07661

UK office: 57 Shelton Street, Covent Garden, London WC2H 9HE

British Library Cataloguing-in-Publication Data
A catalogue record for this book is available from the British Library.

PARTICLE PHYSICS PHENOMENOLOGY
Proceedings of IV International Workshop

Copyright © 1999 by World Scientific Publishing Co. Pte. Ltd.

All rights reserved. This book, or parts thereof, may not be reproduced in any form or by any means, electronic or mechanical, including photocopying, recording or any information storage and retrieval system now known or to be invented, without written permission from the Publisher.

For photocopying of material in this volume, please pay a copying fee through the Copyright Clearance Center, Inc., 222 Rosewood Drive, Danvers, MA 01923, USA. In this case permission to photocopy is not required from the publisher.

ISBN 981-02-3817-7

Printed in Singapore.

PREFACE

The Fourth International Workshop on Particle Physics Phenomenology was held at the National Sun Yat-Sen University, Kaohsiung, from June 18 to June 21, 1998. About 85 participants attended the Workshop, which lasted for three and a half days. For the first time in this series, we had two evening sessions for contributed talks.

In accordance with traditions of this meeting, we had two pedagogical lectures, one on spin physics and one on SUSY phenomenology. These were complemented by twenty-four invited talks and fifteen contributed talks, which covered many aspects of the latest developments in B physics (in particular, exclusive and inclusive hadronic charmless B decays), CP violation from within the Standard Model and beyond, perturbative QCD and its application to B physics. Due to the special timing of this workshop, we also arranged a special report on new results of the Super-Kamiokande experiment.

The Workshop was supported by the National Center for Theoretical Science of R.O.C., the Ministry of Education of R.O.C., The Asia Pacific Center for Theoretical Physics, and sponsored by National Chiao-Tung University, National Taiwan University, Academia Sinica, National Sun Yat-Sen University, and National Cheng-Kung University.

Hai-Yang Cheng
Wei-Shu Hou
Hsiang-nan Li
Guey-Lin Lin

CONTENTS

Contributed Talks

Lectures

POLARIZED STRUCTURE FUNCTIONS AND HIGH ENERGY SPIN PHYSICS

J. SOFFER

Centre de Physique Théorique, CNRS Luminy, Case 907,
F-13288 Marseille Cedex 09, France
E-mail: soffer@cpt.univ-mrs.fr

We review in these lectures, some aspects of the development of the field of spin phenomena in high energy hadron physics, over the last thirty years or so. We also discuss a few crucial tests accessible to existing or future experimental programs,which will, hopefully, provide the answers to many fundamental questions related to the nucleon spin structure and spin observables in perturbative QCD.

1 Introduction and physics motivations

Our purpose is to emphasize the important role spin is bound to play in high energy hadron physics. Nucleon spin structure is a fundamental issue and more precisely, we ought to know how the nucleon spin is shared among its basic constituents. One way to attack this problem is to use polarized Deep Inelastic Scattering (DIS) and considerable progress have been made over the last ten years or so, by various fixed polarized targets experiments at CERN, DESY and SLAC, in determining the polarized parton distributions [1]. Some theoretical ambiguities remain and the restricted Q^2 range, accessible by the data, makes also rather difficult, sensible tests of the Q^2 evolution, predicted by recent higher order QCD calculations. More generally, it is clear that the spin sector of perturbative QCD has not been tested in polarized hadronic collisions and this remains to be done, to firmly establish the theory.

The outline of these lectures is as follows. In the next section we give a brief survey of some specific and striking experimental results in *pp* collisions, which are known for several years, but whose theoretical interpretation is not yet fully understood. Then we consider some aspects of polarized deep inelastic phenomena, which has been extensively studied, for about a quarter of a century, in order to elucidate the spin structure of the nucleon. Several theoretical interpretations are still compatible with the most recent experimental results on spin-dependent structure functions. We next turn to present the spin programme at RHIC-BNL which will, very likely, clarify several questions and also uncover unexpected features of this field. Finally, we give our concluding remarks.

2 Rapid survey of experimental results in pp collisions

Several transverse spin effects have been observed in exclusive two-body reactions, in fixed polarized targets experiments, with a rather low incident beam energy, say up to 200GeV/c or so. This was done in the seventies and one was using Regge phenomenology as a theoretical framework, since it concerns essentially the small scattering angle region [2]. According to a widespread prejudice, spin effects were expected to die off very quickly as the energy increases. An illustration of this is given in Fig.1, where we see that in pp elastic scattering, the analyzing power A_N (or single transverse spin asymmetry), at fixed momentum transfer t, exhibits a strong fall off for increasing beam energy [4]. However, the spin dependence does not necessarily follow this simple rule and, as shown in Fig.2, spin effects do not vanish at large p_T^2. It is also worthwhile to recall the intriguing behavior of A_{NN}, the transverse spin-spin correlation parameter at a fixed value of θ_{cm}, the center of mass scattering angle, $\theta_{cm} = 90^0$. This is displayed in Fig.3, versus p_{lab}, and we observe that A_{NN} is falling off, after a dramatic structure below 2GeV/c and then has a sharp rise up to 60% at the highest energy ever measured. We don't really know what to expect for $p_{lab} > 13GeV/c$, an oscillating pattern or a continuous rise?

Let us now turn to the consideration of spin effects in inclusive reactions. There is a large amount of significant polarization data for inclusive hyperon production as shown in Figs.4,5. Sizeable effects and some regular behavior have been observed which may help to uncover the underlying mechanism for particle production. This was first discovered in 1976 at FNAL by studying hyperons produced by 300GeV/c protons on a beryllium target. More specifically the Λ's produced in the beam fragmentation region have a large polarization perpendicular to the scattering plane. The effect grows with p_T and has almost no energy dependence, since it still remains at ISR energy, a much higher energy (see Fig.5). At FNAL energy, one observes an effect of opposite sign for $\Sigma^\pm$ and same sign for Ξ^- and Ξ^0 (see Fig.4). By using the fact that this Λ polarization is transfered to the proton in the decay Λ to $p + \pi^-$, the E704 Collaboration has developed a 200GeV/c polarized proton beam. They also had a polarized $\bar{p}$ beam, a unique facility in the world. The intensities were too low for experiments at large p_T but at moderate p_T, say $p_T \simeq 1GeV/c$ or so, some interesting results were obtained. They find a large analyzing power A_N in pp collisions which has opposite signs for π^+ and π^- as shown in Fig.6. The signs are reversed in the case of $\bar{p}p$ collisions. We will not try to enter into the discussion about their possible theoretical interpretation, but all this is surely a clear evidence that spin effects are large at high energy.

3 Polarized structure functions from polarized Deep Inelastic Scattering

After a brief introduction on polarized deep inelastic phenomena, we make a simple analysis of the spin-dependent structure functions, by means of an isospin decomposition. We then present the results of a fit to the data, whose lack of accuracy in a sufficiently broad kinematic range, prevents a unique theoretical interpretation. Finally, we discuss an interesting sum rule which has been investigated experimentally, only recently.

3.1 Preamble

The EMC experiment on $g_1^p(x)$ [10] was giving for the first moment of the spin dependent proton structure function $g_1^p(x)$ at $<Q^2>= 10\,GeV^2$

$$\Gamma_1^p = \int_0^1 g_1^p(x)\; dx = 0.126 \pm 0.010 \pm 0.015, \tag{1}$$

a value much smaller than the prediction of the Ellis-Jaffe sum rule [11] for the proton (we use for the nonsinglet couplings $a_3 = G_A/G_V = 1.2573 \pm 0.0028$ and $a_8 = 3\,F - D = 0.579 \pm 0.025$ [12]),

$$\Gamma^p_{1\,EJ} = F/2 - D/18 = 0.185 \pm 0.003. \tag{2}$$

We recall that

$$g_1^p(x) = 1/2 \sum_i e_i^2[\Delta q_i(x) + \Delta\bar{q}_i(x)]\;, \tag{3}$$

the $\Delta q(x)$'s (same for antiquarks) are defined as $\Delta q(x) = q_+(x) - q_-(x)$, where $q_\pm$ are the quark distributions in a polarized proton with helicity parallel (+) or antiparallel (−) to that of the proton. This has stimulated a considerable experimental activity in measuring the polarized structure functions $g_1^p(x)$, $g_1^n(x)$ ($g_1^{^3He}$) and $g_1^d(x)$ [13,14,15].

From the theoretical point of view, the hypothesis has been formulated [16] that a relevant isoscalar contribution, related, in the chiral-invariant factorization scheme [17], to a large gluon polarization, contributes to Eq.(2) at $O(\alpha_s)$, such as

$$\Gamma_1^p(Q^2) = \frac{1}{12}\left(a_3 + \frac{a_8}{3} + \frac{4}{3}a_0(Q^2)\right)\left(1 - \frac{\alpha_s(Q^2)}{\pi}\right) - \frac{\alpha_s(Q^2)}{6\,\pi}\Delta G(Q^2)\;. \tag{4}$$

This would explain the defect in the Ellis-Jaffe sum rule for the proton, without affecting the more theoretically founded Bjorken sum rule [18],

$$(\Gamma_1^p - \Gamma_1^n)_{LO} = \frac{a_3}{6} = \frac{1}{6}\frac{G_A}{G_V}. \tag{5}$$

QCD corrections play an important role in modifying the r.h.s. of Eq.(5) which, strictly speaking, is valid in the limit $Q^2 \to 0$, and reads

$$(\Gamma_1^p - \Gamma_1^n)(Q^2) = \frac{1}{6}\frac{G_A}{G_V}\left(1 - \alpha_s(Q^2)/\pi\right), \tag{6}$$

where only the $O(\alpha_s)$ has been retained. The validity of the Bjorken sum rule is so universally accepted that, in the analysis of the experimental data on polarized structure functions [19,20,21], it is almost always assumed, rather than tested. There are few exceptions to this attitude, as for example the paper by Altarelli *et al.* [22], where a_3 is evaluated to be 1.19 ± 0.09, and the l.h.s. of Eq.(6) at $Q^2 = 4\,GeV^2$ is 0.177 ± 0.014, in agreement with the predicted value at $O(\alpha_s)$, 0.189 ± 0.002.

An alternative interpretation [23] of the defect shown by Eq.(1) has been given by relating the defects in the Ellis-Jaffe sum rule for the proton and the Gottfried sum rule [24] for the isovector unpolarized structure functions of the nucleons,

$$\int_0^1 \frac{F_2^p(x) - F_2^n(x)}{x}\, dx = 1/3 + 2/3 \int_0^1 [\bar{u}(x) - \bar{d}(x)]\, dx = 1/3, \tag{7}$$

to be compared with the measurement performed by the NMC experiment [25], namely 0.235 ± 0.026. The defect in the Gottfried sum rule may be ascribed to a flavor asymmetry in the sea of the proton, which does not mean the breaking of the isospin symmetry since the proton is not an isoscalar. A flavor asymmetry of the parton sea in the proton was advocated many years ago [26], in connection with the Pauli principle and an essential asymmetry in the proton which contains two valence u and one valence d quark. The sign of the asymmetry is just the right one to account for the defect in the Gottfried sum rule, $\bar{d} - \bar{u} > 0$, while to reproduce the value given by NMC one should have

$$\bar{d} - \bar{u} = \int_0^1 [\bar{d}(x) - \bar{u}(x)]\, dx = 0.15 \pm 0.04. \tag{8}$$

The conservation of the SU(3) non-singlet weak currents gives

$$\Delta u_{val} = 2\,F, \qquad \Delta d_{val} = F - D. \tag{9}$$

Since in the proton $u^\uparrow$ is the valence parton with the highest first moment (the quark model sum rules would imply $u^\uparrow = 1 + F = 1.459 \pm 0.006$), it is reasonable to assume, inspired by the Pauli principle and Eq.(8), that it is just this parton which receives less contribution from the sea and to write

$$\Delta u_{val} = 2F + \bar{u} - \bar{d}. \tag{10}$$

This implies a defect of about 0.033 (0.008) in the Ellis-Jaffe sum rule for the proton (neutron) and, what would be very relevant, a defect of about 0.025 in the Bjorken sum rule. In this framework one should expect

$$\Gamma_1^p = F/2 - D/18 + 2/9(\bar{u} - \bar{d}) = 0.152 \pm 0.010,$$
$$\Gamma_1^n = F/3 - 2D/9 + 1/18(\bar{u} - \bar{d}) = -0.033 \pm 0.004, \tag{11}$$

in reasonable agreement with experiment.

These two interpretations can be tested, in principle, by using the world data, which extend on some Q^2 range and therefore requiere a next-to-leading order (NLO) analysis. Although several parametrizations of the polarized structure functions have been proposed in the literature, we will present in section 3.3 the results of a NLO fit based on an expansion in terms of Jacobi polynomials. Before going to that, in the next section, we perform an isospin decomposition of the proton and neutron SLAC data and we discuss what can be learned from such a simple method.

3.2 Isospin decomposition

The very accurate measurement of the neutron spin-dependent structure function g_1^n [13], whose results have been presented recently, possess two remarkable properties. First, the values of the neutron structure function at $Q^2 = 5GeV^2$ are rather large and negative in the region of moderately low x. Second, the data can be rather accurately fitted by the power function $x^{-0.8}$ and this power seems to be unexpectedly large. It is not obvious, where this large number, for which there is no indication in the proton data, is coming from, so it may affect the extrapolation to $x = 0$ and cast some doubts on the validity of Bjorken sum rule. By performing the isospin decomposition of the data for both proton and neutron, we conclude that the isovector contribution is well approximated by the power behavior found earlier by an elaborate method based on a NLO QCD evolution [27]. It may be also interpreted as the manifestation of $ln^2 x$ terms [28]. For the isoscalar part one has the signature for a rather singular singlet contribution, which is compatible with the similar behavior predicted by QCD [29]. This may imply a significant gluon polarization in the nucleon in this range of x.

This new neutron data [13] is large and negative $g_1^n \sim -g_1^p$, and has been measured with a good accuracy up to small x, say $x \sim 0.01$. The SLAC proton data [13] at $Q^2 = 3GeV^2$ are positive and of roughly the same magnitude, but on the contrary, rather flat in this region. This could be understood, qualitatively, as a result of the interplay of a negative contribution at low x, responsible for the singular behavior of the neutron, and a positive contribution at larger x. To check this assumption it is instructive to consider the isovector contribution. Since there is no clear evidence of scaling violations between $Q^2 = 3GeV^2$ and $Q^2 = 5GeV^2$ in any polarized DIS experiment, we may neglect, for the time being, the effects of QCD evolution, because we are only interested in the results, provided by the data at the present level of statistical accuracy. It is then possible, by combining the SLAC neutron data with the SLAC proton data, to determine the quantity $g_1^{p-n} \equiv g_1^p - g_1^n$, entering the Bjorken sum rule, with a higher accuracy, than for proton alone. This is because the neutron data are fortunately negative, so the difference is larger in magnitude than g_1^p, while the errors are practically the same for g_1^{p-n} and g_1^p, given the quality of the neutron data.

Performing this analysis of the data, we see immediately a behavior less sharp than that of neutron. Since phase space effects, as powers of $(1-x)$, are not so important in the region under consideration, we were looking for a simple power parametrization of the data, implied by Regge pole behavior, and we found that this works rather well, for $0.016 \leq x \leq 0.125$ with the following expression

$$g_1^{p-n}(x) = 0.147x^{-0.45}, \tag{12}$$

as shown in Fig.7.

The power we obtain is significantly smaller than the expected contribution of the $a_1(1260)$ meson trajectory (~ 0.14). As a result, the contribution to the Bjorken integral from the region $0 \leq x \leq 0.125$ is large

$$\int_0^{0.125} dx g_1^{p-n}(x) = 0.085. \tag{13}$$

The region of higher x corresponds to a neutron contribution much smaller than that of the proton, the latter also providing a large contribution to the above integral, which is 0.09 for $0.125 < x < 1$. The total contribution to the Bjorken sum rule, obtained in this way, i.e. 0.175, appears to be in the fair agreement with the theoretical value. Note that when this neutron data will be associated to the final very precise proton data, it will allow a more serious analysis, taking also into account the effects of the QCD evolution.

At the present moment we would like to stress, that by combining the current neutron and proton data, one is led to good agreement with Bjorken sum rule. Let us stress that this power is compatible with the one obtained using NLO fit [27] to all previous data (-0.56 ± 0.21). It is consistent with the result of ln^2x summation [28].

Since we found, that the sharp neutron structure function is not seen in the difference between proton and neutron, it should be attributed to the isoscalar channel. Also, the partial cancellation, we suspect to be at the origin of a flat proton structure function, should be manifested in this channel as well. To check this, we calculated the quantity $g_1^{p+n} \equiv g_1^p + g_1^n$. It really shows a rather flat structure for $x \geq 0.035$ which should be related to the interplay of the negative sharp contribution, showing itself in the fit $x^{-0.8}$ to the neutron data [13], and a positive contribution with a smaller power, dominating at larger x. Since there is no counterpart for such a sharp behavior in the conventional Regge analysis, we make a strong, but natural assumption, that it manifested itself in the $SU(3)$-singlet channel. It is in this channel that a strong mixing between polarized quarks and gluons provides the anomalous gluon contribution to the first moment of g_1 [30]. At low x the quark-gluon mixing provides a strong correction to the subleading behavior [29], producing a power close to 1. For a first estimate we find that the data are well described by

$$g_1^{p+n}(x) = 0.145x^{-0.45} - 0.03x^{-0.87}, \tag{14}$$

as seen in Fig.8.

This formula is suggesting that the isoscalar contribution is approximately equal to the isovector one, which is not so surprising, in order to have the neutron structure function, dominated by the most singular power only. Note that Eq.(14) gives also a fair description of the g_1^d data [13] in this kinematical region and further details on this result are given in ref.[31].

3.3 An example of NLO fits

For the polarized parton distributions, we take the following standard parametrization (we assume a flavour symmetric sea, $\Delta u_{sea} = \Delta\bar{u}_{sea} = \Delta d_{sea} = \Delta\bar{d}_{sea} = \Delta s_{sea} = \Delta\bar{s}_{sea} \equiv \Delta q_{sea}$):

$$x\Delta u_{val}(x, Q_0^2) = \eta_u A_u x^{a_v}(1-x)^{b_u}(1+\gamma_v x), \tag{15}$$
$$x\Delta d_{val}(x, Q_0^2) = \eta_d A_d x^{a_v}(1-x)^{b_d}(1+\gamma_v x), \tag{16}$$
$$x\Delta q_{sea}(x, Q_0^2) = x\Delta\bar{q}_{sea}(x, Q_0^2) = \eta_s A_s x^{a_s}(1-x)^{b_s}(1+\gamma_s x), \tag{17}$$
$$x\Delta G(x, Q_0^2) = \eta_G A_G x^{a_G}(1-x)^{b_G}(1+\gamma_G x), \tag{18}$$

at $Q_0^2 = 4GeV^2$, where η_q ($q = u, d, s, G$) are the first moments of the distributions and $A_q = A_q(a_q, b_q, \gamma_q)$ is given by

$$A_q^{-1} = \int_0^1 dx x^{a_q-1}(1-x)^{b_q}(1+\gamma_q x) = \left(1 + \gamma_q \frac{a_q}{a_q+b_q+1}\right) \frac{\Gamma(a_q)\Gamma(b_q+1)}{\Gamma(a_q+b_q+1)}, \tag{19}$$

in such a way that

$$\int_0^1 dx A_q x^{a_q-1}(1-x)^{b_q}(1+\gamma_q x) = 1. \tag{20}$$

We fix $\gamma_s = \gamma_G = 0$, since it is not possible to fit the values of these parameters with sufficient accuracy.

We fit the asymmetries

$$A_1^{p(n)}(x,Q^2) = \frac{g_1^{p(n)}(x,Q^2)}{F_1^{p(n)}(x,Q^2)} = \frac{g_1^{p(n)}(x,Q^2)}{F_2^{p(n)}(x,Q^2)} 2x(1+R^{p(n)}(x,Q^2)), \tag{21}$$

which are just the quantities experimentally measured. The choice of A_1 would also allow one to minimize the higher twist contribution (that should partly cancel in the ratio in Eq.(21)), instead of cutting the data at low Q^2. In constructing the ratio in Eq.(21), we use the MRS parameterization [32] at $Q_0^2 = 4GeV^2$ for the unpolarized distributions. The deuteron asymmetry, $A_1^d(x,Q^2)$, is given by

$$A_1^d(x,Q^2) = \frac{g_1^d(x,Q^2)}{F_1^d(x,Q^2)}, \tag{22}$$

with

$$g_1^d(x,Q^2) = \frac{1}{2}\left(1 - \frac{3}{2}\omega_D\right)(g_1^p(x,Q^2) + g_1^n(x,Q^2)), \tag{23}$$

where $\omega_D = 0.058$ takes into account the small D-wave component in the deuteron ground state [33] . We only include in the fit the data obtained by the SLAC experiments [13]. The HERMES data [15] are not used since their lower precision does not change the results. We will compare the results with the SMC proton and deuteron data [14], by evolving the distributions to the Q^2 of the CERN SMC experiment.

In all the cases we will fix

$$\eta_d = F - D = -0.339 \pm 0.013, \tag{24}$$

as expected from the conservation of the SU(3) non singlet weak currents. The two different interpretations mentioned above in section 3.1, will be characterized, for the first option with $\eta_G \neq 0$ (see below, fit A), by the additional constraint

$$\eta_u = 2F = 0.918 \pm 0.013, \tag{25}$$

in such a way to obey the Bjorken sum rule with the first order correction, Eq.(6), consistently with the NLO approximation. For the second option, we take $\eta_G = 0$ (see below, fit D), η_u will be a free parameter and the Bjorken sum rule will departure from the theoretical expectation. Further limitations on the parameters are imposed, in order to satisfy, in particular, the positivity constraints with respect to the unpolarized distributions. One possible method to solve the QCD evolution equations is to express the structure functions by means of orthogonal polynomials, for example Jacobi polynomials, which is a suitable choice [34]. For a complete explanation, we refer the reader to ref.[34], where the values of all the parameters are also given. The main results are shown on Figs.9,10 and 11. As a result of our analysis, we can conclude that the interpretation related to the Pauli principle is not contradicted by the rather precise SLAC data which, however, slightly favor the generally accepted solution with the Bjorken sum rule obeyed and a sea contribution to the proton spin, in the gauge-invariant factorization scheme. Obviously, a better determination of the small x region could help to clarify further the present situation.

3.4 The generalized GDH sum rule

The generalized Gerasimov-Drell-Hearn (GDH) sum rule [35,36] is just being tested experimentally [37] and the proton data are in the good agreement with our prediction [38,39], making use of its relation to the Burkhardt-Cottingham sum rule. They also agree with a new estimate of the contributions from low-lying resonances [40]. We should stress, that such a similarity is by no means surprising, since the dominant magnetic form factor of $\Delta(1232)$ is contributing entirely through the structure function g_2 [39], which is the key ingredient of our approach. Given the fact that the experimental data [37] are also available for a neutron target (although with a poor accuracy), we present here the quantitative prediction for the neutron case.

Let us consider the Q^2-dependent integral

$$I_1(Q^2) = \frac{2M^2}{Q^2} \int_0^1 g_1(x, Q^2)dx \ , \tag{26}$$

where M is the nucleon mass. It is defined for *all* Q^2, and g_1 is the obvious generalization for all Q^2 of the standard scale–invariant $g_1(x)$. Note that the elastic contribution at $x = 1$ is not included in the above sum rule. One recovers then at $Q^2 = 0$ the GDH sum rule

$$I_1(0) = -\frac{\mu_A^2}{4}, \tag{27}$$

where μ_A is the nucleon anomalous magnetic moment in nuclear magnetons. While the $I_1(0)$ is always negative, its value at large Q^2 is determined by the

Q^2-independent integral $\int_0^1 g_1(x)dx$, which is positive for proton and negative for neutron.

It is possible to decompose I_1 as the difference of the contributions from two form factors I_{1+2} and I_2

$$I_1 = I_{1+2} - I_2, \tag{28}$$

where

$$I_{1+2}(Q^2) = \frac{2M^2}{Q^2}\int_0^1 g_{1+2}(x)dx, \quad I_2(Q^2) = \frac{2M^2}{Q^2}\int_0^1 g_2(x)dx, \tag{29}$$

$$g_{1+2} = g_1 + g_2. \tag{30}$$

There are solid theoretical arguments to expect a strong Q^2-dependence of I_2. It is the well-known Burkhardt-Cottingham(BC) rule[41], derived independently by Schwinger[42], using a rather different method. It states that

$$I_2(Q^2) = \frac{1}{4}\mu G_M(Q^2)[\mu G_M(Q^2) - G_E(Q^2)], \tag{31}$$

where μ is the nucleon magnetic moment, G's denoting the familiar Sachs form factors which are dimensionless and normalised to unity at $Q^2 = 0$. For large Q^2 one can neglect the r.h.s. and get

$$\int_0^1 g_2(x)dx = 0. \tag{32}$$

In particular,

$$I_2(0) = \frac{\mu_A^2 + \mu_A e}{4}, \tag{33}$$

e being the nucleon charge in elementary units. To reproduce the GDH value one should have

$$I_{1+2}(0) = \frac{\mu_A e}{4}. \tag{34}$$

Note that I_{1+2} does not differ from I_1 for large Q^2 due to the BC sum rule, but it is *positive* in the proton case. It is possible to obtain a smooth interpolation for $I^p_{1+2}(Q^2)$ between large Q^2 and $Q^2 = 0$[38], namely

$$I^p_{1+2}(Q^2) = \theta(Q_0^2 - Q^2)\left(\frac{\mu_A}{4} - \frac{2M^2Q^2}{(Q_0^2)^2}\Gamma_1^p\right) + \theta(Q^2 - Q_0^2)\frac{2M^2}{Q^2}\Gamma_1^p, \tag{35}$$

where $\Gamma_1^p = \int_0^1 g_1^p(x)dx$. The continuity of the function and of its derivative is guaranteed with the choice $Q_0^2 = (16M^2/\mu_A)\Gamma_1^p \sim 1GeV^2$, where the integral is given by the world average proton data. It is quite reasonable to distinguish the perturbative and the non-perturbative regions. As a result one obtains, a crossing point at $Q^2 \sim 0.2GeV^2$, below the resonance region [38], while the positive value at $Q^2 = 0.5GeV^2$ is in a good agreement with the E143 data [37].

To generalize this approach to the neutron case, one needs to find a similar parametrization for the neutron. Since the value at $Q^2 = 0$ is equal to zero, it is not sufficient to limit oneself to the simplest linear parametrization and one needs to add two terms, one quadratic and one cubic in Q^2. A simple parametrization providing the continuity of the function and its first and second derivative is

$$I^n_{1+2}(Q^2) = \frac{2M^2}{Q_0^2}\Gamma^n_1 \left[\frac{Q^2}{Q_0^2}\theta(Q_0^2 - Q^2)[6 - 8\frac{Q^2}{Q_0^2} + 3(\frac{Q^2}{Q_0^2})^2] + \theta(Q^2 - Q_0^2)\frac{Q_0^2}{Q^2}\right]. \tag{36}$$

Note that because of the extra terms in the parametrization, Q_0^2 is not determined by the continuity conditions, and we are taking the same value as in the proton case. The parametrization seems to be rather natural in terms of the Q^2-dependent integral $\Gamma^n_{1+2}(Q^2)$, which is equal to its asymptotic value $\Gamma^n_{1+2}(\infty) = \Gamma^n_1$ down to Q_0^2. Below Q_0^2 it is just

$$\Gamma^n_{1+2}(Q^2) = (\frac{Q^2}{Q_0^2})^2\Gamma^n_1[6 - 8\frac{Q^2}{Q_0^2} + 3(\frac{Q^2}{Q_0^2})^2]. \tag{37}$$

To parametrize the elastic contribution to the BC sum rule we need the neutron elastic form factors. While the electric one might be neglected, the magnetic form factor is well decribed by the dipole formula [43]

$$G_M(Q^2) = \frac{1}{(1 + Q^2/0.71)^2}. \tag{38}$$

The plot representing $\Gamma^n_1(Q^2)$ is displayed on Fig.12, where $\Gamma^n_1 = -0.04$ was used. One can observe a behaviour distinct to the case of the resonance model [40], since the minimum occurs at -0.06 for $Q^2 = 0.7GeV^2$. In addition it is worth noting that our prediction is typically below the QCD extrapolation, since we subtract I_2 (which is a sharp positive contribution both for proton and neutron) from smooth I_{1+2}. In the resonance model it is above the QCD extrapolation for the neutron and below for the proton. We hope it will be possible to distinguish between these two predictions, as soon as more accurate data will be available from CEBAF.

The success of our model may be considered as an indirect check of the BC sum rule, but it could be also due to a possible cancellation [39] of non-scaling Regge cuts [45] in I_1, in such a way to preserve our predictions.

4 What to expect from the polarized pp collider at RHIC-BNL?

We first recall some basic definitions of the helicity asymmetries and then go to prompt photon production and jet production, which will allow the first direct determination of the gluon helicity distribution $\Delta G(x,Q^2)$ inside a polarized nucleon. Next we show, how antiquark helicity distributions $\Delta\bar{q}(x,Q^2)$ can be isolated in $W^{\pm}$ production, which leads also to the u-d flavor separation. This has been done, rather inaccurately, in semi-inclusive DIS. From transversely polarized proton beams, it is possible to make the first measurement of the transversity distributions $h_1^{q,\bar{q}}(x,Q^2)$ in Drell-Yan lepton pair production. Finally, we indicate the relevance of the parity violating asymmetry in single jet production. It might provide a clean signature for new physics and, as an example, we will consider the possible effects of a quark-quark contact interaction.

4.1 Basic definitions and helicity asymmetries

Fundamental interactions at short distances which are explored in high energy hadronic collisions, involve hard scattering of quarks, antiquarks and gluons. Let us consider the general hadronic reaction

$$a + b \to c + X \tag{39}$$

where c, in the cases we will consider below, is either a photon, a Z, a $W^{\pm}$ or a single-jet. In the hard scattering kinematic region, the cross section describing (39) reads in the QCD parton model, provided factorization holds, as

$$d\sigma(a+b \to c+X) = \sum_{ij} \frac{1}{1+\delta_{ij}} \int dx_a dx_b \left[f_i^{(a)}(x_a,Q^2) f_j^{(b)}(x_b,Q^2)\, d\hat{\sigma}^{ij} + (i \leftrightarrow j) \right]. \tag{40}$$

The summation runs over all parton configurations which contribute to the hadronic reaction, the $f(x,Q^2)$'s are the parton distributions, directly extracted from DIS for quarks and antiquarks and indirectly for gluons. Here $d\hat{\sigma}_{ij}$ is the cross section for the interaction of two partons i and j which can be calculated perturbatively, some of which, at the lowest order, are given in ref.[47]. If we consider the reaction (39) with *both* initial hadrons, a and b longitudinally polarized, one useful observable is the *double* helicity asymmetry A_{LL} defined as

$$A_{LL} = \frac{d\sigma_{a(+)b(+)} - d\sigma_{a(+)b(-)}}{d\sigma_{a(+)b(+)} + d\sigma_{a(+)b(-)}}\,, \tag{41}$$

when we assume parity conservation, i.e. $d\sigma_{a(\lambda)b(\lambda')} = d\sigma_{a(-\lambda)b(-\lambda')}$. Its explicit expression, assuming factorization, is given by

$$A_{LL}d\sigma = \sum_{ij} \frac{1}{1+\delta_{ij}} \int dx_a dx_b \left[\Delta f_i^{(a)}(x_a, Q^2) \Delta f_j^{(b)}(x_b, Q^2) \hat{a}_{LL}^{ij} d\hat{\sigma}^{ij} + (i \leftrightarrow j) \right], \tag{42}$$

where $d\sigma$ is given by Eq.(40) and $\hat{a}_{LL}^{ij}$ denotes the corresponding subprocess double asymmetry for initial partons i and j. The Δf's are defined as

$$\Delta f(x, Q^2) = f_+(x, Q^2) - f_-(x, Q^2) , \tag{43}$$

where $f_\pm$ are the parton distributions in a polarized hadron with helicity either parallel (+) or antiparallel (−) to the parent hadron helicity. Recall that the unpolarized distributions are $f = f_+ + f_-$ and Δf measures how much the parton f "remembers" the parent hadron helicity. If the subprocess involves parity violating interactions, one can consider another interesting observable which requires only *one* initial hadron polarized, that is the *single* helicity asymmetry A_L, defined as

$$A_L = \frac{d\sigma_{a(-)} - d\sigma_{a(+)}}{d\sigma_{a(-)} + d\sigma_{a(+)}} . \tag{44}$$

In addition, if both a and b are polarized, one can also have two *double* helicity parity violating asymmetries defined as

$$A_{LL}^{PV} = \frac{d\sigma_{a(-)b(-)} - d\sigma_{a(+)b(+)}}{d\sigma_{a(-)b(-)} + d\sigma_{a(+)b(+)}} \quad \text{and} \quad \bar{A}_{LL}^{PV} = \frac{d\sigma_{a(-)b(+)} - d\sigma_{a(+)b(-)}}{d\sigma_{a(-)b(+)} + d\sigma_{a(+)b(-)}} , \tag{45}$$

which can be simply related to A_L [48]. Several sets of polarized parton densities $\Delta f(x, Q^2)$ ($f = q, \bar{q}, G$) have been proposed in the recent literature (see refs.[19-22]). Using some of these parametrizations to calculate helicity asymmetries for various processes, we will show how the RHIC-BNL spin programme will be able, in particular, to pin down the polarized helicity distributions which remain badly constrained by polarized DIS experiments.

4.2 How to pin down $\Delta G(x, Q^2)$?

The cross section for direct photon production on pp collisions at high p_T is considered as one of the cleanest probe of the unpolarized gluon distribution $G(x, Q^2)$. This is partly due to the fact that the photon originates in the hard scattering subprocess and is detected without undergoing fragmentation. Moreover in pp collisions the quark-gluon Compton subprocess

$qG \to q\gamma$ dominates largely and the quark-antiquark annihilation subprocess $q\bar{q} \to G\gamma$ can be neglected. Consequently the double helicity asymmetry A_{LL}^{γ} (see Eq.(42)), which involves in this case only one subprocess, becomes particularly simple to calculate and is expected to be strongly sensitive to the sign and magnitude of $\Delta G(x, Q^2)$. For the Compton subprocess, $\hat{a}_{LL}$ whose expression at the lowest order is given in ref.[47], is always positive and such that $\hat{a}_{LL}(\hat{\theta}_{cm} = 90°) = 3/5$ where $\hat{\theta}_{cm}$ is the center of mass (c.m.) angle in the subprocess.

In Fig.13 we show a complete comparison of the results we obtained for the p_T^{γ} distribution of A_{LL}^{γ}, at pseudo rapidity $\eta = 0$, using the different sets of polarized parton distributions we have mentioned in the previous section, with a leading order Q^2 evolution. Actually we find that the smallest predictions are obtained from the sets refs.[19,20], which have the smallest $\Delta G(x)/G(x)$. The predictions differ substantially at large p_T, which corresponds to the region, say around $x = 0.4$ or so, where the distributions $\Delta G(x)$ have rather different shapes. We have also indicated the expected statistical errors based on an integrated luminosity $L = 800\ \text{pb}^{-1}$ at $\sqrt{s} = 500$ GeV, for three months running time. We have evaluated the event rates in the pseudo-rapidity gap $-1.5 < \eta < 1.5$, assuming a detector efficiency of 100% and for a p_T^{γ} acceptance $\Delta p_T^{\gamma} = 5$ GeV/c. We see that up to $p_T^{\gamma} = 50$ GeV/c or so, A_{LL}^{γ} will be determined with an error less than 5% which therefore will allow to distinguish between these different possible $\Delta G(x)$. For very large p_T^{γ}, the event rate drops too much to provide any sensitivity in the determination of $\Delta G(x)$. The pseudo-rapidity distribution of $A_{LL}^{\gamma}(\eta)$ at a fixed p_T^{γ} value has been also calculated in ref.[52] and it shows a systematic increase of A_{LL}^{γ} for higher η. Inclusive jet production is also a physics area where one can learn a lot about parton densities and, considering the vast amount of unpolarized existing data, it has been regarded as an important QCD testing ground. Event rates are much larger than for prompt photon production, but there is a drawback because many subprocesses are involved, unlike in the previous case. In principle one should take into account gluon-gluon (GG), gluon-quark (Gq) and quark-quark (qq) scatterings. Although these subprocesses cross sections are not so much different, after convolution with the appropriate parton densities (see Eq.(40)), they lead to very distinct contributions to the hadronic spin-average cross section. Here we will restrict ourselves to the double helicity asymmetry A_{LL}^{jet} for single jet production and in order to clarify the interpretation of our results below, let us recall some simple dynamical features. In the very low p_T^{jet} region, say $p_T^{jet} \sim 10$ GeV/c or so, GG scattering dominates by far, but its contribution drops down very rapidly with increasing p_T^{jet}. In the medium p_T^{jet} range, say 20 GeV/c $< p_T^{jet} <$ 80 GeV/c or so, Gq scattering dominates and then decreases for large p_T^{jet}, to be overcome by qq scattering. Of course these are rough qualitative considerations and accurate numerical estimates for the relative fractions of these different contributions depend strongly on the parton densities one uses.Let us look at $A_{LL}^{jet}(\eta = 0)$ and, from the above discussion, we see that in the medium p_T^{jet} range where Gq scattering dominates, $A_{LL}^{jet}(\eta = 0)$ should have a trend similar to $A_{LL}^{\gamma}(\eta = 0)$, with perhaps a magnitude reduced by a factor two, since about half of the jet cross section is due to GG and qq scatterings. This is what we see approximately in Fig.14, where we present the numerical results for $A_{LL}^{jet}(\eta = 0)$ at $\sqrt{s} = 500$ GeV, which should be compared to Fig.13. We have also indicated the statistical errors which are extremely small in this case, because of the huge event rates.

4.3 How to pin down $\Delta q(x,Q^2)$ and $\Delta\bar{q}(x,Q^2)$?

Let us now consider, for the reaction $pp \to W^{\pm}X$, the parity violating single helicity asymmetry A_L defined in Eq.(44). In the Standard Model (SM), the W gauge boson is a purely left-handed object and this asymmetry reads simply for $W^{\pm}$ production

$$A_L^{W^+}(y) = \frac{\Delta u(x_a, M_W^2)\, \bar{d}(x_b, M_W^2) \; - \; (u \leftrightarrow \bar{d})}{u(x_a, M_W^2)\, \bar{d}(x_b, M_W^2) \; + \; (u \leftrightarrow \bar{d})}\,, \tag{46}$$

assuming the proton a is polarized. Here we have $x_a = \sqrt{\tau}e^{y}$, $x_b = \sqrt{\tau}e^{-y}$ and $\tau = M_W^2/s$. For W^- production the quark flavors are interchanged $(u \leftrightarrow d)$. The calculation of these asymmetries is therefore very simple and the results are presented in Figs.15,16 at $\sqrt{s} = 500$ GeV, for different sets of distributions. As first noticed in ref.[48], the general trend of A_L can be easily understood as follows : at $y = 0$ one has

$$A_L^{W^+} = \frac{1}{2}\left(\frac{\Delta u}{u} - \frac{\Delta \bar{d}}{\bar{d}}\right) \quad \text{and} \quad A_L^{W^-} = \frac{1}{2}\left(\frac{\Delta d}{d} - \frac{\Delta \bar{u}}{\bar{u}}\right)\,, \tag{47}$$

evaluated at $x = M_W/\sqrt{s} = 0.164$, for $y = -1$ one has

$$A_L^{W^+} \sim -\frac{\Delta \bar{d}}{\bar{d}} \quad \text{and} \quad A_L^{W^-} \sim -\frac{\Delta \bar{u}}{\bar{u}}\,, \tag{48}$$

evaluated at $x = 0.059$ and for $y = +1$ one has

$$A_L^{W^+} \sim \frac{\Delta u}{u} \quad \text{and} \quad A_L^{W^-} \sim \frac{\Delta d}{d}\,, \tag{49}$$

evaluated at $x = 0.435$. Therefore these measurements will allow a fairly clean flavor separation, both for quarks and antiquarks, for some interesting ranges of x values. We see in Fig.15 that $A_L^{W^+}$, which is driven by the u and $\bar{d}$ polarizations, leads to similar predictions for all cases. This is mainly due to our knowledge of $\Delta u/u$, except for $x \geq 0.3$ where it comes out to be larger for the set proposed in ref.[49]. In Fig.16, for $A_L^{W^-}$ which is sensitive to the d and $\bar{u}$ polarizations, we see that the various predictions lead to the same general trend, with some differences in magnitude due to a large uncertainty in the determination of $\Delta d/d$. Also in ref.[49], one has assumed a larger negative $\Delta\bar{u}/\bar{u}$, which is reflected in the behaviour near $y = -1$. The statistical errors have been calculated with a rapidity resolution $\Delta y = 0.2$ and taking into account only the events from the leptonic decay modes. They are smaller for W^+ production which has larger event rates.

4.4 How to measure $h_1^{q,\bar{q}}(x, Q^2)$?

The existence of $h_1^q(x)$ for quarks ($h_1^{\bar{q}}$ for antiquarks) was first detected in a systematic study of the Drell-Yan process with polarized beams [53] and some of its relevant properties were discussed later in various papers [54,55,56]. Pretty much like $q(x)$ and $\Delta q(x)$, the transversity distributions $h_1^{q,\bar{q}}(x)$ are of fundamental importance for our understanding of the nucleon structure and they are all leading-twist distributions. Due to scaling violations, these quark distributions depend also on the scale Q and their Q^2-behavior is predicted by the QCD evolution equations. They are different in the three cases but, due to lake of time, we will not discuss here this important question[57,59]. We recall that $h_1^q(x, Q^2)$ (or $h_1^{\bar{q}}(x, Q^2)$) are not simply accessible in DIS because they are in fact chiral-odd distributions, contrarely to $q(x, Q^2)$ and $\Delta q(x, Q^2)$ which are chiral-even [56]. However they can be extracted from polarized Drell-Yan processes with two transversely polarized proton beams. For lepton pair production $pp \to \ell^+\ell^- X$ ($\ell = e, \mu$) mediated by a virtual photon γ^*, the double transverse-spin asymmetry $A_{TT}^{\gamma^*}$ defined, similarly to Eq.(42), as $A_{TT} = d\delta\sigma/d\sigma$, reads

$$A_{TT}^{\gamma^*} = \widehat{a}_{TT} \frac{\sum_q e_q^2 h_1^q(x_a, M^2) h_1^{\bar{q}}(x_b, M^2) + (a \leftrightarrow b)}{\sum_q e_q^2 q(x_a, M^2) \bar{q}(x_b, M^2) + (a \leftrightarrow b)}, \tag{50}$$

where $\widehat{a}_{TT}$ is the partonic asymmetry calculable in perturbative QCD and M is the dilepton mass. The rapidity y of the dilepton is $y = x_a - x_b$, and for $y = 0$ one has $x_a = x_b = M/\sqrt{s}$. Note that this is a leading-order expression, which can be used to get a first estimate of $A_{TT}^{\gamma\star}$ from different theoretical results for h_1^q and $h_1^{\bar{q}}$. If the lepton pair is mediated by a Z gauge boson, one has a similar expression for A_{TT}^{Z} [58], namely

$$A_{TT}^{Z} = \frac{\sum_q (b_q^2 - a_q^2) h_1^q(x_a, M_Z^2) h_1^{\bar{q}}(x_b, M_Z^2) + (a \leftrightarrow b)}{\sum_q (b_q^2 + a_q^2) q(x_a, M_Z^2) \bar{q}(x_b, M_Z^2) + (a \leftrightarrow b)}, \tag{51}$$

where a_q and b_q are the vector and axial couplings of the flavor q to the Z. However in the case of $W^{\pm}$ production one expects $A_{TT}^{W} = 0$, because the W gauge boson is a pure left-handed object (*i.e.*, $a_q = b_q$), which does not allow a left-right interference effect associated to the existence of $h_1^{q,\bar{q}}$ [58]. We show in Fig.17 some predictions, at LO and NLO, for A_{TT}, together with some expected statistical errors, where one sees a characteristic effect in the Z mass region. The asymmetry is only a few percents, due to the smallness of $h_1^{\bar{q}}$, but this is a real challenge for the experiment.

4.5 How to uncover new parity violation effects ?

Let us consider again one-jet inclusive production. As discussed in section 4.2, the cross section is dominated by the pure QCD subprocesses GG, Gq and qq scatterings, but the existence of the electroweak (EW) interaction, via the effects of the $W^{\pm}, Z$ gauge bosons, adds to it, a small contribution. Consequently, the parity violating asymmetry A_{LL}^{PV} defined in Eq.(45) and resulting from the QCD-EW interference is non-zero, as shown in Fig.18 (see SM curve) and has a small structure near $E_T = M_{W,Z}/2$. Now if we introduce a new contact interaction, belonging purely to the quark sector and normalized to a certain compositeness scale Λ, under the form

$$\mathcal{L}_{qqqq} = \epsilon \frac{g^2}{8\Lambda^2} \bar{\Psi}\gamma_\mu(1-\eta\gamma_5)\Psi . \bar{\Psi}\gamma^\mu(1-\eta\gamma_5)\Psi \ , \tag{52}$$

where Ψ is a quark doublet and $\epsilon = \pm 1$. If parity is not conserved $\eta = \pm 1$ and we show in Fig.18, how the SM prediction will be affected by such a new interaction assuming $\Lambda = 1.6 TeV$. As expected, the errors are large in the high E_T region, but if the observation of such a signal is confirmed, it will be extremely important.

5 Concluding remarks

The observation of large polarization effects in various areas of medium and high energy particle physics, shows that spin degrees of freedom play a significant role in hadron dynamics. Recent data on polarized DIS have generated new insights in the fundamental problem of the nucleon spin structure which remains to be clarified in the future. We have also stressed the relevance of the polarized *pp* collider at RHIC-BNL, for studying polarized parton distributions and providing the first serious tests of the spin sector of perturbative QCD. This very exciting spin physics programme will also include the measurement of parity violating asymmetries, which may uncover some new physics signals.

Acknowledgments

It is my pleasure to thank H.Y. Cheng and G.L. Lin for their invitation and all the organizers for setting up this excellent workshop in such a pleasant and stimulating atmosphere. I would like also to acknowledge fruitful discussions with all my collaborators on the various topics presented here, in particular, C. Bourrely, F. Buccella, O. Teryaev and J.M. Virey.

References

1. For recent reviews see M. Anselmino, A. Efremov and E. Leader, Phys. Reports **261**, 1 (1995); B. Lampe and E. Reya, hep-ph/9810270, to appear Rev. Mod. Phys. .
2. C. Bourrely, E. Leader and J. Soffer, Phys. Reports **59**, 95 (1980).
3. T.L. Trueman, Proceedings of "Spin 96", 12th Int. Symp. on High Energy Spin Physics, Amsterdam Sept.96, Eds. C.W. de Jager, et al., World Scientific, Singapore, 1997, p.833.
4. This is related to the important issue of the size of the Pomeron flip coupling; see N.H. Buttimore, B.Z. Kopeliovich, E. Leader, J. Soffer and T.L. Trueman, preprint CPT-98/P.3693 (in preparation).
5. D. G. Crabb *et al.*, Phys. Rev. Lett. **65**, 3241 (1990).
6. A. D. Krisch, Z. Phys. **C46**, S133 (1990).
7. C. Wilkinson *et al.*, Phys. Rev. Lett. **58**, 855 (1987).
8. A. M. Smith *et al.*, Phys. Lett. **B185**, 209 (1987).
9. D. L. Adams *et al.*, Phys. Lett. **B264**, 462 (1991).
10. EMC Collaboration: J. Ashman *et al.*, Phys. Lett. **B206**, 364 (1988); Nucl. Phys. **B328**, 1 (1989).
11. J. Ellis and R. L. Jaffe, Phys. Rev. **D9**, 1444 (1974); *Erratum*, ibib **D10**, 69 (1974).
12. Particle Data Group, L. Montanet *et al.*, Phys. Rev. **D50**, 1173 (1994); F. E. Close and R. G. Roberts, Phys. Lett. **B313**, 165 (1993).
13. SLAC-E142 Experiment: P. L. Anthony *et al.*, Phys. Rev. Lett. **71**, 959 (1993); Phys. Rev. **D54**, 6620 (1996). SLAC-E143 Experiment: K. Abe *et al.*, Phys. Rev. Lett. **74**, 346 (1995); ibid**75**, 25 (1995) ; Phys. Lett. **B364**, 61 (1995). SLAC-E154 Experiment: K. Abe *et al.*, Phys. Rev. Lett. **79**, 26 (1997).

14. SMC Collaboration: D. Adams *et al.*, Phys. Lett. **B329**, 399 (1994); B. Adeva *et al.*, Phys. Lett. **B302**, 533 (1993); D. Adams *et al.*, Phys. Lett. **B357**, 248 (1995); Phys. Lett. **B396**, 338 (1997); Phys. Rev. **D56**, 5330 (1997); B. Adeva *et al.*, Phys. Lett. **B412**, 414 (1997).
15. HERMES Collaboration: K. Ackerstaff *et al.*, Phys. Lett. **B404**, 383 (1997).
16. G. Altarelli and G. G. Ross, Phys. Lett. **B212**, 391 (1988); A. V. Efremov and O. V. Teryaev, Dubna preprint **E2-88-287** (1988).
17. For an extensive review on the issue of factorization schemes see H.Y. Cheng, Int. J. Mod. Phys. **A11**, 5109 (1996).
18. J. D. Bjorken, Phys. Rev. **148**, 1467 (1966); Phys. Rev. **D1**, 1376 (1970).
19. M. Glück, E. Reya, M. Stratmann and W. Vogelsang, Phys. Rev. **D53**, 4775 (1996).
20. T. Gehrmann and W.J. Stirling, Phys. Rev. **D53**, 6100 (1996).
21. E. Leader, A. V. Sidorov, and D. B. Stamenov, hep-ph/9708335.
22. G. Altarelli, R.D. Ball, S. Forte and G. Ridolfi, Nucl. Phys. **B496**, 337 (1997).
23. F. Buccella and J. Soffer, Mod. Phys. Lett. **A8**, 225 (1993); F. Buccella and J. Soffer, Europhys. Lett. **24**, 165 (1993).
24. K. Gottfried, Phys. Rev. Lett. **18**, 1154 (1967).
25. NMC Collaboration: M. Arneodo *et al.*, Phys. Rev. **D50**, R1 (1994).
26. R. D. Field and R. P. Feynman, Phys. Rev. **D15**, 2590 (1977).
27. R.D. Ball, S. Forte and G. Ridolfi, Phys. Lett. **B378**, 255 (1996); hep-ph/9608399.
28. J. Bartels, B.I. Ermolaev and M.G. Ryskin, Z. Phys. **C70**, 273 (1996).
29. J. Bartels, B.I. Ermolaev and M.G. Ryskin, Z. Phys. **C72**, 627 (1996).
30. A.V. Efremov, J. Soffer, O.V. Teryaev, Nucl. Phys. **B346**, 97 (1990).
31. J. Soffer and O.V. Teryaev, Phys. Rev. **D56**, 1549 (1997).
32. A. D. Martin, R. G. Roberts, and W. J. Stirling, Phys. Lett. **B354**, 155 (1995).
33. M. Lacombe *et al.*, Phys. Lett. **B101**, 139 (1981).
34. C. Bourrely, F. Buccella, O. Pisanti, P. Santorelli and J. Soffer, Prog. of Theo. Phys., **99**, 1017 (1998).
35. S.B. Gerasimov, Yad. Fiz. **2**, 598 (1965) [Sov.J.Nucl.Phys. **2**, 430 (1966)].
36. S.D. Drell and A.C. Hearn, Phys. Rev. Lett. **16**, 908 (1966).
37. E143 Collaboration: K. Abe *et al.*, Phys. Rev. Lett. **78**, 815 (1997).
38. J. Soffer and O. Teryaev, Phys. Rev. Lett. **70**, 3373 (1993).
39. J. Soffer and O. Teryaev, Phys. Rev. **51**, 25 (1995).

40. V.D. Burkert and B.L. Ioffe, Phys. Lett. **B296**, 223 (1992); JETP **105**, 1153 (1994).
41. H. Burkhardt and W.N. Cottingham, Ann. Phys. (N.Y.) **16**, 543 (1970).
42. J. Schwinger, Proc. Nat. Acad. Sci. U.S.A. **72**, 1559 (1975).
43. A. Lung et al., Phys. Rev. Lett. **70**, 718 (1993).
44. J. Soffer and O. Teryaev, Phys. Rev. **D56**, 7458 (1997).
45. B.L. Ioffe, V.A. Khoze, L.N. Lipatov, *Hard Processes*, (North-Holland, Amsterdam, 1984).
46. See *e.g.*, FNAL-E704 Collaboration: A. Bravar *et al*, Phys. Rev. Lett. **77**, 2626 (1996) and references therein.
47. C. Bourrely, J. Soffer, F.M. Renard and P. Taxil, Phys. Reports **177**, 319 (1989), and the errata preprint CPT - 87/P.2056 (January 1991); see also P. Taxil, Riv. Nuovo Cimento, Vol.16, No. 11, 1 (1993).
48. C. Bourrely and J. Soffer, Phys. Lett. **B314**, 132 (1993).
49. C. Bourrely and J. Soffer, Nucl. Phys. **B445**, 341 (1995).
50. C. Bourrely, F. Buccella, O. Pisanti, P. Santorelli and J. Soffer, preprint CPT-96/PE.3327, DSF-T96/17, to appear in the proceedings of the Fundamental Structure of Matter, Ouranoupoulis (Greece) May 28-31 (1997) (Editor A. Nicolaidis).
51. H.Y. Cheng and H.H. Liu, Phys. Rev. **D53**, 2380 (1996).
52. J. Soffer and J.M. Virey, Nucl. Phys. **B509**, 297 (1998).
53. J.P. Ralston and D.E. Soper, Nucl. Phys. **B152**, 109 (1979).
54. X. Artru and M. Mekhfi, Z. Phys. **C45**, 669 (1990).
55. J.L. Cortes, B. Pire and J.P. Ralston, Z. Phys. **C55**, 409 (1992).
56. R.L. Jaffe and X. Ji, Phys. Rev. Lett. **67**, 552, (1991) ; Nucl. Phys. **B375**, 527 (1992).
57. C. Bourrely, J. Soffer and O.V. Teryaev, Phys. Lett. **B420**, 375 (1998).
58. C.Bourrely and J. Soffer, Nucl. Phys. **B423**, 329 (1994).
59. O. Martin, A. Schäfer, M. Stratmann and W. Volgelsang, Phys. Rev. **D57**, 3084 (1998).
60. P. Taxil and J.M. Virey, Phys. Lett. **B364**, 181 (1995).

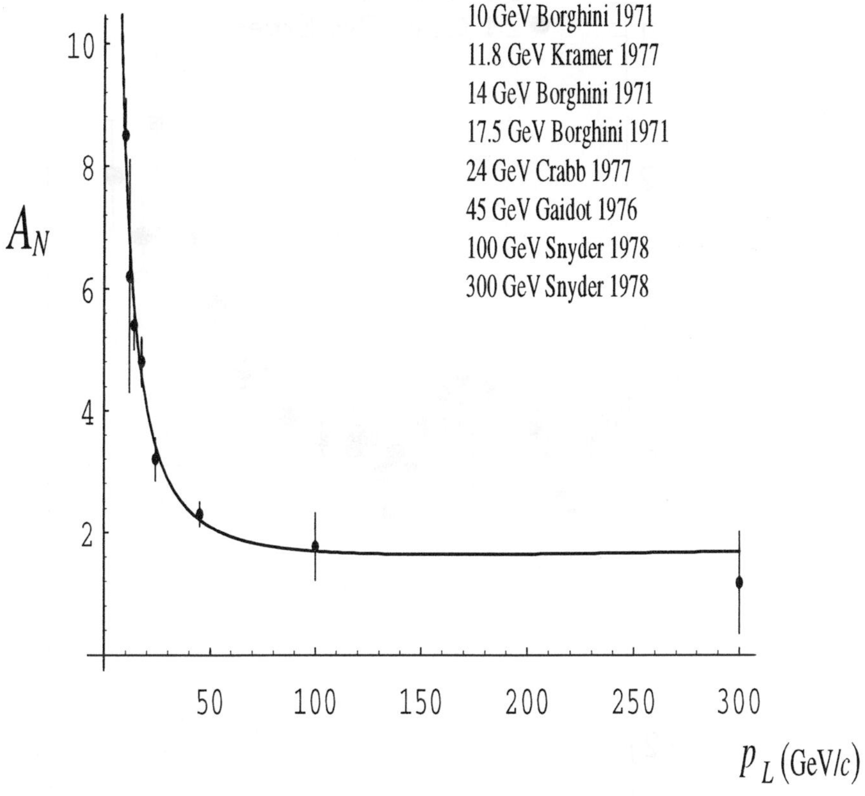

Figure 1: A_N in percent as a function of the lab. momentum p_L at fixed $t = -0.15 GeV^2$ (Taken from ref.[3]).

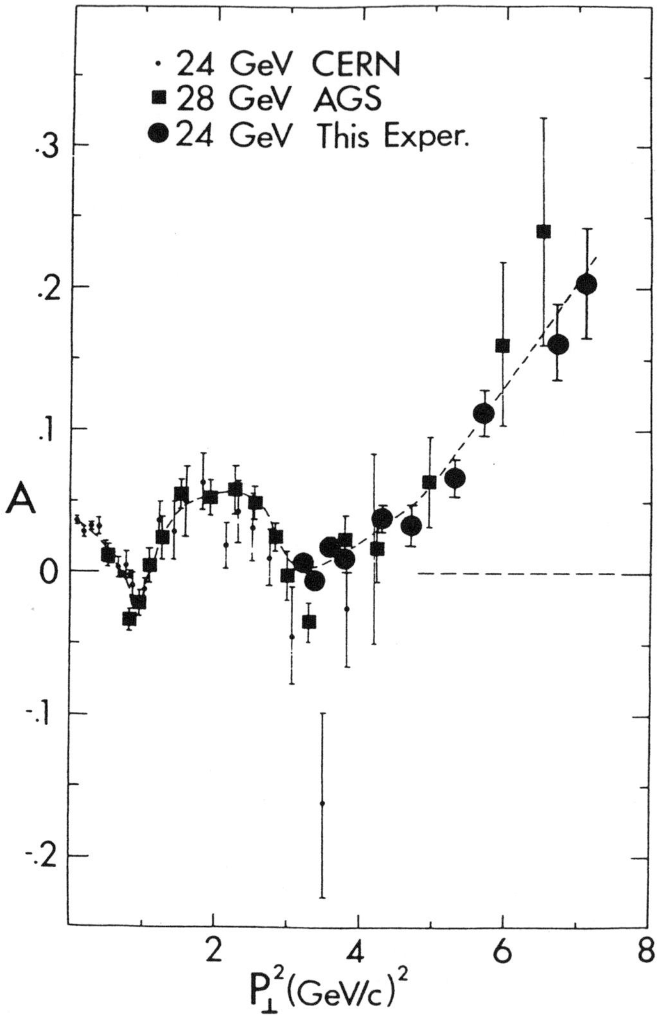

Figure 2: The analyzing power A_N as a function of p_T^2 (Taken from ref.[5]).

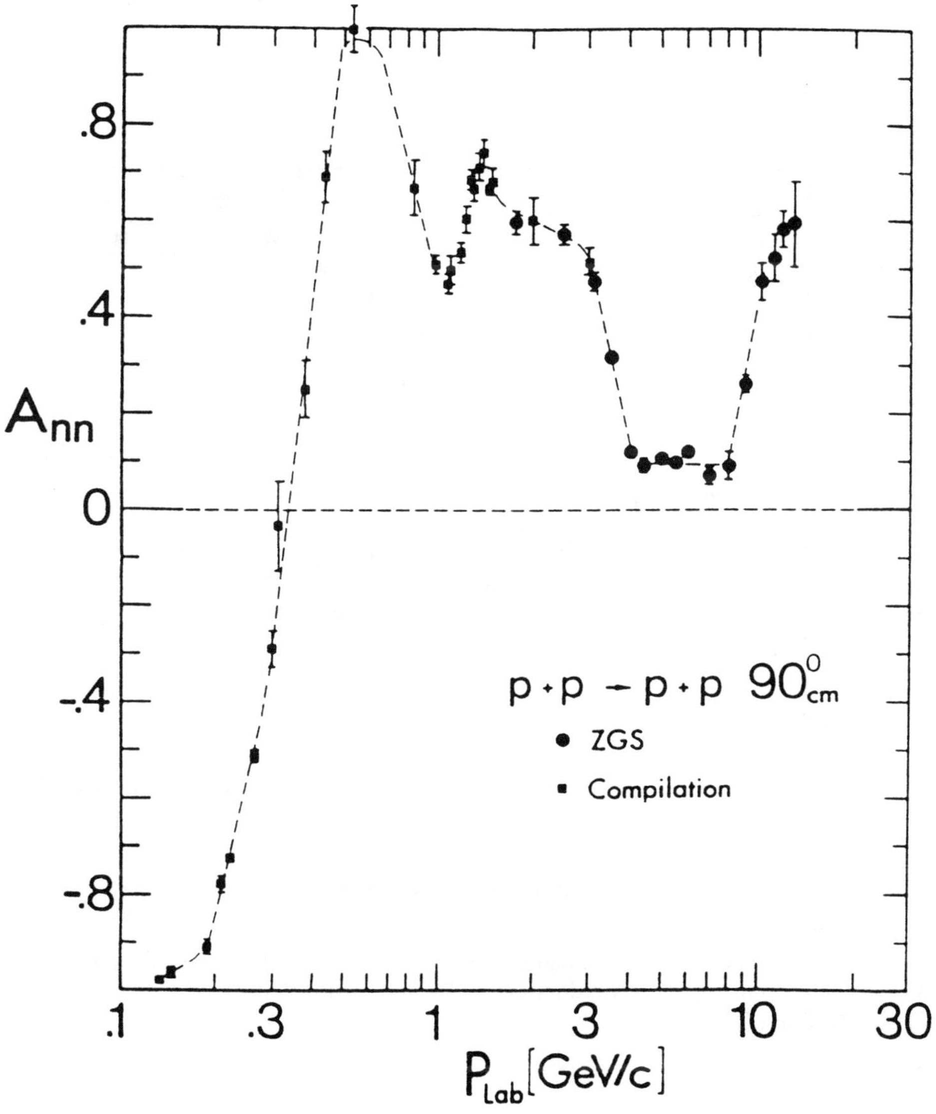

Figure 3: Compilation of the spin-spin parameter A_{NN} at $\theta_{cm} = 90^0$ in pp elastic scattering (Taken from ref.[6]).

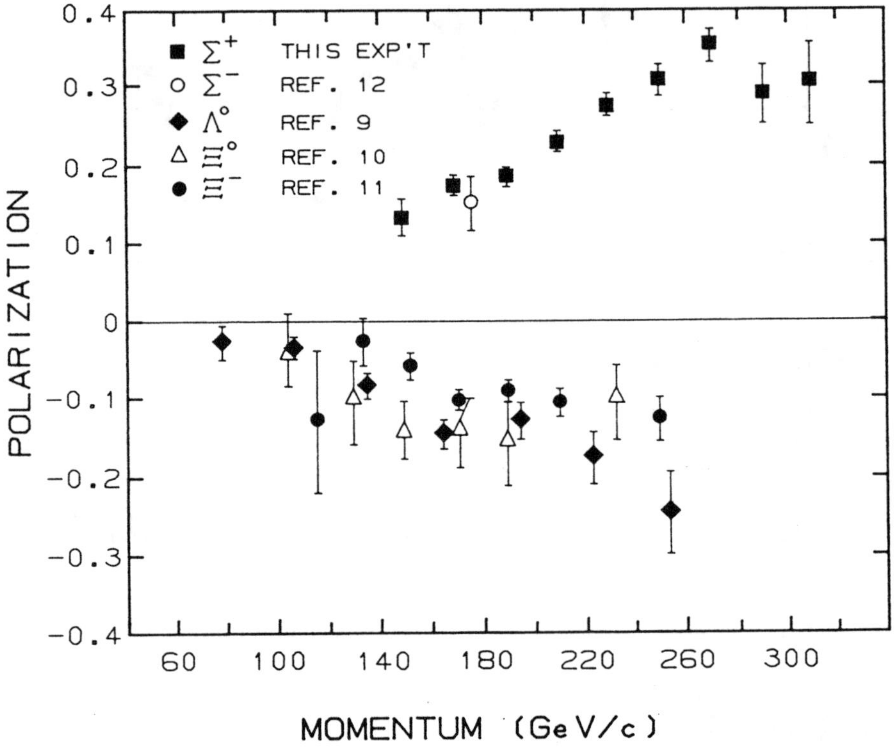

Figure 4: Compilation of the polarization, versus p_T, for hyperons produced by 400GeV/c protons on Be at production angles $\sim$ 5 mrd (7.5 mrd for Σ^-). (Taken from ref.[7]).

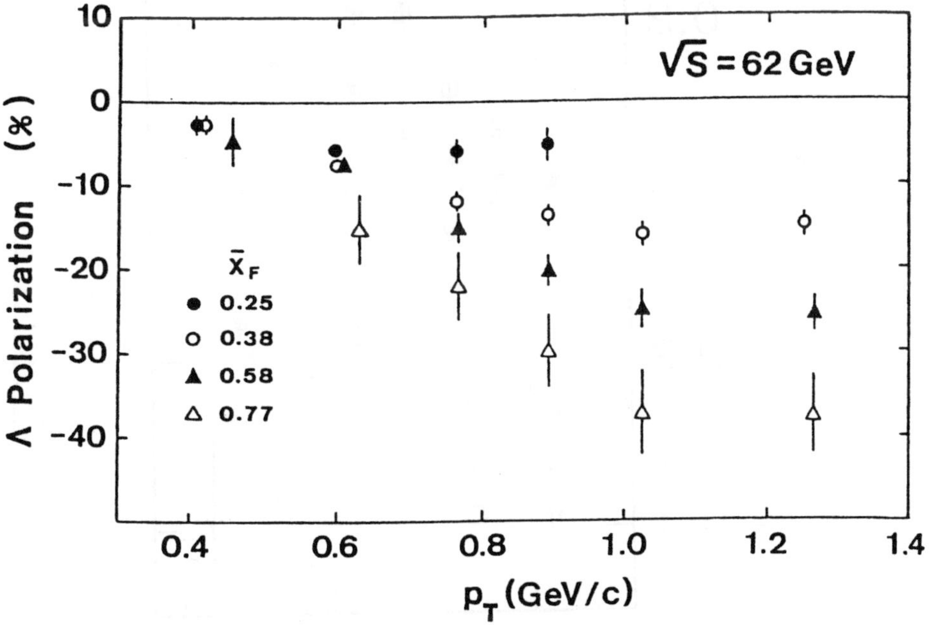

Figure 5: Λ polarization versus p_T for various x_F values at ISR energy (Taken from ref.[8]).

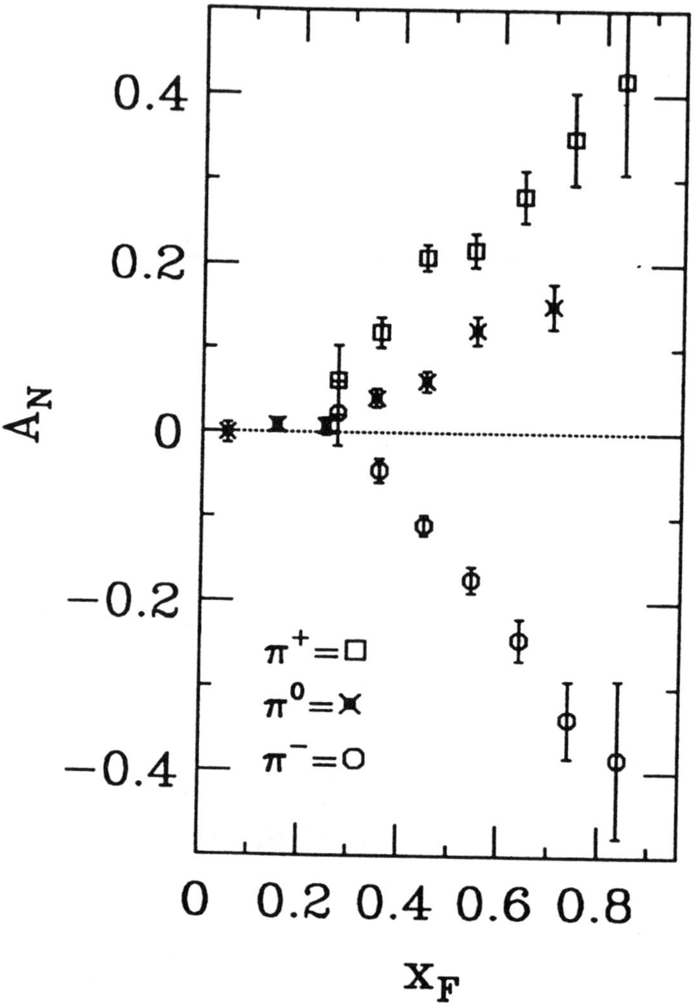

Figure 6: The analyzing power A_N versus x_F for π^+, π^- and π^0 inclusive production at FNAL energy(Taken from ref.[9]).

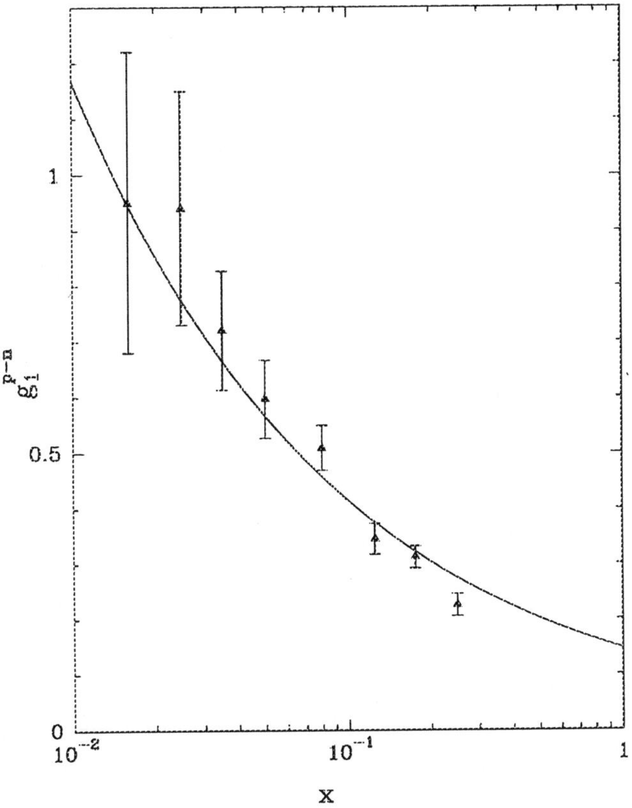

Figure 7: Comparison for g_1^{p-n} between the curve given by Eq.(12) and the SLAC data ref.[13].

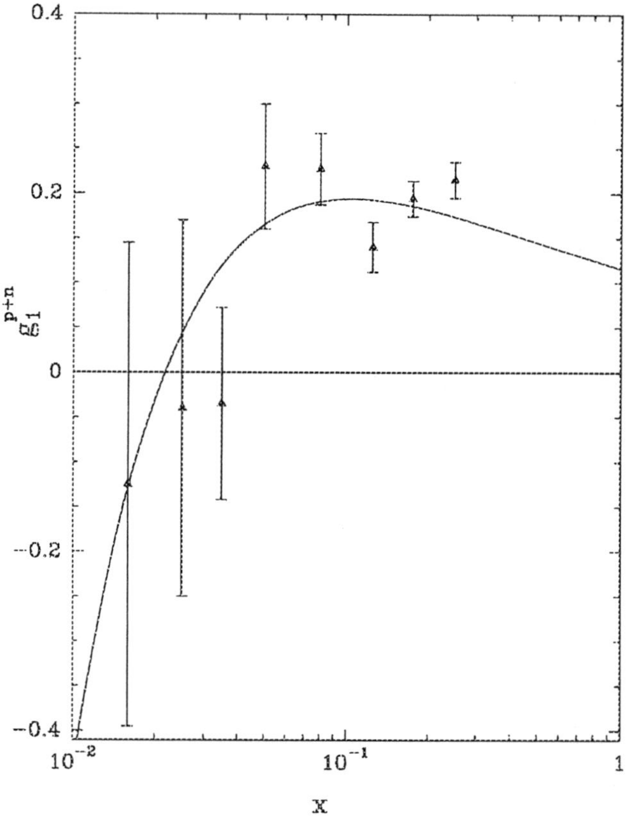

Figure 8: Same as Fig.7 for g_1^{p+n} with the curve given by Eq.(14).

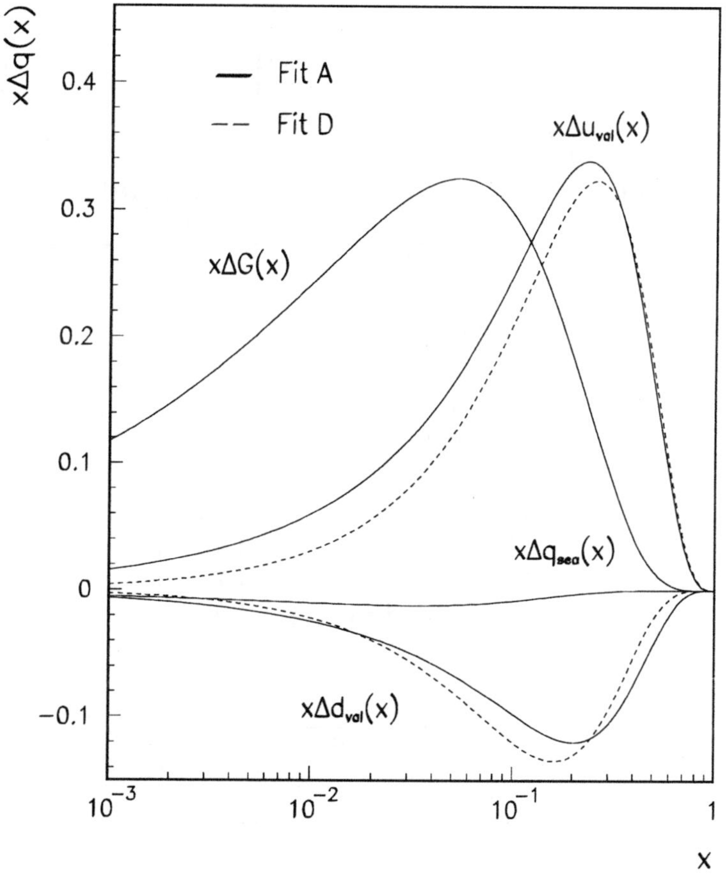

Figure 9: The polarized parton distributions for fits A and D at $Q_0^2 = 4GeV^2$. Note that, in the case of fit D, sea and gluons are vanishing.

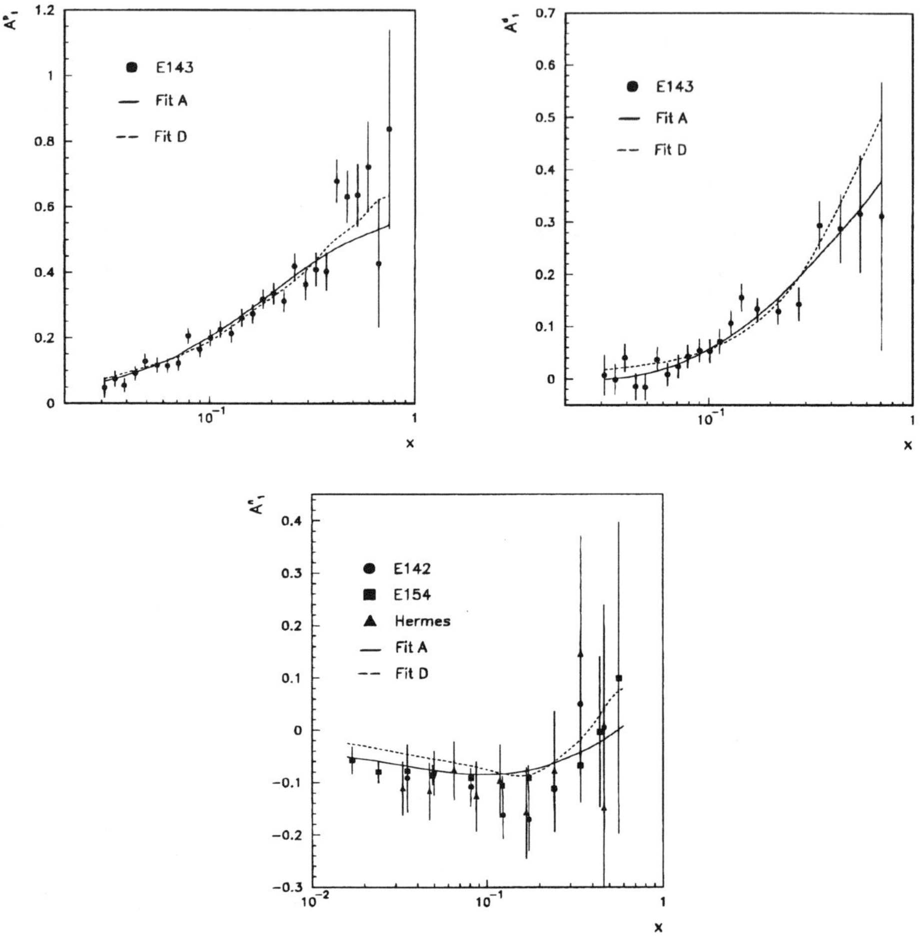

Figure 10: The prediction of fits A and D with data are showed for proton, deuteron, and neutron SLAC asymmetries. The curves correspond to the fits at the Q^2 of the data points for proton and deuteron, and at $Q_0^2 = 4GeV^2$ for neutron. The HERMES data [15] are showed, but they are not used in the fits.

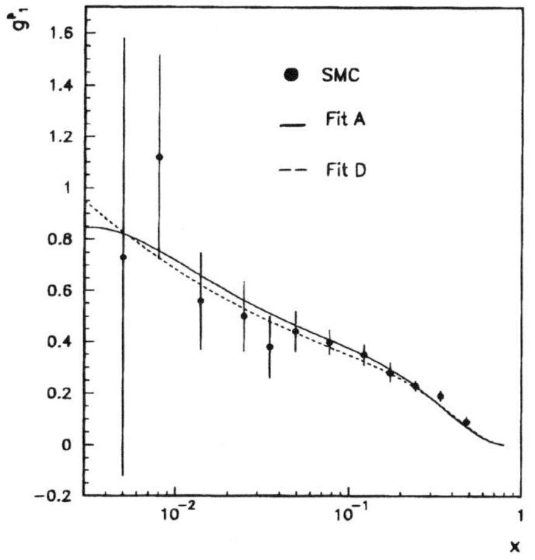

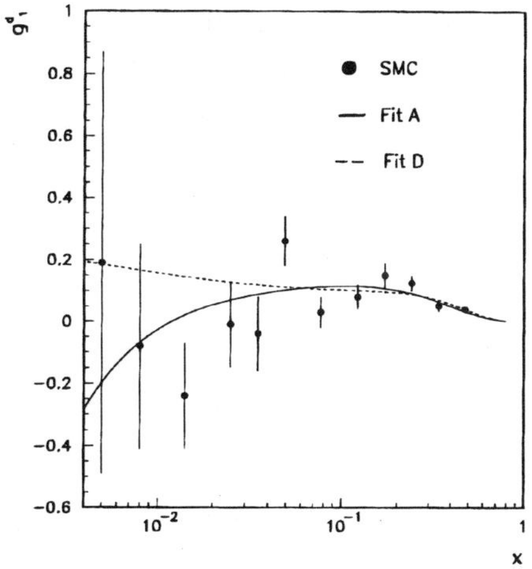

Figure 11: The predictions of fits A and D for the proton and deuteron g_1 structure functions at $<Q^2>= 10\,GeV^2$ are compared with the most recent data obtained by SMC (see Ref.[14]).

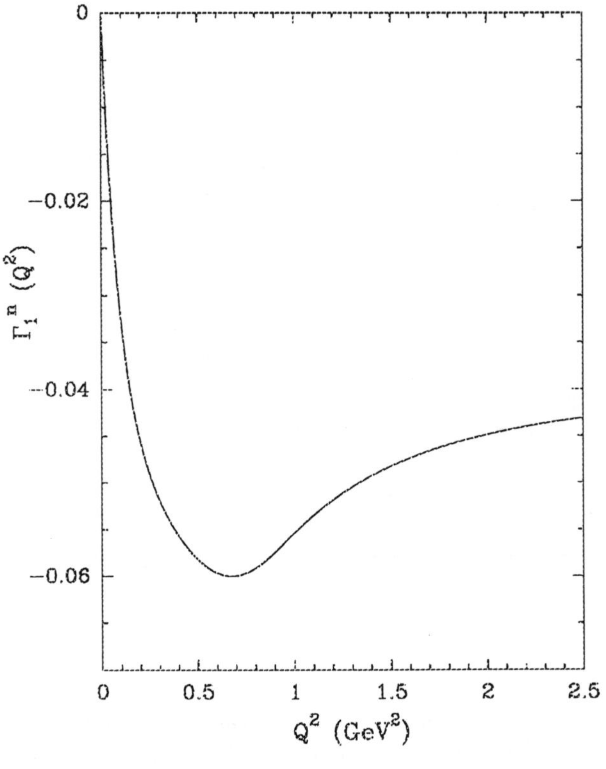

Figure 12: Prediction of the generalized GDH sum rule for the neutron. (Taken from ref.[44])

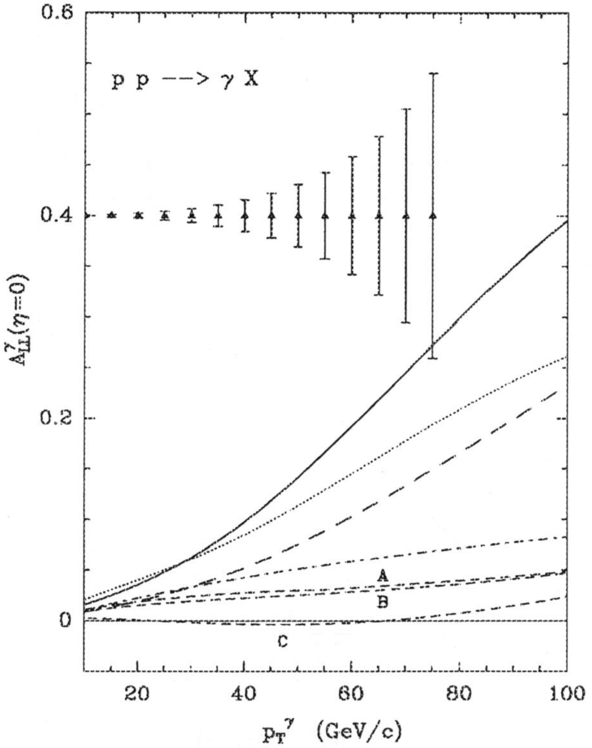

Figure 13: The double helicity asymmetry $A^{\gamma}_{LL}(\eta = 0)$ at $\eta = 0$ versus p^{γ}_T for $\sqrt{s} = 500GeV$ calculated with different parton densities.
(Solid and large dashed curves[49], dotted curve [50], dashed-dotted curve [19], small dashed curves [20]).(Taken from ref.[52])

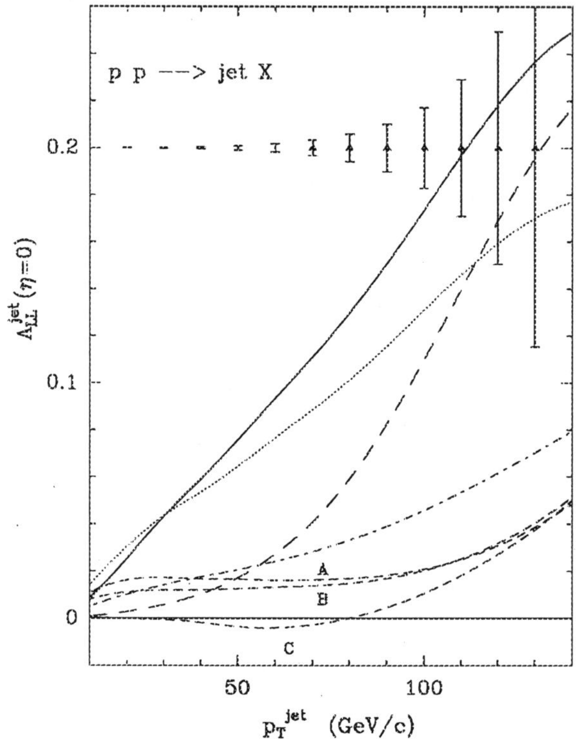

Figure 14: The double helicity asymmetry $A_{LL}^{jet}(\eta=0)$ at $\eta=0$ versus p_T^{jet} for $\sqrt{s}=500GeV$ calculated with different parton densities (Curve labels as in Fig.13).(Taken from ref.[52]).

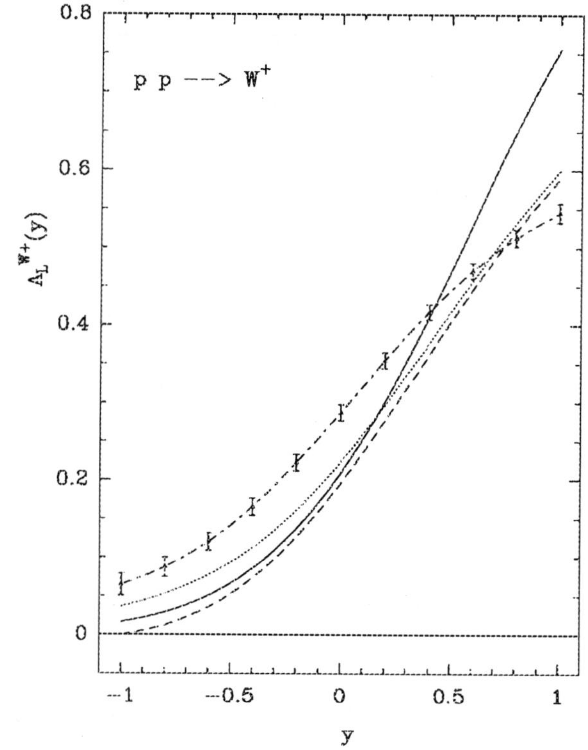

Figure 15: The parity violating asymmetry $A_L^{W^+}(y)$ versus y for $\sqrt{s} = 500 GeV$ calculated with different parton densities (Curve labels as in Fig.13). (Taken from ref.[52]).

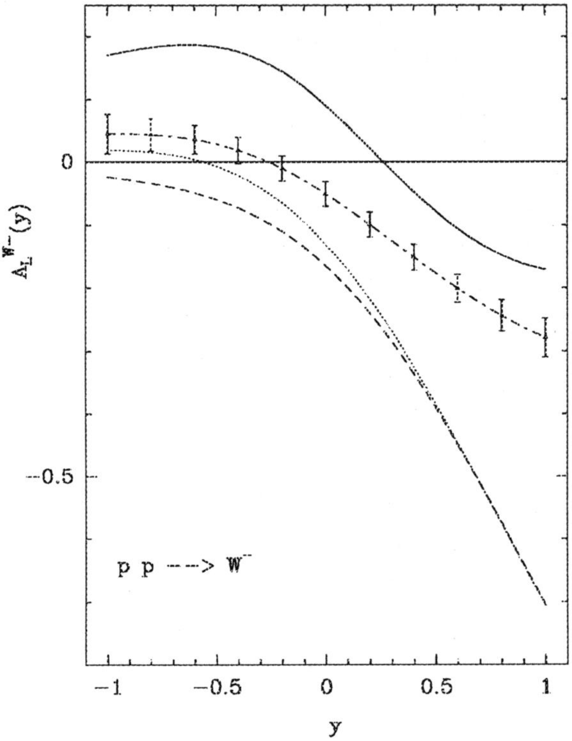

Figure 16: Same as Fig.15 for W^- production.

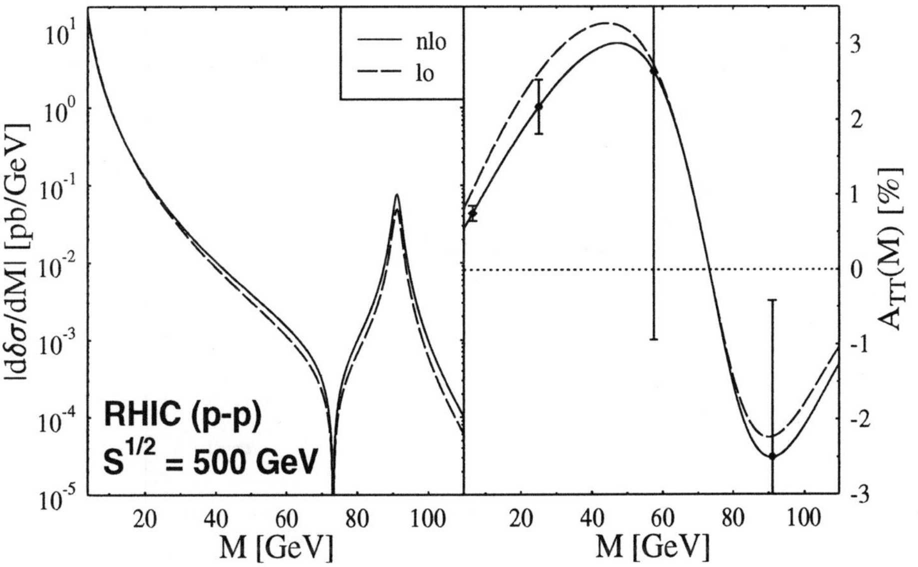

Figure 17: NLO and LO maximal polarized Drell-Yan cross section ($|d\delta\sigma|$) and asymmetry(A_{TT}) for $RHIC$ at $\sqrt{s} = 500GeV$ assuming the maximum luminosity. (Taken from ref.[59]).

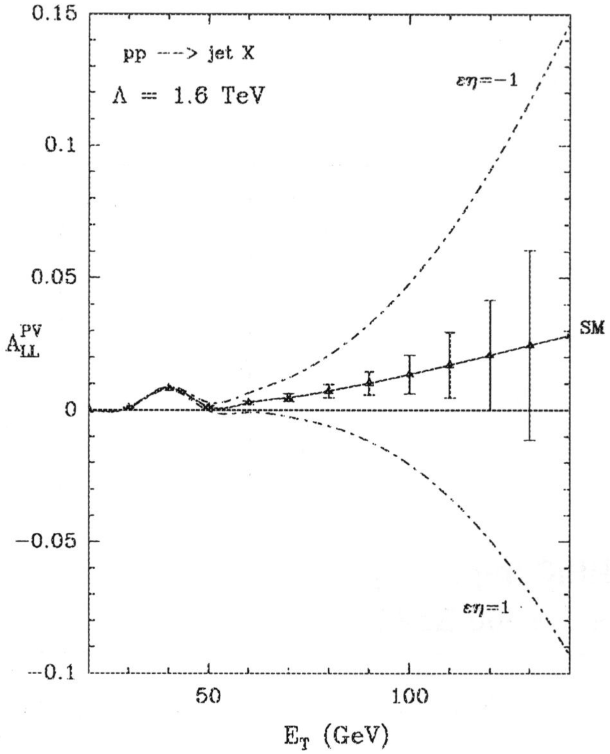

Figure 18: A_{LL}^{PV} versus E_T at RHIC for $\sqrt{s} = 500 GeV$, for $\Lambda = 1.6 TeV$ (Dashed curve corresponds to $\epsilon\eta = -1$, dot-dashed curve to $\epsilon\eta = +1$ and solid curve, SM, is the pure QCD-EW interference). (Taken from ref.[60]).

SUSY Phenomenology

Yasuhiro Okada
KEK, Oho 1-1, Tsukuba, 305-0801 Japan

Three topics on phenomenology in supersymmetric models are reviewed, namely, the Higgs sector in the supersymmetric model, B and K meson decay in the supergravity model and lepton flavor violation processes in supersymmetric grand unified theory.

1 Introduction

Supersymmetry (SUSY) is a symmetry between fermions and bosons and its theoretical development started from the formulation of SUSY invariant Lagrangian in four dimensional field theory in early '70s.[1] Subsequently, SUSY gauge theory, supergravity[2] and string theory with space-time supersymmetry (supersrting)[3] were formulated in '70s.

Phenomenological application of SUSY theory has developed since early '80s in connection with naturalness problem in the Standard Model (SM). One interesting possibility beyond the SM is an idea that various interactions are unified at very high energy scale. Examples are grand unified theory (GUT) and string unification. In such theories the cutoff energy scale for the SM is considered to be close to the Planck scale ($\sim 10^{19}$ GeV) and the quadratic divergence of the Higgs field in the SM is very problematic because very precise fine tuning between the bare mass term and radiative correction is necessary to ensure that the weak scale is smaller by 10^{16} orders than the Planck scale. In SUSY theories this problem can be avoided due to cancellation of the quadratic divergence in scaler mass renormalization. From this point of view SUSY extensions of the SM and the GUT have been extensively studied for more than 15 years.[4]

In the experimental side, the SM has been very successful. LEP and SLAC experiments at the Z pole showed quantitative agreement of various observables with SM predictions. It was also pointed out that three gauge coupling constants determined precisely there are consistent with the SUSY SU(5) GUT assumption although non-SUSY version of the simple SU(5)GUT fails to reproduce the coupling unification.[5] Since then it has become more and more important to know how one can explore SUSY in the present and future high energy experiments.

In this lecture after a brief introduction to the minimal supersymmetric standard model (MSSM) we discuss three topics on SUSY phenomenology, namely the Higgs sector in SUSY models, B and K meson decay in the supergravity model, and lepton flavor violation (LFV) in SUSY GUT.

2 Minimal Supersymmetric Standard Model

2.1 MSSM Lagrangian

Minimal supersymmetric standard model (MSSM) is a minimal SUSY extension of the SM. In this model we introduce a SUSY partner for each particle in the SM. For left- and right-handed quark and lepton complex scalar fields, squark and slepton, are introduced. The superpartner of gauge boson is gauge fermion (or gaugino) and that of the Higgs field is called higgsino. The superpartners of gluon and SU(2) and U(1) gauge fermions are gluino, wino and bino, respectively. After the electroweak symmetry breaking the wino, bino and higgsino mix each other and form two charged Dirac fermions called chargino and four neutral Majorana fermions called neutralino. As for the Higgs sector the SUSY model contains at least two Higgs doublet fields. This is because different Higgs doublet fields should be introduced for mass terms for the up-type quarks and those for the down-type quarks and charged leptons in order to write appropriate Yukawa couplings without conflict with SUSY. Two Higgs doublets are also required from the gauge anomaly cancellation because the higgsino fields generate additional contributions to the anomaly cancellation condition. The particle content of the MSSM is therefore summarized as two Higgs doublet SM with bosonic superpartners (squarks and sleptons) and fermionic superpartners (gluino, charginos and neutralinos).

The MSSM Lagrangian consists of two parts, SUSY invariant part and soft SUSY breaking terms. General rule of writing SUSY invariant Lagrangian is well known.[6] There are two kinds of supermultiplets, *i.e.* gauge supermultiplet which contains gauge fields (A_μ) and gauge fermion (λ) and chiral supermultiplet which consists of a pair of a complex scalar field (ϕ) and a left-handed Weyl fermion (ψ). The SUSY invariant interactions are classified essentially to two kinds, *i.e.* gauge interactions and superpotential interactions.

The gauge interaction is completely determined once we fix gauge group and representation of each chiral supermultiplet. In addition to ordinary gauge interactions between matter fields and gauge fields through covariant derivatives there are two additional interactions specified by the gauge coupling constant. One is gaugino-fermion-scalar coupling and the other is scalar four-point coupling called D term. This kinds of relation between bosonic and fermionic interactions are important feature of SUSY theory.

The superpotential interaction is a set of scalar potential and fermion mass terms or Yukawa type interactions. This is specified by a scalar function called superpotential. In the MSSM we take

$$W = (y_u)_{ij}H_2U_i^cQ_j + (y_d)_{ij}H_1D_i^cQ_j \\ +(y_e)_{ij}H_1E_iL_j + \mu H_1H_2. \quad (1)$$

The SU(3)× SU(2)× U(1) quantum numbers for each superfield are given by $Q(\mathbf{3},\mathbf{2},\frac{1}{6})$, $U^c(\bar{\mathbf{3}},\mathbf{1},-\frac{2}{3})$, $D^c(\bar{\mathbf{3}},\mathbf{1},\frac{1}{3})$, $L(\mathbf{1},\mathbf{2},-\frac{1}{2})$, $E(\mathbf{1},\mathbf{1},1)$, $H_1(\mathbf{1},\mathbf{2},-\frac{1}{2})$, $H_2(\mathbf{1},\mathbf{2},\frac{1}{2})$ and i,j are generation indices. If we only demand gauge invariance we can also include the following coupling in the superpotential.

$$W_{\not R} = \lambda_1 QLD^c + \lambda_2 LEL + \lambda_3 U^cD^cD^c. \quad (2)$$

These couplings in general violate the lepton number and/or baryon number conservation, therefore unless some of these coupling constants are extremely small too fast proton decay is induced. The most popular way to forbid the above terms in the superpotential is to require conservation of a multicative quantum number called R parity which is assigned to be +1 for ordinary particles and -1 for superpartners. The R parity conservation has an important phenomenological consequence. The lightest SUSY particle (LSP) becomes stable. Since LSP should be a neutral particle from cosmological consideration LSP becomes a natural candidate of dark matter of the Universe. The existence of LSP is also important in the direct SUSY search experiment because SUSY particles should be pair-produced and decay to final states including missing energy.

2.2 Soft SUSY Breaking Terms

The soft SUSY breaking terms are defined as those terms in the Lagrangian which violate SUSY invariance but do not induce quadratic divergence. In this way the naturalness is maintained as long as the SUSY breaking scale is close to the weak scale. The general form of the soft SUSY breaking terms are classified.[7] These terms are gaugino mass terms, scalar mass terms and triple and quadratic couplings among scalar components of chiral supermultiplets (or anti-chiral supermultiplets).

$$L_{SUSY Breaking} = -\frac{1}{2}\Sigma M_\alpha \bar{\lambda_\alpha}\lambda_\alpha - \Sigma m_i^2|\phi_i|^2 \\ +(A_uH_2\tilde{u}^c\tilde{q} + A_dH_1\tilde{d}^c\tilde{q} + A_eH_1\tilde{e}\tilde{l}) + h.c. \\ +B\mu H_1H_2 + h.c.. \quad (3)$$

Here we suppress the generation indices and neglect the generation mixing in the scalar mass terms. The role of the SUSY breaking terms is basically to provide mass terms for SUSY partners so that SUSY particles can be heavier than the ordinary particles.

The origin of these soft SUSY breaking terms can be thought of the remnant of spontaneous SUSY breaking. If we consider the MSSM Lagrangian as a low energy effective theory of some unification theory valid at very high energy scale, it is natural to take gravity interaction into consideration. Once gravity is included SUSY becomes local invariance, *i.e.* supergravity theory. Because local invariance is necessary for consistency of the theory the only way to break the symmetry is the spontaneous one. If we want to construct a model based on spontaneous SUSY breaking at the tree level in global SUSY theory, there exists a tree level mass relation which severely constrains the model building.[8] The phenomenologically viable way to generate the soft SUSY breaking terms in the MSSM Lagrangian is therefore to prepare a separate sector where the spontaneous SUSY breaking occurs and to couple this sector with the MSSM through some weak interaction such as gravity or loop effects. As a result appropriate soft SUSY breaking terms are generated in the MSSM Lagrangian.

There are important phenomenological constraints on the SUSY breaking terms from flavor and CP violation physics. The SUSY breaking terms for squarks and sleptons masses can induce too large flavor changing neutral current (FCNC), LFV and neutron and electron electric dipole moments (EDM). This is because squark and slepton mass matrices can be new sources of flavor mixings and CP violation which are not related to the Cabibbo-Kobayashi-Maskawa (CKM) matrix. These flavor and CP problems provide us some clue to consider possible mechanism of SUSY breaking.

For example, if we assume that the SUSY contribution to the $K^0 - \bar{K}^0$ mixing is suppressed because of the cancellation among the squark contributions of different generations, the squarks with the same $SU(3) \times SU(2) \times U(1)$ quantum numbers have to be highly degenerate in masses. When the squark mixing angle is in a similar magnitude to the Cabibbo angle the requirement on degeneracy becomes as

$$\frac{\Delta m_{\tilde{q}}^2}{m_{\tilde{q}}^2} \lesssim 10^{-2} \left(\frac{m_{\tilde{q}}}{100GeV}\right) \tag{4}$$

for at least the first and second generation squarks. Similarly, the $\mu^+ \to e^+\gamma$ process puts a strong constraint on the flavor off-diagonal terms for slepton mass matrices which is roughly given by

$$\frac{\Delta m_{\tilde{\mu}\tilde{e}}^2}{m_{\tilde{l}}^2} \lesssim 10^{-3} \left(\frac{m_{\tilde{l}}}{100GeV}\right)^2. \tag{5}$$

There are several attempts to overcome this problem. One popular assumption is that SUSY breaking mass terms are generated through the coupling of supergravity in a flavor-blind way. In the minimal supergravity model all the scalar fields are assumed to be universal at the Planck scale and therefore there are no FCNC and LFV at this scale. Since physical squark and slepton masses are defined at the weak scale we have to evaluate these mass matrices using the renormalization group equations (RGE) with initial conditions at the Planck scale. As discussed later the squarks for the first and second generations can be highly degenerate at the weak scale in this model and the $K^0 - \bar{K}^0$ mixing from the SUSY loop can be suppressed. Another idea is so called gauge mediated SUSY breaking model.[9] In this case SUSY is assumed to be dynamically broken in a separate sector and the SUSY breaking effect is transmitted to the MSSM sector by loop effects of extra vector-like quarks and leptons. Since this transmission is based on gauge interaction the squarks and sleptons with the same gauge quantum numbers receive the same magnitude of the SUSY breaking effect. Thus the degeneracy necessary to avoid the FCNC problem is guaranteed in this scenario. A general consequence of this scenario is that unlike the gravity-mediated case gravitino becomes LSP. The strategies of SUSY particle search is also quite different in this scenario which is performed in recent collider experiments.[10]

3 Higgs Sector in the MSSM

3.1 The SM Higgs sector

Study of the Higgs sector is not only important to establish the SM but also crucial to search for physics beyond the SM. In this respect the mass of the Higgs boson itself is the most important information. For example, a heavy Higgs boson suggests that the dynamics of the electroweak symmetry breaking is governed by strong interaction. If we assume that fundamental interactions are described by perturbation theory up to the Planck scale or a scale close to it, the Higgs boson is expected to exist below 200 GeV. GUT and SUSY extension of the SM are examples of the latter case.

This situation is more clearly shown with a help of the SM RGE. In the minimal SM the Higgs-boson mass (m_h) and the Higgs-boson self-coupling constant (λ) from the Higgs potential ($V_{Higgs} = m^2|\Phi|^2 + \lambda|\Phi|^4$, ($m^2 \leq 0$)) are related by $m_h^2 = 2\lambda v^2$, where $v(= 246$ GeV) is the vacuum expectation value. Neglecting Yukawa coupling constants except for the one for the top quark, y_t, the RGE for the running self-coupling constant λ at the one-loop level is

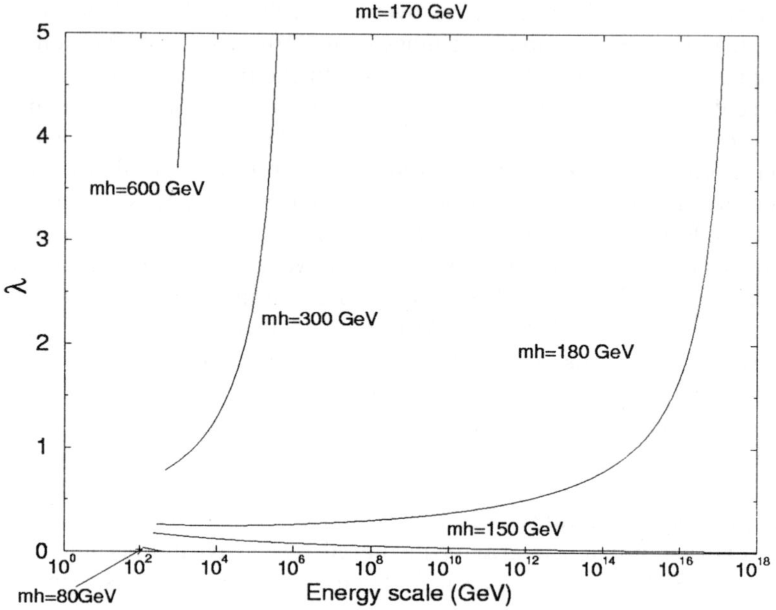

Figure 1: The flow of the Higgs self-coupling constant (λ) in the minimal SM for the several values of the Higgs boson masses. The top quark mass is assumed to be 170 GeV.

written as

$$\frac{d\lambda}{dt} = \frac{1}{(4\pi)^2}\{24\lambda^2 + 12y_t^2\lambda - 6y_t^4 - 12A\lambda + 6B\}, \tag{6}$$

where $A = \frac{1}{4}g_1^2 + \frac{3}{4}g_2^2$, $B = \frac{1}{16}g_1^4 + \frac{1}{8}g_1^2 g_2^2 + \frac{3}{16}g_2^4$ for $U(1)$ and $SU(2)$ gauge coupling constants g_1 and g_2 respectively and $t = \ln\mu$ (μ is the renormalization scale). Using the gauge coupling constants and the top Yukawa coupling constant at the electroweak scale we can draw the flow of the running self-coupling constant for each value of m_h as shown in Figure 1.

From this figure it is clear that possible scenarios for new physics are different for two cases, *i.e.* (i) $m_h \gg m_{top}$ and (ii) $m_h \lesssim m_{top}$. In case (i) the coupling constant becomes very large at a relatively low energy scale, and therefore new physics is indicated well below the Planck scale. On the other hand, the theory can be weakly coupled up to approximately the Planck scale in case (ii), which is consistent with the idea of grand unification. Of course the flow of the coupling constants is different if we change the particle content in the low-energy theory, but the upper bound on the Higgs mass is believed to be about 200 GeV for most GUT models. This is also true for the SUSY GUT. As for the MSSM, however, a stronger bound on the Higgs mass is obtained independently of the assumption of grand unification.

3.2 The Higgs Sector in the MSSM

The most important feature in the MSSM Higgs sector is that the Higgs-self-coupling constant at the tree level is completely determined by the $SU(2)$ and $U(1)$ gauge coupling constants. After electroweak symmetry breaking, the physical Higgs states include two CP-even Higgs bosons (h, H), one CP-odd Higgs boson (A) and one pair of charged Higgs bosons $(H^{\pm})$ where we denote by h and H the lighter and heavier Higgs bosons, respectively. Although at the tree level the upper bound on the lightest CP-even Higgs boson mass is given by the Z^0 mass, the radiative corrections can raise this bound.[11] The Higgs potential is given by

$$\begin{aligned} V_{Higgs} &= m_1^2|H_1|^2 + m_2^2|H_2|^2 - m_3^2(H_1 \cdot H_2 + \bar{H}_1 \cdot \bar{H}_2) \\ &+\frac{g_2^2}{8}(\bar{H}_1\tau^a H_1 + \bar{H}_2\tau^a H_2)^2 + \frac{g_1^2}{8}(|H_1|^2 - |H_2|^2)^2 \\ &+\Delta V, \end{aligned} \tag{7}$$

where ΔV represents the contribution from one-loop diagrams. Since the loop correction due to the top quark and its superpartner, the stop squark, is proportional to the fourth power of the top Yukawa coupling constant the Higgs self-coupling constant is no longer determined only by the gauge coupling constants. The upper bound on the lightest CP-even Higgs mass (m_h) can significantly increase for a reasonable choice of the top-quark and stop-squark masses. Taking account of only one-loop effects of top-quark and stop-squark, the upper bound is given by

$$m_h^2 \leq m_Z^2 \cos^2 2\beta + \frac{6}{(2\pi)^2}\frac{m_t^4}{v^2}\ln\frac{m_{stop}^2}{m_t^2}, \tag{8}$$

where β is the angle determined by the ratio of two Higgs vacuum expectation values ($\tan\beta = \frac{<H_2^0>}{<H_1^0>}$). Figure 2 shows the upper bound on m_h as a function of top-quark mass for several choices of the stop mass and $\tan\beta$. We can see that, in the MSSM, at least one neutral Higgs-boson should exist below 130 - 150 GeV depending on the top and stop masses.

The physical reason for the increase of the upper bound can be easily understood if we take the SUSY scale much larger than the weak scale.[12,13]

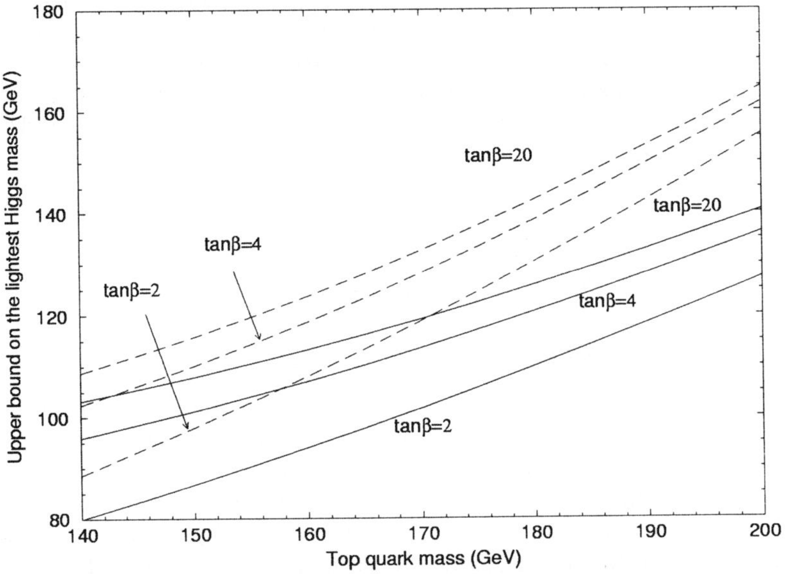

Figure 2: The upper bound on the lightest CP-even Higgs mass in the MSSM as a function of the top quark mass for various $\tan\beta$ and two large stop mass scales. The solid (dashed) line corresponds to m_{stop}=1 (10) TeV without left-right mixing of two stop states. These masses are calculated by the method with the renormalization group equation.[12]

Suppose, for example, that all the SUSY particles as well as physical Higgs states other than the lightest CP-even Higgs boson exist around 1 TeV. Then the effective theory which describes physics below 1 TeV scale is just the minimal SM with one Higgs doublet. On the other hand at the energy scale above 1 TeV the MSSM Lagrangian should be recovered. Information of the SUSY theory can be included in the matching condition between two theories at the 1 TeV scale. Because SUSY relation should be a good approximation above 1 TeV the Higgs self-coupling constant λ in the effective theory is given by $\lambda = \frac{1}{8}(g_1^2 + g_2^2)\cos^2 2\beta$ at 1 TeV. The value of λ changes according to the SM SM RGE below 1 TeV and the Higgs boson mass is determined by λ at the weak scale. Because of the y_t^4 term in Eq.(6) λ becomes larger toward low energy and this results in the increase of the Higgs boson mass. Thus the logarithmic dependence to the stop mass in Eq. (8) can be understood as a renormalization effect between the SUSY scale and the top scale and the increase of the upper bound can be interpreted as a correction to the tree-level SUSY relation due to the mass difference of top quark and stop squark.

Other Higgs states, namely the $H, A, H^\pm$, are also important to clarify the structure of the model. Their existence alone is a proof of new physics beyond the SM, but we may be able to distinguish the MSSM from a general two-Higgs model through the investigation of their masses and couplings. In the MSSM the Higgs sector is essentially determined by three independent parameters if we neglect mass difference and mixing of left- and right-handed stop states. For the free parameters we can take the mass of the CP-odd Higgs boson (m_A), $\tan\beta$, the stop mass (m_{stop}). Note that the stop masses enter through radiative corrections to the Higgs potential. In Figure 3, the masses for the H, A, and $H^\pm$ are shown as a function of m_A for several choices of $\tan\beta$ and m_{stop}=1 TeV. We can see that, in the limit of $m_A \to \infty$, m_h approaches a constant value which corresponds to the upper bound in Figure 1. Also in this limit the H, A and $H^\pm$ become degenerate in mass.

The neutral Higgs-boson couplings to gauge bosons and fermions are determined by $\tan\beta$ and the mixing angle α of the two CP-even Higgs particles defined as

$$
\begin{aligned}
ReH_1^0 &= \frac{1}{\sqrt{2}}(v\cos\beta - h\sin\alpha + H\cos\alpha) \\
ReH_2^0 &= \frac{1}{\sqrt{2}}(v\sin\beta + h\cos\alpha + H\sin\alpha). \qquad (9)
\end{aligned}
$$

In the MSSM α is a function of above three independent parameters. The various coupling constants and therefore properties of Higgs bosons are specified by these two angles. Let us consider the Higgs-boson production in e^+e^- linear

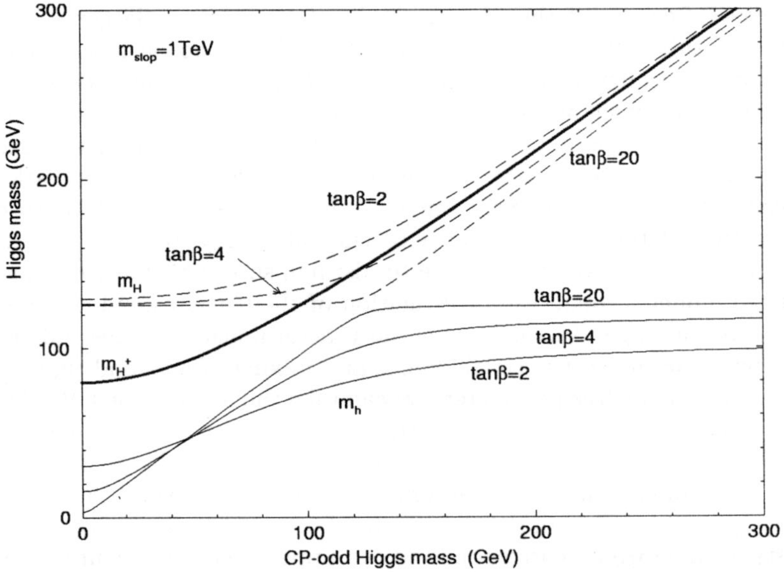

Figure 3: The light (h), heavy (H) CP-even Higgs masses and the charged Higgs ($H^{\pm}$) mass as a function of the CP-odd Higgs (A) mass. The top and stop masses are taken as $m_t = 170$ GeV and $m_{stop} = 1$ TeV.

collider experiments. In this case the Higgs-bremsstrahlung process $e^+e^- \to Zh$ or ZH and the associated production $e^+e^- \to Ah$ or AH play complimentary roles. Namely $e^+e^- \to Zh$ (ZH) is proportional to $\cos(\beta-\alpha)(\sin(\beta-\alpha))$, and $e^+e^- \to Ah$ (AH) is proportional to $\sin(\beta-\alpha)(\cos(\beta-\alpha))$, so at least one of the two processes has a sizable coupling. It is useful to distinguish the following two cases, namely (i) $m_A \lesssim 150$ GeV, (ii) $m_A \gg 150$ GeV. In case (i), the two CP-even Higgs bosons can have large mixing, and therefore the properties of the neutral Higgs boson can be substantially different from those of the minimal SM Higgs. On the other hand, in case (ii), the lightest CP-even Higgs becomes a SM-like Higgs, and the other four states, $H, A, H^\pm$ behave as a Higgs doublet orthogonal to the SM-like Higgs doublet. In this region, $\cos(\beta-\alpha)$ approaches unity and $\sin(\beta-\alpha)$ goes to zero so that $e^+e^- \to Zh$ and $e^+e^- \to AH$ are the dominant production processes.

Scenarios for the Higgs physics at a future e^+e^- linear collider are different for two cases. In case (i) it is possible to discover all Higgs states with $\sqrt{s} = 500$ GeV, and the production cross-section of the lightest Higgs boson may be quite different from that of the SM so that it may be clear that the discovered Higgs is not the SM Higgs. On the other hand, in case (ii), only the lightest Higgs may be discovered at the earlier stage of the e^+e^- experiment, and we have to go to a higher energy machine to find the heavier Higgs bosons. Also, since the properties of the lightest Higgs boson may be quite similar to those of the SM Higgs boson we need precision experiments on the production and decay of the particle in order to investigate possible deviations from the SM.

3.3 The Higgs sector in extended versions of the SUSY SM

Although the MSSM is the most widely studied model, there are several extensions of the SUSY version of the SM. If we focus on the structure of the Higgs sector, the MSSM is special because the Higgs self-couplings at the tree level are completely determined by the gauge coupling constants. It is therefore important to know how the Higgs phenomenology is different for models other than the MSSM.

A model with a gauge-singlet Higgs field is the simplest extension.[14] This model does not destroy the unification of the three gauge coupling constants since extra particles do not carry the SM quantum numbers. Moreover, we can include a term $W_\lambda = \lambda N H_1 H_2$ in the superpotential where N is a gauge singlet superfield. Since this term induces $\lambda^2|H_1 H_2|^2$ in the Higgs potential, the tree-level Higgs-boson self-coupling depends on λ as well as the gauge coupling constants. Including the one loop correction the upper bound of the lightest CP-even Higgs-boson mass is given by

$$m_h^2 \leq \frac{\lambda^2 v^2}{2} \sin^2 2\beta + m_Z^2 \cos^2 2\beta + \frac{6}{(2\pi)^2} \frac{m_t^4}{v^2} \ln \frac{m_{stop}^2}{m_t^2}. \tag{10}$$

There is no definite upper-bound on the lightest CP-even Higgs-boson mass in this model unless a further assumption on the strength of the coupling λ is made. If we require all dimensionless coupling constants to remain perturbative up to the GUT scale we can calculate the upper-bound of the lightest CP-even Higgs-boson mass.[15] Numerical analysis shows that the upper bound is about 130 GeV in case that the stop mass is 1 TeV and the top mass is 170 to 180 GeV. This is not very much different from the MSSM case because for these large top masses we can show that the coupling λ cannot be so large at the weak scale as a result of the RGE analysis.

This result means that the lightest Higgs boson is at least kinematically accessible at an e^+e^- linear collider with $\sqrt{s} \sim 300 - 500$ GeV. This does not necessarily mean that the lightest Higgs boson is detectable. In this model the lightest Higgs boson is composed of one gauge singlet and two doublets, and if it is singlet-dominated its couplings to the gauge bosons are significantly reduced, hence its production cross-section is too small. In such a case the heavier neutral Higgs bosons may be detectable since these bosons have a large enough coupling to gauge bosons. In fact we can put an upper-bound on the mass of the heavier Higgs boson when the lightest one becomes singlet-dominated. By quantitative study of the masses and the production cross-section of the Higgs bosons in this model, we can show that at least one of the three CP-even Higgs bosons has a large enough production cross-section in the $e^+e^- \to Zh_i^0$ $(i = 1, 2, 3)$ process to be detected at an e^+e^- linear collider with $\sqrt{s} \sim 300 - 500$ GeV.[16] For this purpose we define the minimal production cross-section, σ_{min}, as a function of $\sqrt{s}$ such that at least one of these three h_i^0 has a larger production cross section than σ_{min} irrespective of the parameters in the Higgs mass matrix. The σ_{min} turns to be given by one third of the SM production cross-section with the Higgs boson mass equal to the upper-bound value. In Figure 4 we show σ_{min} as a function of $\sqrt{s}$ for $m_{stop} = 1$ TeV . From this figure we can conclude that the discovery of at least one neutral Higgs boson is guaranteed at an e^+e^- linear collider with an integrated luminosity of 10 fb^{-1}.

If we further extend the model we can increase the upper bound of the lightest CP even Higgs boson mass. One such possibility is to include extra matter fields such as extra families. Then loop effect of extra matter fields can raise the upper bound if Yukawa coupling constants associated to the extra families are large. If we require that all the dimensionless coupling constants

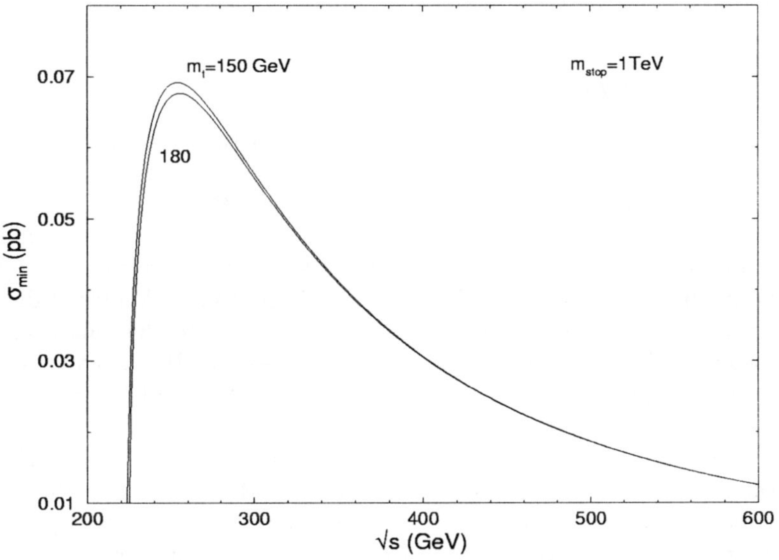

Figure 4: Minimal production cross section, σ_{min}, for the SUSY SM with a gauge singlet Higgs for the top mass m_t=150 and 180 GeV and m_{stop}=1 TeV.

remain perturbative up to the GUT scale we can restrict possible content of extra matter fields and also the magnitude of relevant Yukawa coupling constants. This kind of analysis was performed for models with extra matter multiplets which keep the SU(5)gauge coupling unification.[17] It was shown that the upper bound can be as large as 180 GeV for 1 TeV SUSY scale in the model with extra $\mathbf{10}+\bar{\mathbf{10}}$ or $\mathbf{10}+\bar{\mathbf{5}}+\bar{\mathbf{10}}+\mathbf{5}$ representations of SU(5) gauge group.

Another possibility is to consider triplet Higgs fields instead of a gauge singlet field.[18] A analysis similar to the model with a gauge singlet Higgs field leads to the maximum value of the lightest CP-even Higgs boson mass is about 180 GeV if we require that all the dimensionless coupling constants remain perturbative below the GUT scale.

3.4 Phenomenology of MSSM Higgs searches

The Higgs boson in the intermediate mass region such as the lightest Higgs boson in the MSSM is a target of the future collider experiments both at LHC and e^+e^- linear colliders. In the experiment at LEP II the SM Higgs boson is expected to be discovered if its mass is below 105 GeV and the upgraded Tevatron experiment also cover a similar mass region. Since the upper bound exceeds the discovery limit of LEP II it is very important to know the discovery potential of the SUSY Higgs in the LHC experiment. In this mass range the main decay mode of the SM Higgs boson is $h \to b\bar{b}$. But because of QCD backgrounds of this mode the LHC Higgs boson search mainly relies on the two photon mode whose branching ratio is $O(10^{-3})$. In the SUSY case its branching ratio can be even smaller. However in some parameter space, especially for large $\tan\beta$, $H, A, h \to \tau^+\tau^-$ modes are useful. Recent study shows that it is possible to observe at least one mode of the MSSM Higgs signals in all parameter space after several years of running at high luminosity.[19]

On the other hand an e^+e^- linear collider with $\sqrt{s} \sim 300 - 500$ GeV is a suitable place to study the Higgs boson in this mass region. Here we can not only discover the Higgs boson easily but also measure various quantities, *i.e.* production cross sections and branching ratios related to the Higgs boson.[20,21,22,23] These measurements are very important to clarify nature of the discovered Higgs boson and distinguish the SM Higgs boson from Higgs particles associated with some extensions of the SM like the MSSM. In Figure 5 the parameter region of the MSSM is shown where some deviation from the SM is expected to be observable at e^+e^- linear-collider. In this figure we take $m_{stop} = 1$ TeV and $m_t = 175$ GeV. There are three lines in this figure which corresponds to: (1)The total width of the lightest CP even Higgs boson mass is larger than 200 MeV which is expected to be measurable using leptonic decay

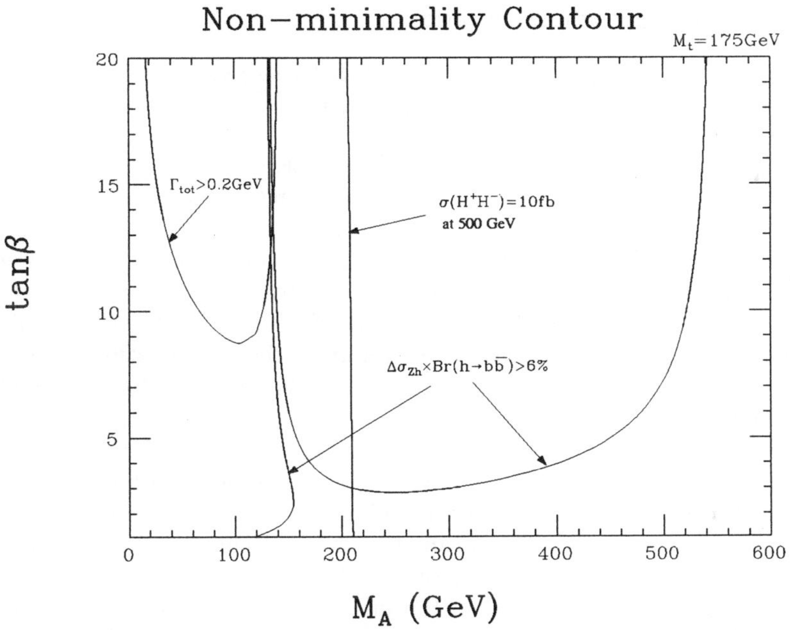

Figure 5: Parameter space in the MSSM where deviation from the SM is expected to be observable at e^+e^- linear-collider. Three lines corresponds to: (1) The total width of the lightest CP even Higgs boson mass is larger, (2) $\sigma(e^+e^- \to Zh)Br(h \to b\bar{b})$ deviates from the SM value more than 6% for $\sqrt{s} = 300$ GeV, (3) $\sigma(e^+e^- \to H^+H^-) > 10$ fb for $\sqrt{s} = 500$ GeV. This figure is provided by A. Miyamoto.

mode of Z boson in the $e^+e^- \to Zh$ reaction. (2) The number of $e^+e^- \to Zh$, $h \to b\bar{b}$ events deviates from the SM prediction more than 6% for e^+e^- linear-collider with $\sqrt{s} = 300$ GeV. (3) The discovery limit of charged Higgs pair production at e^+e^- linear-collider with $\sqrt{s} = 500$ GeV. We can see that there are sizable deviation from the SM in a large part of the parameter space by measuring the production cross multiplied by the $h \to b\bar{b}$ branching ratio.

Another important information on the MSSM parameter is obtained measuring various branching ratio of the Higgs boson. We show in particular the ratio of the following branching ratios,

$$R_{br} \equiv \frac{Br(h \to c\bar{c}) + Br(h \to gg)}{Br(h \to b\bar{b})}, \tag{11}$$

is useful to constrain the heavy Higgs mass scale if we assume that only one CP-even Higgs boson is discovered at the first stage of the e^+e^- linear-collider experiment [24].

If we assume that the lightest CP-even Higgs boson mass (m_h) is precisely known we can solve for one of the three free parameters m_A, $\tan\beta$ and m_{stop} in terms of the other parameters. For example, if we solve for $\tan\beta$ the unknown parameters become m_A and m_{stop}. In Figure 6 the above ratio R_{br} is shown as a function m_A for different values of m_{stop}. We can see that R_{br} is almost independent of m_{stop}.

This simple dependence can be explained in the following way. In the MSSM, each of the two Higgs doublets couples to either up-type or down-type quarks. Thus the ratio of the Higgs couplings to up-type quarks and to down-type quarks is sensitive to the parameters of the Higgs sector, *i.e.* the angles α and β. Since the gluonic width of the Higgs boson is also generated by a one-loop diagram with an internal top-quark, the Higgs-gluon-gluon coupling is essentially proportional to the Higgs-top coupling. Thus R_{br} is proportional to square of the ratio of the up-type and down-type Yukawa coupling constants which depend on two angles as $(\tan\alpha\tan\beta)^{-2}$. By examining the neutral Higgs boson mass matrix in the MSSM at one loop level we can derive approximate relation for R_{br} in the MSSM, normalized by R_{br} in the SM as

$$\frac{R_{br}(MSSM)}{R_{br}(SM)} \approx \left(\frac{m_h^2 - m_A^2}{m_Z^2 + m_A^2}\right)^2 \tag{12}$$

for $m_A \gg m_h \sim m_Z$. This is actually a very good approximation so that R_{br} is almost independent of m_{stop}.

Measuring this quantity to a good accuracy is therefore important for constraining the scale of the heavy Higgs mass and in choosing the beam energy

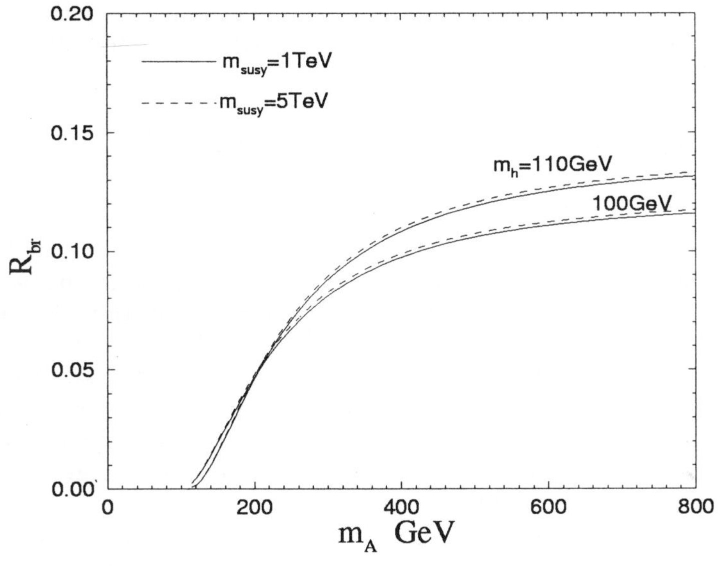

Figure 6: $R_{br} \equiv \frac{(Br(h \to c\bar{c}) + Br(h \to gg))}{Br(h \to b\bar{b})}$ as a function of m_A for $m_{susy} = 1, 5$ TeV and $m_h = 100, 110$ GeV.[24] The following parameters are used for the calculation of the branching ratios: $m_t = 170$ GeV, $\bar{m}_c(m_c) = 1.2$ GeV, $\bar{m}_b(m_b) = 4.2$ GeV, $\alpha_s(m_Z) = 0.12$.

of the e^+e^- linear-collider at th second stage when only one Higgs boson is discovered at the first stage. Note that R_{br} approaches the SM value rather slowly in the large m_A limit. We can see that R_{br} is reduced by 20% even for $m_A = 400$ GeV. By simulation study for e^+e^- linear collider experiments it is shown that the sum of the charm and gluonic branching ratios can be determined reasonably well. The statistical error in the determination of R_{br} after two years at an e^+e^- linear collider with $\sqrt{s} = 300$ GeV is 17%.[25] We also need to know the theoretical ambiguity of the calculation of the branching ratios in $h \to b\bar{b}, c\bar{c}, gg$. At the moment the theoretical error in the calculation of R_{br} is estimated to be rather large ($\sim 20\%$) due to uncertainties in α_s and m_c.[24,26] But these uncertainties can be reduced in future from both theoretical and experimental improvements.

4 B and K decay in the supergravity model

4.1 *Flavor mixing in the supergravity model*

In the minimal SM various FCNC processes and CP violation in B and K decays are determined by the CKM matrix. The constraints on the parameters in the CKM matrix elements V_{ij} can be conveniently expressed in terms of the unitarity triangle which is based on the unitarity relation $V_{ud}V_{ub}^* + V_{cd}V_{cb}^* + V_{td}V_{tb}^* = 0$ in the complex plane. With CP violation at B factory as well as rare K decay experiments we will be able to check consistency of the unitarity triangle and at the same time search for effects of physics beyond the SM. In order to distinguish possible new physics effects it is important to identify how various models can modify the SM predictions.

As we discussed in section 2.2 the squark and the slepton mass matrices become new sources of flavor mixings in the SUSY model and generic mass matrices would induce too large FCNC and LFV effects if the superpartners' masses are in the 100 GeV region. In the SUSY model based on the supergravity these flavor problems can be avoided by setting SUSY breaking mass terms universal at the very high energy scale. In fact all the scalar fields are assumed to have the same SUSY breaking mass at the Planck scale in the minimal supergravity model and therefore there are no FCNC and LFV at this scale. Physical squark and slepton masses defined at the weak scale are determined through RGE. General consequences are: (i) Squarks for the first and second generations remain highly degenerate so that the constraint from the $K^0 - \bar{K}^0$ mixing can be safely satisfied. (ii) Due to the effect of large top Yukawa coupling constant the stop and the sbottom can be significantly lighter than other squarks. This will induce sizable contributions to FCNC processes such as $b \to s\gamma$[27,28], $b \to sl^+l^-$ [27,29], ΔM_B[27,30], ϵ_K[30] and $K \to \pi\nu\bar{\nu}$. (iii) In the SUSY

GUT the large top Yukawa coupling constant also induces the flavor mixing in the slepton sector so that LFV processes such as $\mu^+ \to e^+\gamma$, $\mu^+ \to e^+e^+e^-$ and $\mu^- - e^-$ conversion in atoms receive large SUSY contributions [31]. In this section we deal with the processes listed in (ii) following recent update of the calculation [32] and the LFV processes are discussed in the next section.

In the minimal supergravity model we can introduce four new complex phases in SUSY breaking terms and the μ term, of which only two are physically independent phases. These phases in general induce too large electron and neutron EDMs if the phase is $O(1)$ and the squark and slepton masses are a few hundred GeV.[33] Here we assume that all these parameters are real so that the source of the CP violation is only in the CKM matrix element. In this case we can show by evaluating the RGE that the SUSY loop contributions to FCNC amplitudes approximately have the same dependence on the CKM elements as the SM contributions. In particular, the complex phase of the $B^0 - \bar{B}^0$ mixing amplitude does not change even if we take into account the SUSY and the charged Higgs loop contributions. In terms of the unitarity triangle this means that the angle measurements through CP asymmetry in B decays determine the CKM matrix elements as in the SM case. On the other hand the length of the unitarity triangle determined from ΔM_B and ϵ_K can be modified.

If we consider the supergravity model with SUSY CP phases we have to take into account the constraints from EDM explicitly. Even in such case it is shown that the the phase of the $B^0 - \bar{B}^0$ mixing amplitude cannot be much different from the SM amplitude.[34]

There is important difference between two classes of the processes in constraining the SUSY parameter space, namely,(i) $b \to s\gamma$ and $b \to sl^+l^-$ and (ii) ΔM_B, ϵ_K and $K \to \pi\nu\bar{\nu}$. While the class (i) processes only depend on CKM parameters which is already well known, the class (ii) processes depend on V_{td} element or (ρ, η) in the Wolfenstein parameterization. Present constraint on (ρ, η) which is independent of ΔM_B and ϵ_K only comes from the $b \to u$ transition, thus there are still large uncertainty. While the class (i) processes, especially the $b \to s\gamma$ process, already provide important constraint on possible SUSY loop effect, the class (ii) processes become useful constraint after new information on the unitarity triangle is obtained at B factories.

4.2 $b \to s\gamma$, $b \to sl^+l^-$

First we discuss the branching ratios of $b \to s\gamma$, $b \to sl^+l^-$ in supergravity model.[32] In the calculation in this and the next subsection we have used updated results of various SUSY search experiments at LEP2 and Tevatron as well as the next-to-leading QCD corrections in the calculation of the $b \to s\gamma$ branching ratio. We calculated the SUSY particle spectrum based on two different assumptions on the initial conditions of RGE. The minimal case corresponds to the minimal supergravity where all scalar fields have a common SUSY breaking mass at the GUT scale. In the second case shown as "all" in the figures we enlarge the SUSY parameter space by relaxing the initial conditions for the SUSY breaking parameters, namely all squarks and sleptons have a common SUSY breaking mass whereas an independent SUSY breaking parameter is assigned for Higgs fields.

The Br($b \to s\gamma$) and Br($b \to sl^+l^-$) can be calculated using the weak effective Hamiltonian

$$H_{eff} = \sum_{i=1}^{10} C_i O_i + h.c.. \tag{13}$$

The relevant operator for the calculation of $b \to s\gamma$ process is O_7 and the $b \to sl^+l^-$ process also depends on O_9 and O_{10}. These operators are defined as

$$O_7 = \frac{e}{16\pi^2} m_b (\bar{s}_L \sigma_{\mu\nu} b_R) F^{\mu\nu}, \tag{14}$$

$$O_9 = \frac{e^2}{16\pi^2} (\bar{s}_L \gamma_\mu b_L)(\bar{l}_L \gamma^\mu l_L), \tag{15}$$

$$O_{10} = \frac{e^2}{16\pi^2} (\bar{s}_L \gamma_\mu b_L)(\bar{l}_L \gamma^\mu \gamma_5 l_L). \tag{16}$$

The effect of SUSY particle and charged Higgs loop can be taken into account in the calculation of the coefficient functions for these operators at the weak scale. By numerical calculation it was shown that only the coefficient of O_7 receives sizable correction in this model. Moreover, although the charged Higgs boson loop contribution has the same sign as the SM one the SUSY loop effect can interfere with the SM amplitude either destructively and constructively.

In Figure 7 the $b \to s\gamma$ branching ratio is shown as a function of charged Higgs mass for $\tan\beta = 2$. The dark points correspond to the minimal case and the light points to the enlarged parameter space. We can see that unlike the type II two Higgs doublet model the charged Higgs boson mass less than 400 GeV is allowed because of destructive interference. Since only the coefficient of O_7 is significantly affected by the SUSY contribution there is a strong correlation between the $b \to s\gamma$ and $b \to sl^+l^-$ branching ratios. Figure 8 shows this correlation for $\tan\beta = 30$. Here $b \to s\mu^+\mu^-$ branching ratio is integrated over lepton invariant spectrum below the J/ψ threshold to avoid the large resonance peak. We can see that there are two fold ambiguity which corresponds to the

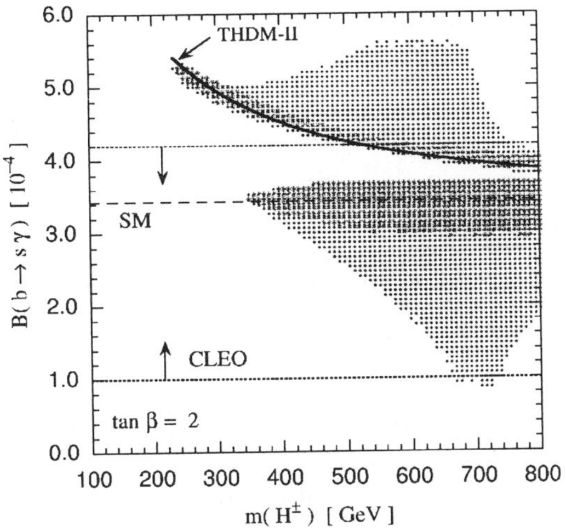

Figure 7: Br($b \to s\gamma$) in the supergravity model as a function of the charged Higgs boson mass for $\tan\beta$=2. The solid line shows the branching ratio in the two Higgs doublet model (Model II). The dark points represent the results for the minimal supergravity and the light points correspond to the extended parameter space where SUSY soft breaking parameters at the GUT scale are different for squark/slepton and Higgs fields. Also the SM prediction and the CLEO 95 bound [35] are shown.

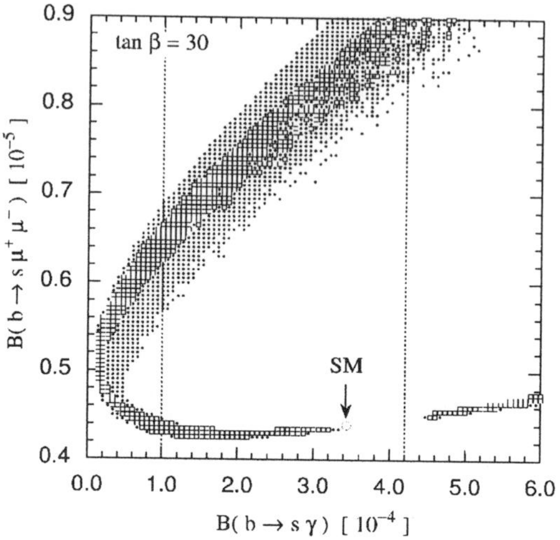

Figure 8: Correlation between $\mathrm{Br}(b \to s\gamma)$ and $\mathrm{Br}(b \to sl^+l^-)$ in the supergravity model for $\tan\beta = 30$. Here $\mathrm{Br}(b \to s\mu^+\mu^-)$ is obtained by integrating in the range $2m_\mu < \sqrt{s} < m_{J/\psi} - 100$ MeV where $\sqrt{s}$ is the invariant mass of $\mu^+\mu^-$ pair. The vertical lines correspond to the CLEO 95 upper and lower bounds.

sign of the coefficient of O_7. When the SUSY and charged Higgs contribution is (-2) times the SM contribution the $b \to s\gamma$ branching ratio is the same as the SM while the $b \to sl^+l^-$ branching ratio can be enhanced by about a factor of two. In the supergravity model this situation arises only for large $\tan\beta$ case. The lepton forward and backward asymmetry can also show sizable deviation from the SM in this case.

4.3 ΔM_B, ϵ_K, $K_L \to \pi^0\nu\bar{\nu}$

We have calculated the ΔM_B, ϵ_K and branching ratios of $K_L \to \pi^0\nu\bar{\nu}$ and $K^+ \to \pi^+\nu\bar{\nu}$ in the SUSY model based on supergravity[32]. In Figures 9 and 10 we present ΔM_{B_d} and $\mathrm{Br}(K_L \to \pi^0\nu\bar{\nu})$ in the present model normalized by the same quantities calculated in the SM as the function of the $b \to s\gamma$ branching ratio. Note that these ratios are essentially independent of the CKM parameters because the SUSY and the charged Higgs boson loop contributions have the same dependence on the CKM parameters. Although we only present the results for ΔM_{B_d} and $\mathrm{Br}(K_L \to \pi^0\nu\bar{\nu})$, ϵ_K and $\mathrm{Br}(K^+ \to \pi^+\nu\bar{\nu})$ provide the same constraints on the SUSY parameters respectively because these quantities are almost equal if normalized by corresponding quantities in the SM.

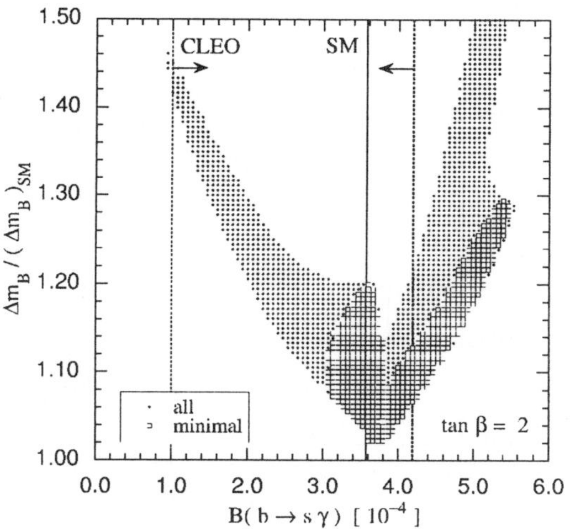

Figure 9: ΔM_{B_d} normalized by the SM value for $\tan\beta = 2$ as a function of $b \to s\gamma$ branching ratio. The dark (light) points correspond to the minimal (enlarged) parameter space of the supergravity model. The vertical lines correspond to the CLEO 95 upper and lower bounds.

From Figures 9 and 10 we can conclude that the ΔM_{B_d} (and ϵ_K) can be enhanced by up to 40% and $\mathrm{Br}(K_L \to \pi^0 \nu\bar{\nu})$ (and $\mathrm{Br}(K_+ \to \pi^+ \nu\bar{\nu})$)is suppressed by up to 10% for extended parameter space and the corresponding numbers for the minimal case are 20% and 3%. The ratio of two Higgs vacuum expectation value, $\tan\beta$, is 2 for these figures and the deviation from the SM turns out to be smaller for large value of $\tan\beta$.

These deviations may be evident in future when B factory experiments provide additional information on the CKM parameters. It is expected that the one of the three angles of the unitarity triangle is determined well through the $B \to J/\psi K_S$ mode. Then assuming the SM, one more physical observable can determine the CKM parameters or (ρ, η) in the Wolfenstein parameterization. New physics effects may appear as inconsistency in the determination of these parameters from different inputs. For example, the ρ and η parameters determined from CP asymmetry of B decay in other modes, $\frac{\Delta M_{B_s}}{\Delta M_{B_d}}$ and $|V_{ub}|$ can be considerably different from those determined through ΔM_{B_d} ϵ_K and $\mathrm{Br}(K \to \pi\nu\bar{\nu})$ because ΔM_{B_d} ϵ_K are enhanced and $\mathrm{Br}(K \to \pi\nu\bar{\nu})$'s are suppressed in the present model. It turns out that there is strong correlation between deviation of two quantities, namely the latters are most suppressed where the formers have the largest deviation from the SM. The pattern of these

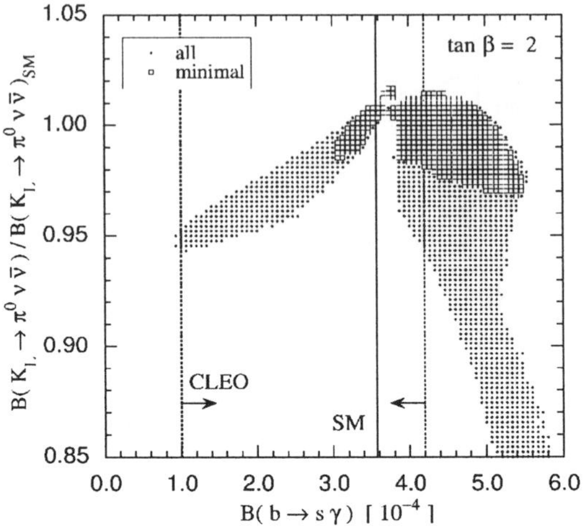

Figure 10: $\mathrm{Br}(K_L \to \pi^0 \nu\bar{\nu})$ normalized by the SM value for $\tan\beta = 2$ as a function of $b \to s\gamma$ branching ratio.

deviations from the SM will be a key to distinguish various new physics effects. We also note from Figure 9 and 10 that, although the new results reported at ICHEP98 ($2.0 \times 10^{-4} < \mathrm{Br}(b \to s\gamma) < 4.5 \times 10^{-4}$)[36] does not change the situation very much, future improvement on the $b \to s\gamma$ branching will give great impacts on constraining the size of possible deviation from the SM in FCNC processes.

5 LFV in the SUSY GUT

5.1 $\mu^+ \to e^+\gamma$ in SUSY GUT

In the minimal SM with massless neutrinos the lepton number is conserved separately for each generation. This is not necessarily true if we consider physics beyond the SM. LFV can easily occur in many extension of the SM. Experimentally, LFV searches are continued in various processes and especially strong bounds are obtained for muon processes such as $\mu^+ \to e^+\gamma$, $\mu^+ \to e^+e^+e^-$ and $\mu^- - e^-$ conversion in atoms. The experimental upper bound on these processes quoted in PDG 98 are $\mathrm{Br}(\mu^+ \to e^+\gamma) \le 4.9 \times 10^{-11}$, $\mathrm{Br}(\mu^+ \to e^+e^+e^-) \le 1.0 \times 10^{-12}$ and $\frac{\sigma(\mu^- Ti \to e^- Ti)}{\sigma(\mu^- Ti \to caputure)} \le 4.3 \times 10^{-12}$. Recently there are considerable interests on these processes because predicted branching ratios turn out to be close to the upper bounds in the SUSY GUT.[31]

As discussed in previous sections no LFV is generated at the Planck scale in the context of the minimal supergravity model. In the SUSY GUT scenario, however, the LFV can be induced through renormalization effects on slepton mass matrix because the GUT interaction breaks lepton flavor conservation. In the minimal SUSY SU(5) GUT, the effect of the large top Yukawa coupling constant results in the LFV in the right-handed slepton sector. This induces $\mu^- \to e_R^- \gamma$ $(\mu^+ \to e_L^+ \gamma)$ decay. The branching ratio can be calculated based on this model and it was pointed out that there is unfortunate cancellation between different diagrams so that $\text{Br}(\mu^+ \to e^+\gamma)$ is below 10^{-13} level for most of the parameter space.[37]

In the SO(10) SUSY GUT model, on the other hand, both left- and right-handed sleptons induce LFV because of different structure of Yukawa coupling constants at the GUT scale. More importantly, there are diagrams which enhance the branching ratio by a factor $(\frac{m_\tau}{m_\mu})^2$ compared to the minimal SU(5) SUSY GUT model and predicted branching ratio is at least larger by two order of magnitudes.[38]

A similar enhancement can occur also in the context of the SUSY SU(5) model for large value of $\tan\beta$ once we take into account effects of higher dimensional operators to explain realistic fermion masses.[39] In the minimal case the Yukawa coupling is given by the superpotential $W = (y_u)_{ij}\mathbf{T_i}\cdot\mathbf{T_j}\cdot\mathbf{H(5)} + (y_d)_{ij}\mathbf{T_i}\cdot\bar{\mathbf{F}}_\mathbf{j}\cdot\bar{\mathbf{H}}(\mathbf{5})$ where $\mathbf{T_i}$ is 10 dimensional and $\bar{\mathbf{F}}_\mathbf{j}$ is 5 dimensional representation of SU(5). This superpotential alone cannot explain the lepton and quark mass ratios for the first and second generations although the m_b/m_τ ratio is in reasonable agreement. One way to obtain realistic mass ratios are to introduce higher dimensional operators such as $\frac{f_{ij}}{M_{Planck}}\mathbf{\Sigma(24)}\cdot\mathbf{T_i}\cdot\bar{\mathbf{F}}_\mathbf{j}\cdot\bar{\mathbf{H}}(\mathbf{5})$. We investigated how inclusion of these terms changes prediction of the branching ratio. It turns out that the branching ratio is quite sensitive to the details of these higher dimensional operators. Firstly, once we include these terms the slepton mixing matrix elements $\lambda_\tau \equiv V_{\tilde{e}31}^* V_{\tilde{e}32}$ which appear in the formula of the $\mu^+ \to e^+\gamma$ amplitude is no longer related to the corresponding CKM matrix elements. More importantly, for large value of $\tan\beta$, the left-handed slepton also induces the LFV and the predicted branching ratio becomes enhanced by two order of magnitudes as in the SO(10) case.[40] The destructive interference among the different diagrams also disappear. We show one example of such calculation in Figure 11 where $\text{Br}(\mu^+ \to e^+\gamma)$ can be close to 10^{-11} level for large values of $\tan\beta$ in the non-minimal case.

Another example of possible large LFV is a SUSY model with the see-saw type neutrino mass.[41] If we include the right-handed neutrino supermultiplet at high energy scale, we can include large Yukawa coupling constant among lepton doublet, right-handed neutrino and Higgs doublet field. The renormalization

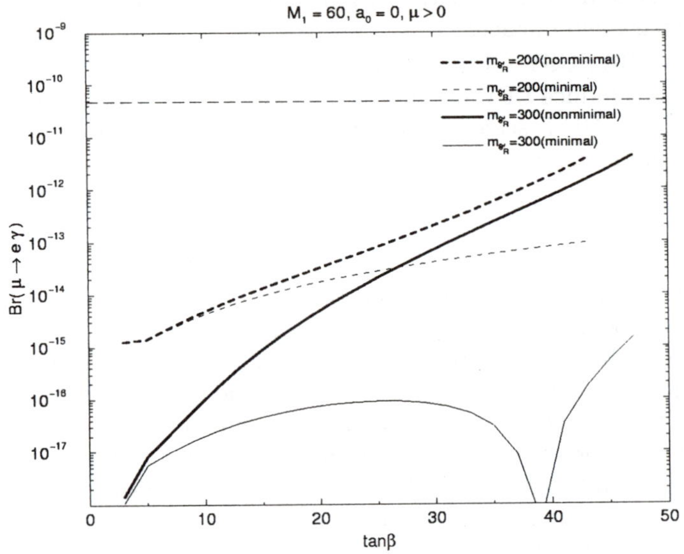

Figure 11: Dependence of the branching ratio of $\mu \to e\gamma$ on $\tan\beta$ for the right-handed selectron mass 200GeV (dashed lines) and 300GeV (solid lines). The thick lines are for the non-minimal case that $V_{\bar{e}}$ and V_l are the same as $V_{\rm KM}$, and the thin lines are for the minimal case in which $V_{\bar{e}} = V_{\rm KM}$ and $V_l = \mathbf{1}$. In this figure we choose the bino mass 60GeV, the coefficient of the A term a_0 =0, the higgsino mass positive. The long-dashed line is the experimental upper bound.

effect can induce the LFV in the left-handed slepton sector and in this case $\mu^- \to e_L^- \gamma$ ($\mu^+ \to e_R^+ \gamma$) decay can occur. The branching ratio depends on, among other things, the magnitude of the Yukawa coupling constant which is a function of the neutrino mass and the Majorana mass scale. Therefore even if we fix the neutrino mass motivated by atmospheric and solar neutrino problem the branching ratio strongly depends on the right-handed neutrino mass scale. If the Yukawa coupling constant is assumed to be as large as the top Yukawa coupling constant the predicted the branching ratio can reach the present experimental upper bound. Note that without SUSY neutrino mass can only induce negligible effect to the LFV processes for μ and τ decay. It is therefore interesting that the same source of LFV can be responsible for the neutrino oscillation and the $\mu \to e\gamma$ decay in the SUSY model.

5.2 LFV process with polarized muons

Finally we would like to discuss usefulness of the polarized muons in search for LFV. A highly polarized muon beam is available in μ^+ decay experiments. Muons from π^+ decay stopped near the surface of pion production target is 100% polarized opposite to the muon momentum and this muon is called surface muon.

The first obvious merit of polarized muons in $\mu^+ \to e^+\gamma$ is that we can distinguish $\mu^+ \to e_R^+\gamma$ and $\mu^+ \to e_L^+\gamma$ by the angular distribution of the decay products with respect to the muon polarization direction. For example, the positron from the $\mu^+ \to e_R^+\gamma$ decay follows the $(1 - P\cos\theta)$ distribution where θ is the angle between the polarization direction and the positron momentum and P is the the muon polarization. In the previous examples the SU(5) SUSY GUT for small $\tan\beta$ predicts $\mu^+ \to e_L^+\gamma$ because LFV is induced only in the right-handed slepton sector. On the other hand the SO(10) SUSY GUT generates both $\mu^+ \to e_L^+\gamma$ and $\mu^+ \to e_R^+\gamma$. If LFV is induced by the right-handed neutrino Yukawa coupling constant, only $\mu^+ \to e_R^+\gamma$ should be observed.

Polarized muons are also useful to suppress background processes for the $\mu^+ \to e^+\gamma$ search.[42] In this mode the experimental sensitivity is limited by appearance of the background processes. There are two major background processes. The first one is physics background process which is a tail of radiative muon decay. If neutrino pair carries out only little energy in the $\mu^+ \to e^+\nu\bar{\nu}\gamma$ process, we cannot distinguish this from the signal process. The second background process is an accidental background process where detections of 52.8 MeV positron and 52.8 MeV photon from different muon decays coincide within time and angular resolutions for selection of signals. The source of the 52.8

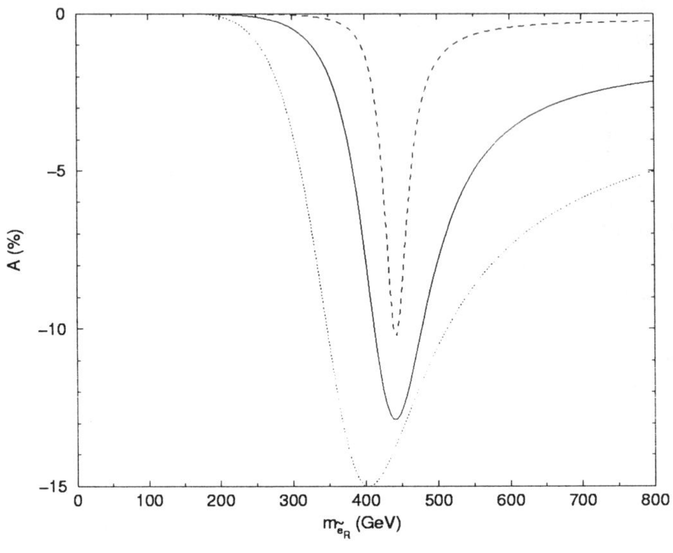

Figure 12: T-odd asymmetry **A** as a function of the right-handed selectron mass in $\mu \to 3e$ precess in the SUSY SU(5) GUT. We fix the SUSY parameters as $M_2 = 200$ GeV, $A_X = i$, $\mu > 0$ and $\tan\beta = 3$ (dotted line), 10 (solid line), 30 (dashed line) and top quark mass as 175 GeV.

MeV positron is the ordinary $\mu^+ \to e^+\nu\bar{\nu}$ decay whereas the 52.8 MeV photon mainly comes from a tail of the radiative muon decay. In the experiment with polarized muons angular distribution is useful to suppress both backgrounds. Suppression of the accidental background turns out to be more important in the next generation experiment. In this case both positrons and photons follow the $(1 + P\cos\theta)$ distribution. Since the signal is back-to-back the background suppression works independently of the signal distribution. If we use 97% polarized muons we can expect to reduce the accidental background by one order of magnitude. This looks promising for search of $\mu^+ \to e^+\gamma$ at the level of 10^{-14} branching ratio.

Last example of usefulness of polarized muons is that we can measure T violating asymmetry in the $\mu^+ \to e^+e^+e^-$ decay. Using polarization of initial muons we can define T odd triple vector correlation $< \vec{P} \cdot (\vec{p_1} \times \vec{p_2}) >$ where $\vec{P}$ is muon polarization and $\vec{p_1}$ and $\vec{p_1}$ are two independent momenta of decay particles.[43] We have investigated T odd asymmetry in the SU(5) SUSY GUT.[44] In order to have this asymmetry we need to introduce a CP violating phase other than the KM phase. This phase can be provided by the complex phases in the SUSY breaking terms, for example, the phase in the triple scalar coupling constant (A term). Since this phase also induces electron and neutron EDMs, we have calculates the T odd asymmetry in the $\mu^+ \to e^+e^+e^-$ taking into account EDM constraints. In Figure12 we show one example of such calculation for the T odd asymmetry. By examining SUSY parameter space we found that the asymmetry up to 20% is possible. The branching ratio for $\mu^+ \to e^+e^+e^-$ turns out to crucially depend on the slepton mixing element λ_τ which is an unknown parameter once we take into account the higher dimensional operators for the Yukawa coupling constants. For $\lambda_\tau = 10^{-2}$ we can show that the branching ratio of 10^{-14} is possible with 10% asymmetry which can be reached in future experiment with sensitivity of 10^{-16} level.

6 Conclusions

I have reviewed some aspects of the Higgs physics and flavor physics in SUSY theories. An important feature of SUSY models is existence of light Higgs boson which is a target of LEP II, Tevatron and LHC experiments. I have also shown that a future e^+e^- linear collider is an ideal place to study the SUSY Higgs sector. At earlier stage of the experiment with $\sqrt{s} \sim 300$ - 500 GeV, it is easy to find a light Higgs boson predicted in SUSY standard models. In particular, both in the MSSM and the SUSY SM with a gauge singlet Higgs, at least one of neutral Higgs bosons is detectable. More importantly, detailed study on properties of the Higgs boson is possible at an e^+e^- linear collider through measurements of various production cross-sections and branching ratios. As an example we show that the measurement of Higgs couplings to $c\bar{c}/gg/b\bar{b}$ gives us important information on the Higgs sector of the MSSM. Combining expected results at LHC we may be able to clarify the Higgs sector of the SM and explore physics beyond the SM such as the SUSY SM in future e^+e^- linear collider experiments.

We have considered various flavor changing processes in the supersymmetric standard model based on the supergravity. We have seen that the branching ratio of $b \to s\gamma$ is very sensitive to SUSY effects and further improved measurement of this process is important. Flavor changing neutral current processes in B and K decays such as $B^0 - \bar{B}^0$ mixing, ϵ_K and branching ratio of $K \to \pi\nu\bar{\nu}$ are calculated and it is shown that the deviation from the SM becomes as large as 40 % for $B^0 - \bar{B}^0$ mixing and ϵ_K but somewhat smaller for $K \to \pi\nu\bar{\nu}$ processes. We also investigated LFV in the SU(5) SUSY GUT. It is pointed out that the $\mu \to e\gamma$ branching ratio can be enhanced for large $\tan\beta$ if we take into account the higher dimensional operators in the Yukawa coupling constants at the GUT scale. The T odd triple vector correlation is also calculated for the $\mu^+ \to e^+e^+e^-$ process and it is shown that the asymmetry up to 20% is possible due to the CP violating phases in the supersymmetry breaking terms. Experiments on B, K and LFV processes in near future, therefore, will provide very important opportunities to investigate into the structure of the SUSY breaking sector.

References

1. J. Wess and B. Zumino, *Nucl. Phys.* B **70**, 39 (1974); D.V. Volkov and V.P. Akulov, *Phys. Lett.* B **46**, 109 (1973).
2. D.F. Freedman, P. van Nieuwenhuizen and S. Ferrara, *Phys. Rev.* D **13**, 3714 (1976).
3. F. Gliozzi, J. Scherk and D.I. Olive, *Phys. Lett.* B **65**, 282 (1973); *Nucl. Phys.* B **122**, 253 (1977) .
4. For review, see H.P. Nilles, *Phys. Rep.* **110**, 1 (1984).
5. P. Langacker and M. Luo,*Phys. Rev.* D **44**, 817 (1991); U. Amaldi, W. de Beor and H. Fürstenau,*Phys. Lett.* B **260**, 447 (1991) .
6. For example, J. Wess and J. Bagger *Supersymmetry and Supergravity* (Princeton University Press, Princeton, 1992).
7. L. Girardello and M.T. Grisaru, *Nucl. Phys.* B **194**, 65 (1982).
8. S. Ferrara, L. Girardello and F. Palumbo, *Phys. Rev.* D **20**, 403 (1979).
9. M. Dine and A. Nelson, *Phys. Rev.* D **48**, 1277 (1993); M. Dine, A. Nelson, and Y. Shirman, *Phys. Rev.* D **51**, 1362 (1995); M. Dine, A. Nelson, Y. Nir, and Y. Shirman, *Phys. Rev.* D **53**, 2658 (1996).
10. For example, ALEPH collaboration. Barate *et al.*, *Phys. Lett.* B **429**, 201 (1998); *Phys. Lett.* B **433**, 176 (1998).
11. Y. Okada, M. Yamaguchi and T. Yanagida, *Prog. Theor. Phys.* **85**, 1 (1991); J. Ellis, G. Ridolfi and F. Zwirner, *Phys. Lett.* B **257**, 83 (1991); H.E. Haber and R. Hempfling, *Phys. Rev. Lett.* **66**, 1815 (1991).
12. Y. Okada, M. Yamaguchi and T. Yanagida, *Phys. Lett.* B **262**, 54 (1991).
13. R. Barbieri, M. Frigeni, and F. Caravaglios, *Phys. Lett.* B **258**, 167 (1991).
14. M. Drees, *Int. J. Mod. Phys.* A **4**, 3635 (1989) 3635; J. Ellis, J.F. Gunion, H.E. Haber, L. Roszkowski and F. Zwirner, *Phys. Rev.* D **39**, 844 (1989); L. Durand and J.L. Lopez, *Phys. Lett.* B **217**, 463 (1989); J.R. Espinosa and M. Quirós, *Phys. Lett.* B **279**, 92 (1992).
15. P.Binétruy and C.A. Savoy, *Phys. Lett.* B **277**, 453 (1992); T. Moroi

and Y. Okada, *Phys. Lett.* B **295**, 73 (1992); G. Kane, C. Kolda and J. Wells, *Phys. Rev. Lett.* **70**, 2686 (1993); W.T.A. ter Veldhuis, Purdue preprint, PURD-TH-92-11; J.R. Espinosa and M. Quirós, *Phys. Lett.* B **302**, 51 (1993); U. Ellwanger,*Phys. Lett.* B **303**, 271 (1993); P.N. Pandita,*Phys. Lett.* B **318**, 338 (1993); *Z. Phys.* C **59**, 575 (1993); T. Elliot, S.F. King and P.L. White, *Phys. Lett.* B **305**, 71 (1993); *Phys. Rev.* D **49**, 4435 (1994).

16. J. Kamoshita, Y. Okada and M. Tanaka, *Phys. Lett.* B **328**, 67 (1994); S.F. King and P.L. White, *Phys. Rev.* D **53**, 4049 (1996).
17. T. Moroi and Y. Okada, *Phys. Lett.* B **295**, 73 (1992).
18. J.R. Espinosa and M. Quirós, *Phys. Rev. Lett.* **81**, 516 (1998).
19. ATLAS Technical Proposal (1994); CMS Technical Proposal (1994); E. Richter-Was, D. Froidevaux, F. Gianotti, L. Poggioli, D. Cavalli, and S. Resconi, *Int.J. Mod. Phys.* A **13**, 1371 (1998).
20. P. Janot in *Proceedings of the Workshop on Physics and Experiments with Linear e^+e^- Colliders*, Waikola, Hawaii, 1993 edited by F. Harris *et al.* (World Scientific, Singapore, 1993).
21. K. Kawagoe in *Proceedings of the Workshop on Physics and Experiments with Linear e^+e^- Colliders*, Waikola, Hawaii, 1993 edited by F. Harris *et al.* (World Scientific, Singapore, 1993).
22. JLC-I, KEK Report 92-16, December 1992.
23. M.D. Hildreth, T.B. Barklow and D.L. Burke, *Phys. Rev.* D **49**, 3441 (1994).
24. J. Kamoshita, Y. Okada and M. Tanaka, in *Proceedings of the Workshop on Physics and Experiments with Linear Colliders*, Morioka-Appi, Iwate, Japan 1995 edited by A. Miyamoto *et al.* (World Scientific, Singapore, 1996); *Phys. Lett.* B **391**, 124 (1997).
25. I. Nakamura and K. Kawagoe, in *Proceedings of the Workshop on Physics and Experiments with Linear Colliders*, Morioka-Appi, Iwate, Japan 1995 edited by A. Miyamoto *et al.* (World Scientific, Singapore, 1996); *Phys. Rev.* D **54** , 3634 (1996).
26. A. Djouadi, M. Spira and P.M. Zerwas, *Z. Phys.* C **70**, 427 (1996).
27. S. Bertolini, F. Borzumati, A. Masiero and G. Ridolfi, *Nucl. Phys.* B **353**, 591 (1991).
28. T. Goto and Y. Okada, *Prog. Theor. Phys.* **94**, 407 (1995) and references therein.
29. A. Ali, G. Giudice and T. Mannel, *Z. Phys.* C **67**, 417 (1995); P. Cho, M. Misiak and D. Wyler, *Phys. Rev.* D **54**, 3329 (1996); T. Goto, Y. Okada, Y. Shimizu and M. Tanaka, *Phys. Rev.* D **55**, 4273 (1997); J. Hewett, J.D. Wells, *Phys. Rev.* D **55**, 5549 (1997).

30. T. Goto, T. Nihei and Y. Okada, *Phys. Rev.* D **53**, 5233 (1996); *Erratum, ibid.* **D54**, 5904 (1996).
31. R. Barbieri and L.J. Hall, *Phys. Lett.* B **338**, 212 (1994).
32. T. Goto, Y. Okada, and Y. Shimizu, *Phys. Rev.* D **58**, 094006 (1998).
33. J. Ellis, S. Ferrara and D.V. Nanopoulos, *Phys. Lett.* B **114**, 231 (1982); W. Buchmüller and D. Wyler, *Phys. Lett.* B **121**, 321 (1983); J. Polchinski and M. Wise, *Phys. Lett.* B **125**, 393 (1983); F. del Aguila, M. Gavela, J. Grifols and A. Mendez, *Phys. Lett.* B **126**, 71 (1983); D.V. Nanopoulos and M. Srednicki, *Phys. Lett.* B **128**, 61 (1983); M. Dugan, B. Grinstein and L. Hall, *Nucl. Phys.* B **255**, 413 (1985); Y. Kizukuri and N. Oshimo, *Phys. Rev.* D **45**, 1806 (1992); *Phys. Rev.* D **46**, 3025 (1992); T. Falk and K.A. Olive, *Phys. Lett.* B **375**, 196 (1996); T. Ibrahim and P. Nath, *Phys. Lett.* B **418**, 98 (1998); *Phys. Rev.* D **57**, 478 (1998).
34. T. Nihei, *Prog. Theor. Phys.* **98**, 1157 (1997).
35. CLEO Collaboration, M.S. Alam *et al.*, *Phys. Rev. Lett.* **74**, 2885 (1995).
36. CLEO Collaboration, CLEO CONF 98-17.
37. J. Hisano, T. Moroi, K. Tobe and M. Yamaguchi, *Phys. Lett.* B **391**, 341 (1997); *Erratum, ibid.* **B 397**, 357 (1997).
38. R. Barbieri, L. Hall, and A. Strumia, *Nucl. Phys.* B **445**, 219 (1995).
39. J. Hisano, D. Nomura, Y. Okada, Y. Shimizu and M. Tanaka, *Phys. Rev.* D **58**, 116010 (1998).
40. N. Arkani-Hamed, H. Cheng, and L. Hall, *Phys. Rev.* D **53**, 413 (1996).
41. F. Borzumati and A. Masiero, *Phys. Rev. Lett.* **57**, 961 (1986); J. Hisano, T. Moroi, K. Tobe, M. Yamaguchi and T. Yanagida, *Phys. Lett.* B **357**, 579 (1995); J. Hisano, T. Moroi, K. Tobe and M. Yamaguchi, *Phys. Rev.* D **53**, 2442 (1996); J. Hisano, D. Nomura, and T. Yanagida, *Phys. Lett.* B **437**, 351 (1998); J. Hisano and D. Nomura KEK-TH-602, hep-ph/9810479.
42. Y. Kuno and Y. Okada *Phys. Rev. Lett.* **77**, 434 (1996); Y. Kuno, A. Maki, and Y. Okada *Phys. Rev.* D **55**, 2517 (1997).
43. S.B. Treiman, F. Wilczek and A. Zee, *Phys. Rev.* D **16**, 1977 (152).
44. Y. Okada, K. Okumura and Y. Shimizu, *Phys. Rev.* D **58**, 051901 (1998).

Invited Talks

RARE NON-HADRONIC b DECAYS

SHELDON STONE
Physics Department, 201 Physics Building, Syracuse Univerisity, Syracuse, NY 13244-1130, USA
E-mail: stone@suhep.phy.syr.edu

This paper summarizes current results for rare non-hadronic b decay processes. The world average $\mathcal{B}(b \to s\gamma) = (3.14 \pm 0.48) \times 10^{-4}$ is in agreement with the standard model prediction. Upper limits on $b \to s\ell^+\ell^-$ and $b \to \ell^+\ell^-$ are also given. Finally $B^- \to \ell^-\overline{\nu}$ upper limits are presented as well as the world average value of $(255 \pm 21 \pm 28)$ MeV for f_{D_s} from $D_s^+ \to \ell^+\nu$ decay rate measurements.

1 Introduction

This paper will describe current experimental results on both inclusive and exclusive B meson decays to charmless final states containing one photon or two leptons. Specifically the topics include $b \to s\gamma$ and $d\gamma$, $b \to s\ell^+\ell^-$, $b \to s\nu\overline{\nu}$, the associated exclusive reactions for these final states, and decays to dileptons and diphotons. I will also discuss the annihlation processes $B^- \to \ell^-\overline{\nu}$ and $D_s^+ \to \mu^+\nu$.

These processes, with the exception of the lepton-neutrino final state, proceed through higher order weak interactions involving loops, which are often called "Penguin" processes, for unscientific reasons.[1] A Feynman loop diagram is shown in Fig. 1 that describes the transition of a b quark into a charged -1/3 s or d quark, which is effectively a neutral current transition. The dominant charged current decays change the b quark into a charged +2/3 quark, either c or u.

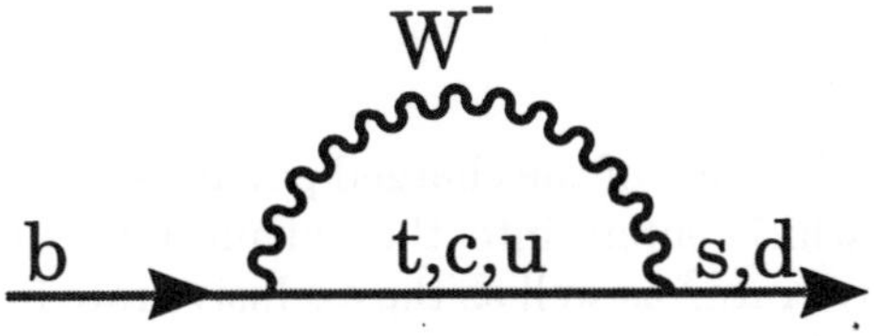

Figure 1: Loop or "Penguin" diagram for a $b \to s$ or $b \to d$ transition.

The intermediate quark inside the loop can be any charge +2/3 quark. The relative size of the different contributions arises from different quark masses and CKM elements. In terms of the Cabibbo angle (λ=0.22), we have for t:c:u - λ^2:λ^2:λ^4. The mass dependence favors the t loop, but the amplitude for c processes can be quite large ≈30%. Moreover, as pointed out by Bander, Silverman and Soni,[2] interference can occur between t, c and u diagrams and lead to CP violation. In the standard model it is not expected to occur when $b \to s$, due to the lack of a CKM phase difference, but could occur when $b \to d$. In any case, it is always worth looking for this effect; all that needs to be done, for example, is to compare the number of $K^{*-}\gamma$ events with the number of $K^{*+}\gamma$ events.

There are other possibilities for physics beyond the standard model to appear. For example, the W^- in the loop can be replaced by some other charged object such as a Higgs; it is also possible for a new object to replace the t.

2 Standard Model Theory

In the Standard Model the effective Hamiltonian for the intermediate t quark is given by

$$H_{eff} = -\frac{4G_F}{\sqrt{2}} V_{tb} V_{ts}^* \sum_{i=1}^{10} C_i(\mu) O_i(\mu) \quad . \tag{1}$$

Some of the operators are

$$O_1 = \bar{s}_L^i \gamma_\mu b_L^j \bar{c}_L^j \gamma^\mu c_L^i, \qquad O_7 = \frac{e}{16\pi^2} m_b \bar{s}_L^i \sigma_{\mu\nu} b_R^j F^{\mu\mu} \quad . \tag{2}$$

The matrix elements are evaluated at the scale $\mu = M_W$ and then evolved to the b mass scale using renormalization group equations, which mixes the operators:

$$C_i(\mu) = \sum_j U_{ij}(\mu, M_W) C_j(M_W) \quad . \tag{3}$$

3 $b \to s\gamma$

This process occurs when any of the charged particles in Fig. 1 emits a photon. The only operator which enters into the calculation is $C_7(\mu)$. CLEO first measured the inclusive rate,[3] as well as the exclusive rate into $K^*(890)\gamma$.[4] I will report here on an updated CLEO measurement [5] using 1.5 times the original data sample and a new measurement from ALEPH.[6]

The momentum spectrum of the γ peaks close to its maximum value at half the B mass. If we had data with only B mesons, it would be easy to pick out $b \to s\gamma$. We have, however, a large background from other processes. At the $\Upsilon(4S)$, the γ spectrum from the different background processes is shown. The largest is π^o production from continuum e^+e^- collisions, but another large source is initial state radiation (ISR), where one of the beam electrons radiates a hard photon before annihilation. The backgrounds and the expected signal are illustrated in Fig. 2. Similar backgrounds exist at LEP.

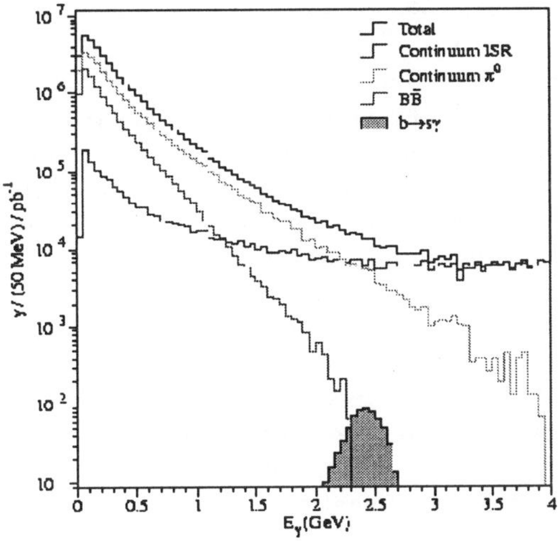

Figure 2: Levels of inclusive photons from various background processes at the $\Upsilon(4S)$ labled largest to smallest at 2.5 GeV/c. Also shown is the expected signal from $b \to s\gamma$.

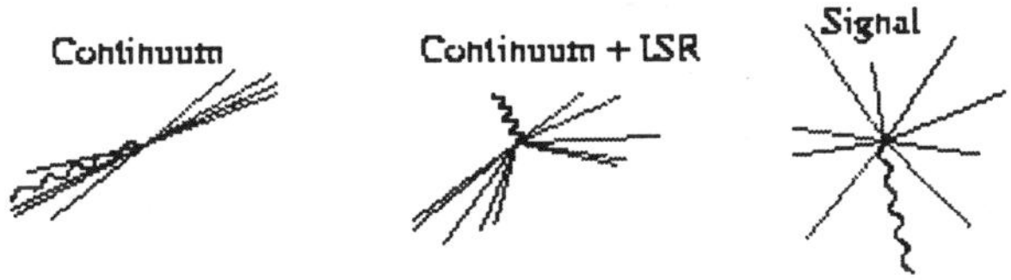

Figure 3: Examples of idealized event shapes. The straight lines indicate hadrons and the wavy lines photons.

To remove background CLEO used two techniques originally, one based on "event shapes" and the other on summing exclusively reconstructed B samples. Examples of idealized events are shown in Fig. 3. CLEO uses eight different shape variables described in Ref. [3], and defines a variable r using a neural network to distinguish signal from background. The idea of the B reconstruction analysis is to find the inclusive branching ratio by summing over exclusive modes. The allowed hadronic system is comprised of either a $K_s \to \pi^+\pi^-$ candidate or a $K^\mp$ combined with 1-4 pions, only one of which can be neutral.

The restriction on the number and kind of pions maximizes efficiency while minimizing background. It does however lead to a model dependent error. For all combinations CLEO evaluates

$$\chi_B^2 = \left(\frac{M_B - 5.279}{\sigma_M}\right)^2 + \left(\frac{E_B - E_{beam}}{\sigma_E}\right)^2, \tag{4}$$

where M_B is the measured B mass for that hypothesis and E_B is its energy. χ_B^2 is required to be < 20. If any particular event has more than one hypothesis, the solution which minimizes χ_B^2 is chosen. For events with a reconstructed B candidate CLEO also considers the angle between the thrust axis of the B and the thrust axis of event with the B candidate removed, $\cos(\theta_t)$. This is highly effective in removing continuum background.

Another neural network is used to combine r, χ_B^2, $\cos(\theta_t)$ into a new variable r_c and events are then weighted according to their value of r_c. This method maximizes the statistical potential of the data.[5] Fig. 4 shows the photon energy spectrum of the inclusive signal, compared with the model of Ali and Greub.[7] A fit to the model over the photon energy range from 2.1 to 2.7 GeV/c gives the branching ratio result shown in Table 1, where the first error is statistical and the second systematic.

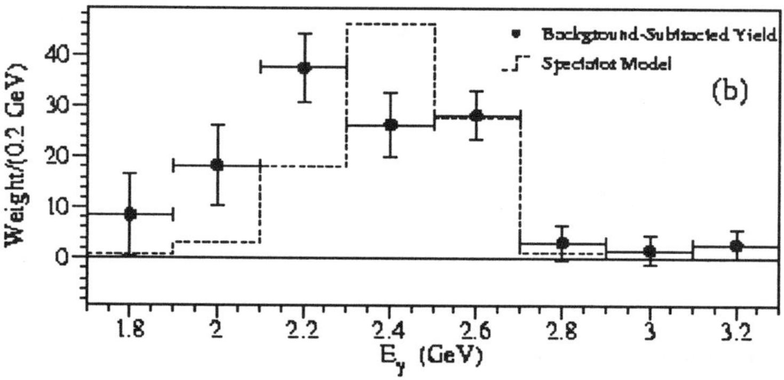

Figure 4: The background subtracted photon energy spectrum from CLEO. The dashed curve is a spectator model prediction from Ali and Greub.

ALEPH reduces the backgrounds by weighting candidate decay tracks in a $b \to s\gamma$ event by a combination of their momentum, impact parameter with respect to the main vertex and rapidity with respect to the b-hadron direction.[6] There result is shown in Table 1, and the resulting photon energy spectrum in Fig. 5. The world average value experimental value is also given, as well as the theoretical prediction.

Table 1: Experimental results for $b \to s\gamma$

Sample	branching ratio
CLEO	$(3.15 \pm 0.35 \pm 0.41) \times 10^{-4}$
ALEPH	$(3.11 \pm 0.80 \pm 0.72) \times 10^{-4}$
Average	$(3.14 \pm 0.48) \times 10^{-4}$
Theory[8]	$(3.28 \pm 0.30) \times 10^{-4}$

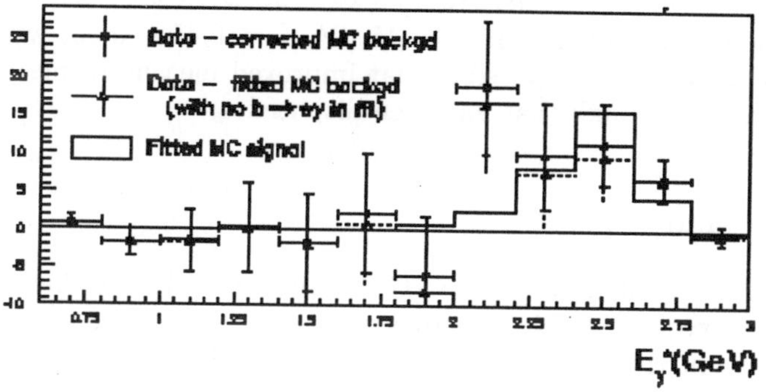

Figure 5: The average photon energy in the rest frame of the reconstructed b jet from ALEPH.

The standard model prediction is in good agreement with the data. It is evaluated by including u, c and t loops. The t loop has the CKM factors $V_{tb}V_{ts}^*$. The c and u can be combined, since by unitarity

$$V_{cb}V_{cs}^* + V_{ub}V_{us}^* = -V_{tb}V_{ts}^* \quad . \tag{5}$$

Thus, comparing the model prediction with the data in Table 1 gives a measurement of $\left|\frac{V_{ts}}{V_{cb}}\right| = 0.98 \pm 0.28$, consistent with the expection of unity.[9]

The consistency with standard model expectation has ruled out many models. Hewett has given a good review of the many minimal supergravity models which are excluded by the data.[10]

Triple gauge boson couplings are of great interest in checking the standard model. If there were an anomalous $WW\gamma$ coupling it would serve to change the standard model rate. $p\bar{p}$ collider experiments have also published results limiting such couplings.[11] In a two-dimensional space defined by $\Delta\kappa$ and λ, the D0 constraint appears as a tilted ellipse and and the $b \to s\gamma$ as nearly vertical bands. In the standard model both parameters are zero.

4 The Exclusive Decays $K^*\gamma$ and $\rho\gamma$

The CLEO measurements have not as yet been updated. The exclusive branching ratio is far more difficult to predict than the inclusive. CLEO measures $\mathcal{B}(B \to K^*(890)\gamma) = (4.2 \pm 0.8 \pm 0.6) \times 10^{-5}$, with this exclusive final state comprising $(18 \pm 7)\%$ of the total rate.[12]

CLEO also limits $\mathcal{B}(B \to \rho\gamma) < 1.2 \times 10^{-5}$ at 90% confidence level.[12] This leads to a model dependent limit on $|V_{td}/V_{ts}|^2 < 0.45 - 0.56$, which is not very significant. It may be possible that improved measurements can find a meaningful limit, although that has been disputed.[13]

5 $b \to s\ell^+\ell^-$ and $b \to s\nu\bar{\nu}$

The diagrams that contribute to $b \to s\ell^+\ell^-$, where ℓ refers to either an electron or muon are shown in Fig. 6. The diagrams for $b \to s\nu\bar{\nu}$ are similar, with ν replacing ℓ, except that only the Z^o contributes in the left-hand diagram.

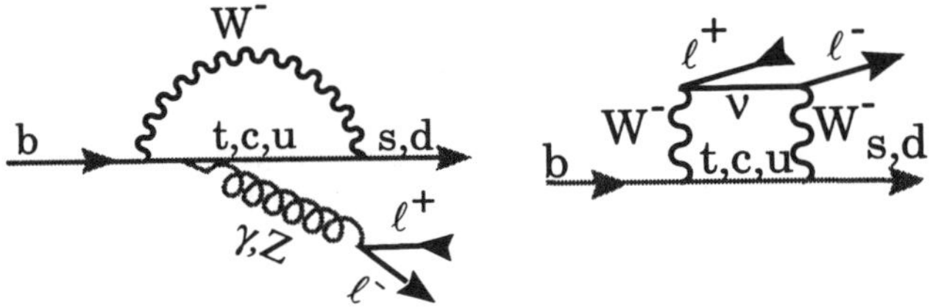

Figure 6: Loop or "Penguin" diagram for a $b \to s\ell^+\ell^-$ transistion.

The operator structure is more complicated as C_9 and C_{10} contribute along with C_7. CP violation can be looked at in both the branching ratios and the polarization of the lepton pair.[15] No signals have been seen as yet in any inclusive or exclusive modes. Current upper limits are given in Table 2 and Table 3.

Table 2: Upper limits at 90% c.l. for $b \to s\ell^+\ell^-$ and $b \to s\nu\bar{\nu}$ inclusive decays

Sample	$b \to s\mu^+\mu^-$	$b \to se^+e^-$	$b \to s\nu\bar{\nu}$
SM theory[14]	$(0.8 \pm 0.2) \times 10^{-5}$	$(0.6 \pm 0.1) \times 10^{-5}$	$(3.8 \pm 0.8) \times 10^{-5}$
CLEO[16]	$< 5.7 \times 10^{-5}$	$< 5.7 \times 10^{-5}$	
ALEPH[17]			$< 7.7 \times 10^{-4}$

Table 3: Upper limits at 90% c.l. for exclusive $b \to s\ell^+\ell^-$ and $b \to s\nu\overline{\nu}$ decays

Mode	CLEO[19]	CDF[20]	DELPHI[21]	Theory[18]
$K^-\mu^+\mu^-$	$< 1.0 \times 10^{-5}$	$< 1.0 \times 10^{-5}$		$0.2\text{-}1.0\times10^{-6}$
$K^-e^+e^-$	$< 1.1 \times 10^{-5}$	$< 2.5 \times 10^{-5}$		$0.2\text{-}1.0\times10^{-6}$
$K^o\mu^+\mu^-$	$< 2.1 \times 10^{-5}$			$0.2\text{-}1.0\times10^{-6}$
K^oe^+e-	$< 1.7 \times 10^{-5}$			$0.2\text{-}1.0\times10^{-6}$
$K^{*o}\mu^+\mu^-$	$< 1.1 \times 10^{-5}$			$0.8\text{-}4.2\times10^{-6}$
$K^{*o}e^+e^-$	$< 1.4 \times 10^{-5}$			$1.1\text{-}6.0\times10^{-6}$
$K^{*o}\nu\overline{\nu}$			$< 10^{-3}$	$1\text{-}\times10^{-5}$
$B_s \to \phi\nu\overline{\nu}$			$< 5.4 \times 10^{-3}$	

6 $b \to \ell^+\ell^-$ or $b \to \gamma\gamma$

The standard model diagrams for neutral b decays into two leptons are shown in Fig. 7. (I have not shown the $\nu\overline{\nu}$ final state.) Larger rates are expected for B_s decays than B_d decays since V_{ts} is larger than V_{td}.

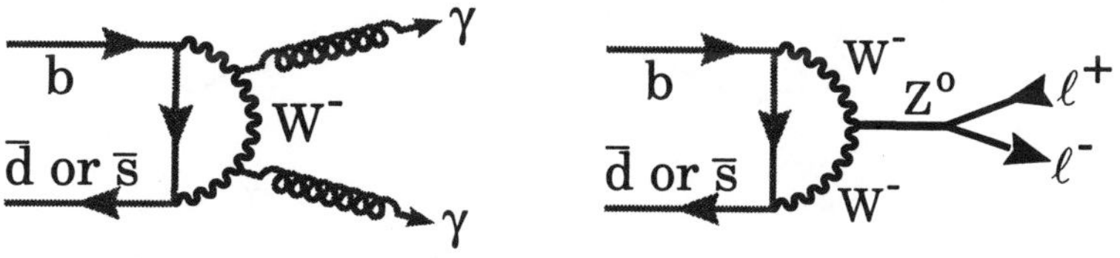

Figure 7: Diagrams for a $b \to \ell^+\ell^-$ and $b \to \gamma\gamma$.

Only upper limits have been determined in these modes. The limits are orders of magnitude higher than the predictions. The best limit in each mode is given in Table 4.

Table 4: Upper limits for exclusive $b \to \ell^+\ell^-$ and $b \to \gamma\gamma$ decays

B_d mode	Exp.	$\mathcal{B} \times 10^6$	confidence	Theory
$\gamma\gamma$	L3[22]	<38	90%	10^{-8}
e^+e^-	CLEO[23]	<5.9	90%	10^{-15}
$\mu^+\mu^-$	CDF[24]	<0.86	95%	10^{-10}
$\tau^+\tau^-$			10^{-8}	
B_s mode	**Exp.**	**$\mathcal{B} \times 10^6$**	**confidence**	**Theory**
$\gamma\gamma$	L3[22]	<148	90%	10^{-7}
e^+e^-	L3[25]	<54	10^{-14}	
$\mu^+\mu^-$	CDF[24]	<2.6	95%	10^{-9}
$\tau^+\tau^-$				10^{-7}

7 The decay $B^- \to \ell^- \overline{\nu}$

8 $B^- \to \ell^- \overline{\nu}$

This reaction proceeds via the annihilation of the b quark with the $\overline{u}$ into a virtual W^- which materializes as $\ell^- \overline{\nu}$ pair as illustrated in Fig. 8. The decay rate for this process can be written as

$$\Gamma(B^- \to \ell^- \overline{\nu}) = \frac{G_F^2}{8\pi} f_B^2 m_\ell^2 M_B \left(1 - \frac{m_\ell^2}{M_B^2}\right)^2 |V_{ub}|^2 \quad , \tag{6}$$

where f_B is the so called "decay constant," a parameter that can be calculated theoretically or determined by measuring the decay rate. This formula is the same for all pseudoscalar mesons using the appropriate CKM matrix element and decay constant.

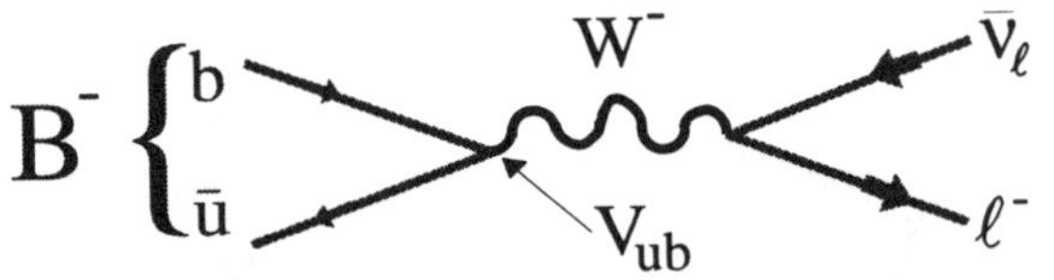

Figure 8: Diagram for a $B^- \to \ell^- \overline{\nu}$ decay.

Knowledge of f_B is important because it is used to determine constraints on CKM matrix elements from measurements of neutral B mixing. Since the decay is helicity suppressed, the heavier the lepton the larger the expected rate. Thus looking for the $\tau^- \overline{\nu}$ has its advantages. The big disadvantage is that there are least two missing neutrinos in the final state. The most stringent limit has been set by L3 of $< 5.7 \times 10^{-4}$ at 90% confidence level, using a missing energy technique.[26] This is still one order of magnitude higher than what is expected. Other limits are poorer.[27]

Since f_B is so difficult to measure, models, especially lattice gauge models, are used.[28] However, it is prudent to test these models. $D_s^+ \to \mu^+ \nu$ can be used; it is Cabibbo favored and the predicted branching ratio is close to 1%. Several groups have made measurements. The results are shown in Table 5. I have changed the values of f_{D_s} according to the new PDG D_s decay branching fractions for the normalization modes,[34] and have corrected the old CLEO result by using the new fake rates determined in their updated analysis. In addition, there are new results using the $D_s^+ \to \tau^+ \nu$ decay from the L3 collaboration [26] of $(309 \pm 58 \pm 33 \pm 38)$ MeV, and (330 ± 95) MeV from the DELPHI collaboration.[27] The world average value for f_{D_s} is $(255 \pm 21 \pm 28)$ MeV, where the common systematic error is due the error on the absolute branching ratio for $D_s^+ \to \phi\pi^+$. These numbers are consistent with C. Bernard's world average for lattice theories of (221±25) MeV.[28]

Table 5: Measured values of f_{D_s} from experimental values of $\Gamma(D_s^+ \to \mu^+\nu)$

Collaboration	Observed Events	Published f_{D_s} value (MeV)	Corrected f_{D_s} value (MeV)
CLEO (old) [29]	39±8	$344 \pm 37 \pm 52 \pm 42$	$282 \pm 30 \pm 43 \pm 34$
WA75 [30]	6	$232 \pm 45 \pm 20 \pm 48$	$213 \pm 41 \pm 18 \pm 26$
BES [31]	3	$430^{+150}_{-130} \pm 40$	Same
E653 [32]	$23.2 \pm 6.0^{+1.0}_{-0.9}$	$194 \pm 35 \pm 20 \pm 14$	$200 \pm 35 \pm 20 \pm 26$
CLEO [33]	182±22	-	$280 \pm 19 \pm 28 \pm 34$

9 Conclusions

The study of rare non-hadronic b decays has been very important in confirming the standard model. (Although, we may have wished for non-standard model effects to show up in these decays.) We have seen that the decay rate for $b \to s\gamma$ is consistent with the standard model within errors and many non-standard models have been excluded. Further improvements in the statistical accuracy of the data can be expected, but improvements in the systematic errors will be slow, as will reductions in the theoretical errors.

The next inclusive decay to be studied will be $b \to s\ell^+\ell^-$. CLEO has shown that there is a large background from events where both b's decay semileptonically at the $\Upsilon(4S)$. This background can, in principle, be removed by insisting that the two leptons come from the same vertex.[35] The upcoming higher luminosity e^+e^- experiments Babar, Belle and CLEO III may be able to see this mode. If they are not successful, perhaps hadronic b collider experiments (BTeV, LHC-B) will be. Though some have thought that inclusive measurements are not possible at such machines, the CLEO technique that finds the inclusive rate by summing the exclusive channels can be used here, albeit with no neutral pions rather than the one allowed by CLEO.

Decays to dileptons can only be seen at hadronic machines, because of the small predicted rates. Charged B decays to $\ell^-\overline{\nu}$ are difficult but can possibly be done at $\Upsilon(4S)$ machines with large integrated luminosity.

Exclusive modes are very important for studies of CP violation. Final states such as $K^*\ell^+\ell^-$ can be studied for rate asymmetries and differences in dilepton polarization. The only exclusive mode seen thus far is $K^*\gamma$, but hopefully that will soon be augmented.

Clearly the study of rare b decays has just begun. We look forward to learning much from these reactions.

Acknowledgments

I thank Amajit Soni, Joanne Hewett, Bill Marciano, Jon Rosner, and Tomasz Skwarnicki for interesting conversations about this material.

1. K. Lingel, T. Skwarnicki and J. G. Smith, "Penguin Decays of the b Quark," hep-ex/9804015, submitted to Annual Reviews of Nuclear and Particle Physics (1998).
2. M. Bander, D. Silverman and A. Soni, *Phys. Rev. Lett.* **43**, 242 (1979).
3. M. S. Alam *et al.* (CLEO), *Phys. Rev. Lett.* **74**, 2885 (1995).
4. R. Ammar *et al.* (CLEO), *Phys. Rev. Lett.* **71**, 674 (1993).
5. S. Glenn *et al.* (CLEO), "Improved Measurement of $\mathcal{B}(b \to s\gamma)$," submitted to XXIX Int. Conf. on High Energy Physics, Vancouver, Canada, July 1998 paper ICHEP98 1011 (1998).
6. B. Barate *et al.* (ALEPH), "A Measurement of the Inclusive $b \to s\gamma$, Branching Ratio," CERN-EP/98-044 (1998).
7. A. Ali and C. Greub, *Phys. Lett.* B **259**, 182 (1991). The parameters for this fit are $< m_b >$=4.88 GeV and P_F =250 MeV/c.
8. A. Czarnecki and W. J. Marciano, "Electroweak Radiative Corrections to $b \to s\gamma$," submitted to XXIX Int. Conf. on High Energy Physics, Vancouver, Canada, July 1998 paper ICHEP98 714 (1998); ibid *Phys. Rev. Lett.* **81**, 277 (1998); see also M. Neubert, "Theoretical Status of $b \to X_s\gamma$ Decays," hep-ph/9809377 (1998); A. Ali, "Theory of Rare B Decays," hep-ph/9709507 DESY 97-192 (1997); N. G. Deshpande, "Theory of Penguins in B Decays," in *B Decays Revised 2nd Edition*, ed. by S. Stone, World Scientific, Singapore, (1994).
9. A. Ali, "Flavour Changing Neutral Current Processes and CKM Phenomenology," o be published in the Proceedings of the First APCTP Workshop: Pacific Particle Physics Phenomenology, Oct. 31 - Nov. 2, 1997, Seoul National University, Seoul, South Korea hep-ph/9801270.
10. J. L. Hewett, "*B* Physics Beyond the Standard Model," hep-ph/9803370 (1998).
11. S. Abachi *et al.* (D0), *Phys. Rev.* D **56**, 6742 (1997); F. Abe *et al.* (CDF), *Phys. Rev. Lett.* **78**, 4536 (()1997).
12. R. Ammar *et al.* (CLEO), "Radiative Penguin Decays of the B Meson," CONF 96-5, ICHEP-96 PA05-9 (1996).
13. D. Atwood, B. Blok & A. Soni, *Int. J. Mod. Phys.* **A11**, 3743 (1994) and *Nuovo Cimento* **109A**, 873 (1994); N. Deshpande, X. He & J. Trampetic, Preprint *Phys. Lett.* B **362**, 1996 (;) see also J. M. Soares, *Phys. Rev.* D **53**, 241 (1996); G. Eilam, A. Ioannissian & R. R. Mendel, *Z. Phys.* **C71**, 95 (1995).

14. A. Ali, C. Greub, and T. Mannel, DESY 93-016 to be published in the *Proceedings of the ECFA Workshop on the Physics of the B Meson Factory,* ed. R. Aleksan and A. Ali (1993).
15. S. Fukae, C.S. Kim , T. Morozumi, and T. Yoshikawa, "A Model Independent Analysis of the Rare B Decay $B \to X_s \ell^+ \ell^-$, KEK-TH578, hep-ph/9807254 (1998), and references cited therein.
16. S. Glen *et al.* (CLEO), *Phys. Rev. Lett.* **80**, 228 (1998).
17. ALEPH Collaboration, "Measurements of $\mathcal{B}(b \to \tau\nu X)$ and $\mathcal{B}(b \to \tau\nu D^{*+} X)$ and Upper Limits on $\mathcal{B}(B \to \tau\nu)$ and $\mathcal{B}(b \to s\nu\overline{\nu})$ submitted ICHEP98 Vacouver, paper 982.
18. For $\ell^+\ell^-$ final states I use a range of values from the following papers (including a $\pm\ 1\sigma$ error estimate from the authors). A. Ali and T. Mannel, *Phys. Lett.* B **273**, 505 (1991); C. Greub, A. Ioannisian and D. Wyler, *Phys. Lett.* B **346**, 149 (1994); P. Ball and V. M. Braun, CERN-TH-98-162 (1998); D. Melikhov and N. Nikitin, hep-ph/9609503 (1996). For $K^*\nu\overline{\nu}$ see A. Ali and T. Mannel, *Phys. Lett.* B **264**, 447 (1991); G. Buchalla and A. J. Buras, *Nucl. Phys.* B **400**, 225 (1993).
19. R. Godang *et al.* (CLEO), "Search for Electroweak Penguin Decays $B \to K\ell^+\ell^-$ and $B \to K\ell^+\ell^-$ at CLEO," CLEO CONF 98-22, ICHEP98-1012 (1998).
20. F. Abe *et al.* (CDF), *Phys. Rev. Lett.* **76**, 4675 (1996).
21. W. Adam *et al.* (DELPHI), *Z. Phys.* **C72**, 207 (1996).
22. M. Acciarri *et al.* (L3), *Phys. Lett.* B **363**, 137 (1995).
23. R. Ammar *et al.* (CLEO), *Phys. Rev.* D **49**, 5701 (1994).
24. F. Abe *et al.* (CDF), *Phys. Rev.* D **57**, R3811 (1998).
25. M. Acciarri *et al.* (L3), *Phys. Lett.* B **391**, 474 (1997).
26. M. Acciarri *et al.* (L3 Collaboration), Phys. Lett. B **396**, 327 (1997).
27. A. Sokolov and G.C. Zucchelli (DELPHI), submitted to XXIX Int. Conf. on High Energy Physics, Vancouver, Canada, July 1998 paper ICHEP98 242 (1998); ALEPH collaboration, Paper submitted to Int. Conf on HEP, Warsaw PA10-019 (1996); M. Artuso *et al.*(CLEO), *Phys. Rev. Lett.* **75**, 785 (1985).
28. C. Bernard, "Lattice Calculations of Decay Constants," in proceedings of 7th Int. Symp on Heavy Flavor Physics, Santa Barbara (1997).
29. D. Acosta *et al.*, Phys. Rev. **D49**, 5690 (1994).
30. S. Aoki *et al.*, Progress of Theoretical Physics **89**, 131 (1993). The WA75 value was based on the 1992 PDG value of $\mathcal{B}(D_s^+ \to K^+K^-\pi^+) = (3.9 \pm 0.4)\%$ and $\mathcal{B}(D^o \to \mu\nu X) = (8.8 \pm 2.5)\%$. We scale the WA75 result using PDG [34] values of $\mathcal{B}(D_s^+ \to K^+K^-\pi^+) = (4.4 \pm 1.2)\%$ and $\mathcal{B}(D^o \to e\nu X) = (6.75 \pm 0.29)\%$ which we use for $\mathcal{B}(D^o \to \mu\nu X)$, after

reducing the value by 3% to account for the smaller muon phase space.
31. J. Z. Bai *et al.*, Phys. Rev. Lett. **74**, 4599 (1995).
32. K. Kodama *et al.*, Phys. Lett. **B382** 299 (1996). The published result normalizes to $\mathcal{B}(D_s \to \phi\mu\nu) = (1.88 \pm 0.29)\%$, while the new PDG value is $(2.0 \pm 0.5)\%$.[34]
33. M. Chadha *et al.* (CLEO) *Phys. Rev.* D **58**, 032002-1 (1998).
34. C. Caso *et al.*, The European Physics Journal, **C3** 1, 1998.
35. Xia Xing, Thesis, Syracuse University, (1998) unpublished.

HADRONIC B MESON DECAYS
- Getting Ready for CP Violation -

HITOSHI YAMAMOTO
Department of Physics and Astronomy, The University of Hawaii, Honolulu, HI 96822 USA
E-mail: hitoshi@uhheph.phys.hawaii.edu

We review recent results on hadronic B meson decays including rare as well as non-suppressed modes. The main emphasis is on those channels relevant to measurement of CP violation in the coming B factories. After briefly describing flavor-tagged charm counting, we cover $B \to DK^{(*)}$, $B \to$ charmless 2-body decays, inclusive η' production, and final-state-interaction phases in $B \to \Psi K^*$ and $B \to D^* \rho$.

1 Flavor-tagged charm counting

According to experiments, the semileptonic branching ratio of B mesons, $B_{s.l.}$, and the number of charms generated in a B meson decay, n_c, are (the numbers are averages over B^0 and B^+)

$$B_{s.l.} = 10.41 \pm 0.29\,\%^1\,, \quad n_c = 1.10 \pm 0.05^{\,2}\,, \tag{1}$$

while the corresponding theoretical predictions are [3]

$$B_{s.l} = \begin{cases} 12.0 \pm 1.0\% \\ 10.9 \pm 1.0\% \end{cases}, \; n_c = \begin{cases} 1.20 \pm 0.06\% \\ 1.21 \pm 0.06\% \end{cases}, \quad \text{for } \mu = \begin{cases} m_b \\ m_b/2 \end{cases}, \tag{2}$$

where the errors are largely due to uncertainties in quark masses. We see that the agreement between experiment and theory is within the errors for $B_{s.l.}$ and about two sigmas for n_c; namely, it is inconclusive at present.

Further information can be obtained by counting c and $\bar{c}$ separately for a given b flavor. This is accomplished on the $\Upsilon 4S$ resonance[4] by tagging D^0 and D^+ mesons (denoted simply as D) with leptons from the other B meson. In the sample of D-lepton pairs, however, D and the lepton may come from the same B, in which case D and the lepton tend to be back-to-back. Those coming from different B's would be uncorrelated (B mesons decay nearly at rest). This can be used to statistically separate the two kinds of pairs. [5] Figure 1 shows the measured angular correlations. One can see that the $\bar{D}\ell^-$ (or $D\ell^+$) pairs are mostly from different B's, namely, the directions of D and ℓ are uncorrelated (flat distribution). On the other hand, the $D\ell^-$ (or $\bar{D}\ell^+$) pairs are mostly from the same B (back-to-back); there is, however, a small uncorrelated component.

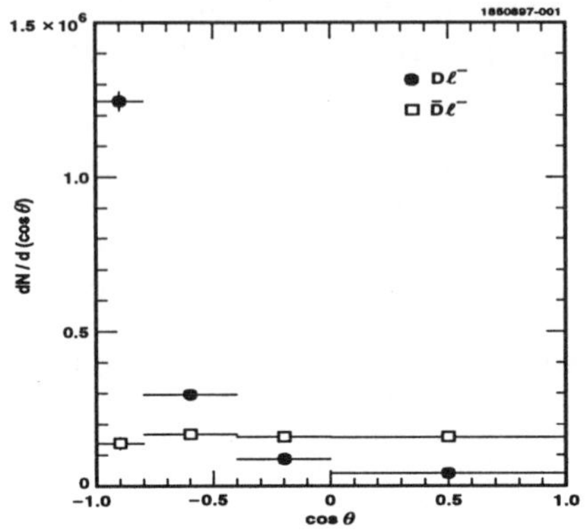

Figure 1: The $\cos\theta_{D-\ell}$ distribution where $\theta_{D-\ell}$ is the angle between D and the lepton. The charm quantum number of D and the charge of the lepton are the same for the open squares and opposite for the solid squares. If D and the lepton are from the same B, they are mostly back-to-back. The lepton momentum is required to be greater than 1.5 GeV/c.

After corrected for the B^0-$\bar{B}^0$ mixing, the latter component represents the 'wrong-sign' D meson production $b \to \bar{D}$ which can be due to the process $b \to cW^-$, $W^- \to \bar{c}s$:

$$\frac{\#(\bar{D}\ell^+)_{opp.B}}{\#(D\ell^+)_{opp.B}} \to \frac{Br(\bar{B} \to \bar{D}X)}{Br(\bar{B} \to DX)} = 0.100 \pm 0.026 \pm 0.016\,. \tag{3}$$

Before this measurement, $\bar{c}$ in $W^- \to \bar{c}s$ was usually assumed to hadronize either as a D_s^-, one of its excited states, or a charmonium. A possibility of sizable wrong-sign D production, however, had been pointed out prior to the measurement [6].

As a byproduct, one can form the ratio of $\#(D\ell^+)_{opp.B}$, which measures $Br(\bar{B} \to DX)$, to $\#(D\ell^-)_{same\ B}$, which measures $Br(\bar{B} \to DX\ell^-\bar{\nu})$, to obtain

$$r \equiv \frac{Br(\bar{B} \to DX)/Br(\bar{B} \to X)}{Br(\bar{B} \to DX\ell^-\bar{\nu})/Br(\bar{B} \to X\ell^-\bar{\nu})} = 0.901 \pm 0.034 \pm 0.015\,, \tag{4}$$

where $Br(\bar{B} \to X) = 1$ by definition, and $Br(\bar{B} \to X\ell^-\bar{\nu})$ is the separately-measured semileptonic branching ratio. Note that the numerator of r is the probability that the b quark will result in a right-sign D, while the denominator is the same probability given that the decay was semileptonic. If we ignore $b \to u$, D_s^+ production from $b \to c$, production of charmed baryon or charmonium, and $b \to$ 'sg', then we expect $r = 1$. Here, $b \to$ 'sg' stands for any quark-level transition $b \to$ (no charm) other than $b \to u$. Correcting for all but $b \to$ 'sg', we get

$$Br(b \to \text{`}sg\text{'}) = 0.2 \pm 3.4 \pm 1.5 \pm 1.7\% \quad \text{or} \quad < 6.8\% \ (90\%\, c.l.) \tag{5}$$

where the first error is statistical, second systematic, and the third is due to models used in the corrections. It should be noted that the ratio r above is insensitive to the uncertainty in the decay branching ratios of D mesons used in the analysis since it largely cancels between the numerator and the denominator.

Another ratio of interest is the denominator of r itself which is expected to be unity if we ignore $\bar{B} \to D_s^+ \ell^- \bar{\nu}$ and $b \to u\ell^-\bar{\nu}$. The D mesons are detected by $D^0 \to K^-\pi^+$ and $D^+ \to K^-\pi^+\pi^+$ where $Br(D^+ \to K^-\pi^+\pi^+)$ is normalized to $Br(D^0 \to K^-\pi^+)$. Thus, by setting the above ratio to unity after the corrections, one can extract $Br(D^0 \to K^-\pi^+)$. The result is

$$Br(D^0 \to K^-\pi^+) = 3.69 \pm 0.11 \pm 0.16 \pm 0.04\,\%\,, \tag{6}$$

which may be compared to the PDG value $Br(D^0 \to K^-\pi^+) = 3.85 \pm 0.09\,\%$ [1].

2 $B^- \to D^0 K^{(*)-}$

Along with $B \to K\pi, \pi\pi$, these are modes that we hope can be used to measure the CKM angle γ. A large CP asymmetry is possible in $B^- \to D_i^0 K^-$ vs. $B^+ \to D_i^0 K^+$ (D_i^0: CP eigensates) through interference of $B^- \to D^0 K^-$ and $B^- \to \bar{D}^0 K^-$ (and its charge conjugate modes). The original method [7] proposed to measure these 4 modes to construct two triangles to extract the angle γ. A difficulty in measuring the 'wrong-sign' D^0 decay due to the doubly-Cabibbo-suppressed D^0 decay was pointed out by Atwood, Dunietz and Soni, [8] who also proposed a solution for the problem even though it required a large data sample. Analyses using only the favored modes were recently proposed by Gronau [9] and Xing. [10] Also, the charged B decays to $DK^{(*)}$ can be combined with their corresponding neutral B to enhance sensitivities. [11,12]

We detect D^0 in $B^- \to D^0 K^-$ in the modes $D^0 \to K^-\pi^+$, $K^-\pi^+\pi^0$, and $K^-\pi^+\pi^-\pi^+$. The major background comes from the Cabibbo-favored counter part $B^- \to D^0\pi^-$. In fact, we expect

$$\frac{Br(B^- \to D^0 K^-)}{Br(B^- \to D^0 \pi^-)} \sim \lambda^2 \left(\frac{f_K}{f_\pi}\right)^2 \sim 0.08\,, \tag{7}$$

where $\lambda \sim 0.225$ is the Cabibbo factor and $f_{K,\pi}$ are decay constants. The $B^- \to D^0\pi^-$ background is suppressed by the ionization loss information in the drift chamber and the fact that miss-assigning kaon mass to the pion results in the total energy greater than the expected B meson energy. There is also substantial background from the continuum events, and they are reduced by cutting on the angle between the sphericity axis of the B candidate and that of the rest of the event, the polar angle of B in the lab frame, and a Fischer discriminant. The Fischer discriminant F is a linear combination of a set of measured variables $\vec{x}$

$$F = \vec{\lambda} \cdot \vec{x} \tag{8}$$

where the coefficients $\vec{\lambda}$ is determined by maximizing the separation between the signal sample and the background sample:

$$S \equiv \frac{(\langle F \rangle_s - \langle F \rangle_b)^2}{\sigma_F^2} = \frac{[\vec{\lambda} \cdot (\langle \vec{x} \rangle_s - \langle \vec{x} \rangle_b)]^2}{\vec{\lambda}^T V \vec{\lambda}} \,. \tag{9}$$

where V is the covariant matrix of $\vec{x}$ and $\langle\ \rangle_{s,b}$ denotes the average over the signal or background sample. Taking $\partial S / \partial \lambda_i = 0$ gives

$$\vec{\lambda} = V^{-1}(\langle \vec{x} \rangle_s - \langle \vec{x} \rangle_b) \,. \tag{10}$$

Here, $\vec{x}$ are the energy flows in 9 cones around the event axis, the second Fox-Wolfram event shape variable, and the polar angle in lab of B candidate event axis.

The B mass is calculated from the total 3-momentum $\vec{P}_{tot}$ of the daughter particles and the known beam energy E_{beam} as

$$M_B \equiv \sqrt{E_{beam}^2 - \vec{P}_{tot}^2} \,, \tag{11}$$

and called the beam-constrained mass. Note that the beam-constrained mass is equivalent to the absolute value of the total momentum of candidate particles. The distribution of M_B after the background rejection cuts is given in Figure 2. There is a significant excess beyond the expected background and we obtain [13]

$$\frac{Br(B^- \to D^0 K^-)}{Br(B^- \to D^0 \pi^-)} = 0.055 \pm 0.014 \pm 0.005 \tag{12}$$

which is consistent with the theoretical expectation.

The large background from $B^- \to D^0\pi^-$ indicates a need for a better particle identification. One way around it is to detect the channel $B^- \to D^0 K^{*-}$, $K^{*-} \to K_S \pi^-$, where the background from the Cabibbo-favored counterpart $B^- \to D^0 \rho^-$ is negligible. Such analysis is currently under way,

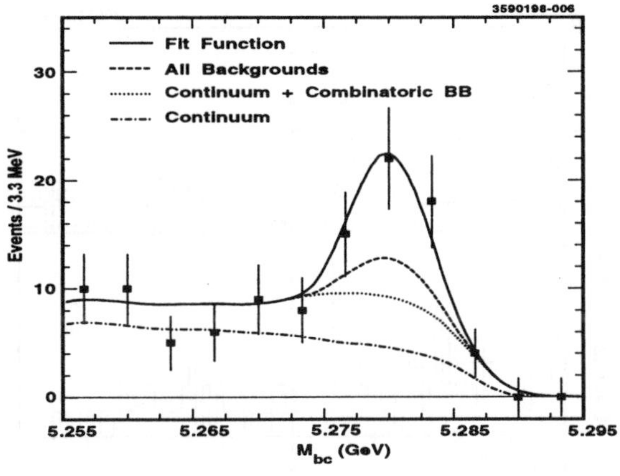

Figure 2: The beam-constrained mass distribution for the $B^- \to D^0K^-$ candidates. The events between the dotted line and the dashed line represent the background from $B^- \to D^0\pi^-$.

3 $B \to$ charmless 2-body decays

The interference of tree and penguin diagrams is expected to cause substantial CP violating decay rate asymmetries in charmless 2-body modes such as $B \to K\pi$ and $\pi\pi$. Many methods have been proposed to extract the CKM phase angles γ and α using these modes. [14]

In the rest frame of $\Upsilon 4S$, the B mesons are nearly at rest ($P_B \sim 350$ MeV/c), and thus the light mesons are nearly monochromatic at $P \sim m_B/2$. Such momentum is the highest that can be created by a B decay, and as a consequence, the background from generic B decays are small. The main background thus comes from the continuum events where the two-jet events $e^+e^- \to q\bar{q}$ can generate particles with momenta all the way up to the beam energy. And the continuum background is suppressed by the same set of parameters as in the D^0K^- analysis discussed above.

As in the case of D^0K^-, it is critical to distinguish charged kaons from charged pions and again we use the ionization loss information (dE/dx) and the B meson energy shift that occurs when the light meson mass is miss-assigned. Figure 3 shows the beam constrained mass distribution for the modes $B^- \to K^-\pi^+$, $h^-\pi^0$, $K_S\pi^-$, [15] where h^- is either K^- or π^-. The variable ΔE is the total energy of the candidate particles minus the beam energy, and expected to peak at zero for the signal when the masses are correctly assigned. When the wrong mass assignment is made (namely, $K \leftrightarrow \pi$ flip), ΔE shifts by about 42 MeV. In the figure, the charged kaons or charged pions are selected

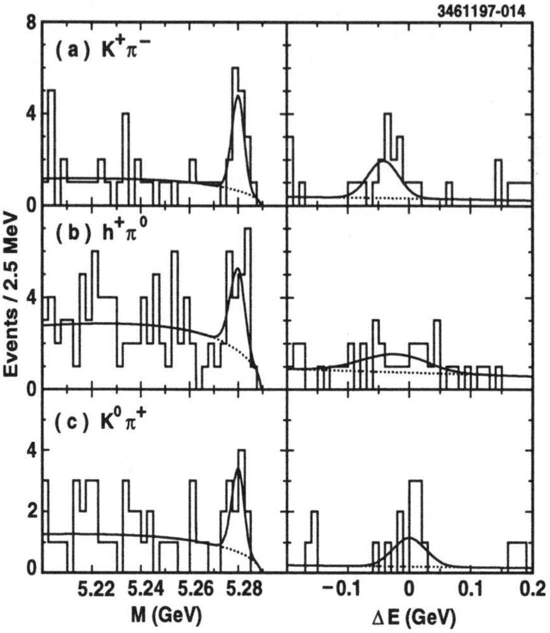

Figure 3: The beam-constrained mass distribution for the modes $B^- \to K^-\pi^+$, $h^-\pi^0$, and $K_S\pi^-$ where h^- is K^- or π^-. All charged tracks are assumed to be pion. The dotted lines show the continuum backgrounds.

by the ionization loss only without using ΔE. All charged tracks are assigned the pion mass and thus one sees that the ΔE peak for $K^-\pi^+$ is shifted. The dE/dx also provides $\sim 1.8\sigma$ separation between K^+ and π^+ at $p = 2.6$ GeV/c (2.0σ for the new dataset taken with a He-based drift chamber gas), and ΔE gives close to 2σ separation for a single $K \leftrightarrow \pi$ flip. Final numbers are obtained by likelihood fit using the beam constrained mass, ΔE, dE/dx and the set of variables used for the continuum suppression. Table 1 gives the results where all numbers are published values [15] except for the column labeled 'ICHEP98' which shows the values updated at the ICHEP 1998 conference. [16]

One sees that we are starting to observe quite a few modes that are important in measuring the CP violating CKM angles. Most of these modes are self-tagging and do not require measurements of decay times (including the neutral B meson modes). Studies to measure decay asymmetries of these and other modes are currently under way.

Table 2 lists the other hadronic rare modes of interest that involve $\eta^{(\prime)}$, ω,

Table 1: Measured branching fractions for $B \to \pi\pi$, $K\pi$, KK.

	N(signal)	signif.	$Br(10^{-5})$(PRL)	$Br(10^{-5})$(ICHEP98)
$\pi^+\pi^-$	9.9	2.2σ	< 1.5	< 0.84
$\pi^+\pi^0$	11.3	2.8σ	< 2.0	< 1.6
$\pi^0\pi^0$	2.7	2.4σ	< 0.93	
$K^+\pi^-$	21.6	5.6σ	$1.5^{+0.5}_{-0.4} \pm 0.1 \pm 0.1$	$1.4 \pm 0.3 \pm 0.2$
$K^+\pi^0$	8.7	2.7σ	< 1.6	$1.5 \pm 0.4 \pm 0.3$
$K^0\pi^+$	9.2	3.2σ	$2.3^{+1.1}_{-1.0} \pm 0.3 \pm 0.2$	$1.4 \pm 0.5 \pm 0.2$
$K^0\pi^0$	4.1	2.2σ	< 4.1	
K^+K^-	0.0	0.0σ	< 0.43	< 0.24
K^+K^0	0.6	0.2σ	< 2.1	< 0.93
K^0K^0	0	-	< 1.7	
$h^+\pi^0$	20.0	5.5σ	$1.6^{+0.6}_{-0.5} \pm 0.3 \pm 0.2$	

or vector mesons. There are upper limits for other modes which can be found in the references.[18,20] Some salient features one notices are: (1) the $\eta' K^-$ rate is large, (2) $\eta' K^- > \eta K^-$, and (3) $\eta' K \gtrsim \eta' K^*$.

Variety of mechanisms have been proposed to explain these features. The large $\eta' K^-$ rate may be due to $c\bar{c}$ content of η',[23] color-octet $c\bar{c}$ contribution,[24] new physics to enhance $b \to sg$,[25,26] $b \to sgg$ followed by $gg \to \eta'$,[27] and gluon fusion where one gluon is from valence quarks.[28,29] They all have something to do with gluon-η' coupling, and many are closely related; in fact, when the gluon creation is dominated by $c\bar{c}$ loop, it may be interpreted as $c\bar{c}$ content (or coupling) to η'. Often, it is said that the $c\bar{c}$ content of η' would result in $\eta' K^*/\eta' K \sim 2$ which is in contradiction with the experiment. This is not necessarily so. With the value of the $c\bar{c}$ coupling needed to explain the $\eta' K$ rate, this ratio is still consistent with the data.[24] The observation $\eta' K^- > \eta K^-$ can be naturally explained by the fact that gluon couples to $u\bar{u}$, $d\bar{d}$, and $s\bar{s}$ with the same amplitude, and the valence quark content of η' and η are roughly (ignoring the mixing) $\eta' = u\bar{u} + d\bar{d} + s\bar{s}$ while $\eta = u\bar{u} + d\bar{d} - 2s\bar{s}$; namely, $u\bar{u} + d\bar{d}$ and $s\bar{s}$ interfere constructively for η' and destructively for η. The feature $\eta' K \gtrsim \eta' K^*$ is predicted by many of the above mechanisms and agrees with the qualitative argument by Lipkin.[30] At the end of the day, however, one should note that the standard 4-quark operators (penguins included) can explain the observed rates within the errors.[31]

Table 2: Measured branching fractions for modes with $\eta^{(\prime)}$ and ω.

	$Br(10^{-5})$	significance	ref.
$\eta' K^-$	$7.4^{+0.8}_{-1.3} \pm 0.9$	12.7σ	17
$\eta' K^0$	$5.9^{+1.8}_{-1.6} \pm 0.9$	7.3σ	17
$\eta' \pi^-$	< 1.2		17
ηh^-	< 0.8		18
$\eta' K^{*0}$	< 9.9		18
ηK^{*0}	< 3.3		18
ωK^-	$1.5^{+0.7}_{-0.6} \pm 0.2$	4.3σ	19
$\omega \pi^-$	$1.1^{+0.6}_{-0.5} \pm 0.3$	2.9σ	19
ϕK^{*} †	$1.3^{+0.7}_{-0.6} \pm 0.2$	3.5σ	19
ϕK^-	< 0.53		19
$\rho^0 K^0$	< 3.6		34
$K^{*+} \pi^-$	< 4.5		34
$K^{*+} K^-$	< 1.1		34

†Average of $B^- \to \phi K^{*-}$ and $B^0 \to \phi K^{*0}$.

4 Inclusive η' production

As mentioned above, the gluonic penguin process ($b \to sg$) plays a critical role in direct CP violation in B decays. One signature of gluonic penguin is excess production of η' which is believed to be 'gluon rich'.

In searching for η' which might be originating from $b \to sg$, one looks for them at the end point of the momentum spectrum in order to suppress η' from $b \to c$ processes. The momentum window is chosen to be $2.0 < p_{\eta'} < 2.7$ GeV/c. The background from continuum is suppressed by the pseudo-B-reconstruction technique which was used previously [21] for the inclusive study of the $b \to s\gamma$ (electro-magnetic penguin) process. Namely, we explicitly reconstruct $B \to \eta' K^{\pm}(n\pi)$ where $n \leq 4$ and at most one neutral pion can be included, then the invariant mass of the decay $\eta' \to \eta\pi^+\pi^-$ ($\eta \to \gamma\gamma$) is plotted when there is at least one B candidate in the signal region of ΔE vs. M_B. The η' mass distributions [22] for the on-resonance data and for the continuum data are shown in Figure 4. The effective luminosity for the continuum data relative to the on-resonance data is 0.524; namely, the continuum background in (a) is about twice the amount seen in (b). The excess in (a) after the continuum background is subtracted is 39.0 ± 11.6.

The continuum-subtracted η' signal is due to B decays in general and not necessarily due to $b \to sg$. Possible backgrounds are

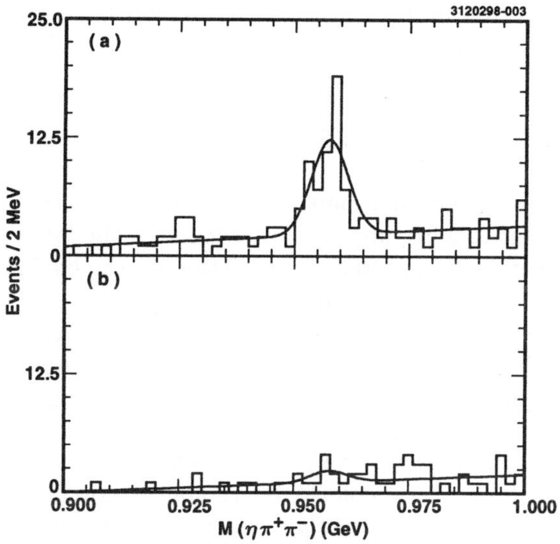

Figure 4: The mass distribution for the η' candidates for $2.0 < p'_\eta < 2.7$ GeV/c when there is at least one B candidate in the signal region (see text). It is plotted for the on-resonance data (a) and for the continuum data (b).

1. $b \to D, D_S \to \eta'$. The rates and spectra of D and D_S in B decay are well-measured and their decays to η' are also experimentally known reliably. The contribution from these sources are estimated to be less than 0.2 events in the signal region.

2. Color-allowed modes $b \to (DX)(\eta' X)$ where the total charge of $(\eta' X)$ is -1 (from W^-). This contribution is modeled by 3-body decay $B \to D\eta'\pi$ with flat phase space and searching for η' at a lower momentum range. Upper limit is set at < 1.4 events in the signal region.

3. $b \to u \to \eta'$. Theoretical estimates [32,24] predicts that $3.5 \sim 7.0\%$ of the signal is due to this source. The uncertainties, however, are large.

4. Color-suppressed modes $B \to D^{(*)0}\eta'$. Theoretical branching fractions predictions for these channels [33] leads to $2.1 \sim 8.6\%$ of the signal being due to this source albeit with large uncertainties. Experimentally, limits can be set by searching for D^0 or D^{*0} in the signal sample which indicates that up to 41% of the signal can be due to this source.

About 10% of the signal is the exclusive mode $\eta' K^-$ and for the rest the recoil mass distribution broadly peaks in the region 1.7 GeV to 2.4 GeV. The fall off at the high mass end is due to the cut $p_{\eta'} > 2.0$ GeV/c. There is no signal seen for $\eta' K^*$. The branching ratio corresponding to the excess is

$$Br(B \to \eta' X_s) = (6.2 \pm 1.6 \pm 1.3^{+0.0}_{-1.5}) \times 10^{-4} \quad (2.0 < p_{\eta'} < 2.7 GeV) \quad (13)$$

where the central number corresponds to the case where the backgrounds discussed above are set to zero, and the lower value of the last error corresponds to 1.5 times the worst case theoretical prediction for the backgrounds discussed above.

If the excess is indeed largely due to $b \to sg$ process, the measured η' production is probably too large compared to the conventional process where the corresponding 4-quark penguin operators are naively coupled to η' meson. [31] Possible enhancement mechanisms proposed are related to those proposed for the large exclusive $\eta' K$ rates: $c\bar{c}$ content of η' which can couple to the Cabibbo-favored transition $b \to c\bar{c}s$, [35] color-octet $c\bar{c}$ contribution, [36] coupling of η' to two gluons through the QCD anomaly resulting in three-body decay $b \to \eta' sg$[32] or two-body decay $b \to \eta' s$, [37] and new physics that enhances $b \to sg$. [25,26] The 2-body decay tends to produce hard η''s (namely, small recoil mass) and seems at odd with the measurement. At this point, there appears to be no strong evidence to force us to introduce new physics.

5 Final-state-interaction phases

Decay rate CP asymmetry in general requires interference of more than one processes having different final-state-interaction (FSI) phases as well as different weak phases. In B decays, final-state-interaction may be looked for in the re-scattering $\bar{B}^0 \to D^+\pi^- \to D^0\pi^0$ where $D^+\pi^-$ is color-favored while $D^0\pi^0$ is color-suppressed, and an enhancement of $D^0\pi^0$ beyond what is expected from naive factorization would indicate re-scattering. Currently, however, only upper limit exists: [38] $Br(\bar{B}^0 \to D^0\pi^0) < 1.2 \times 10^{-4}$ while factorization predicts around 1×10^{-4}. Thus, no evidence for re-scattering is observed in this channel.

Another way is to search phase differences in helicity amplitudes for $B \to VV$ decays such as ΨK^* and $D^*\rho$. Full angular fits has been performed for both modes. The analysis of ΨK^* modes [39] showed no evidence of FSI phase. The same analysis measured the parity content of the ΨK^* final state to be $16 \pm 8 \pm 4\%$ parity $-$. Since the final state ΨK^{*0} ($K^{*0} \to K_S\pi^0$) has C= +, the CP is mostly + which is opposite to the CP of the ΨK_S final state. With full angular analysis, this modes is equivalent to about 1/4 of the ΨK_S sample in measuring the CKM angle β.

The angular distribution for $B \to D^*\rho$ is given in terms of the three helicity amplitudes $H_{\pm,0}$ by

$$\begin{aligned}
&\frac{32\pi}{9\Gamma}\frac{d^3\Gamma}{d\cos\theta_1 d\cos\theta_2 d\chi} \\
&= 4|H_0|^2\cos^2\theta_1\cos^2\theta_2 + (|H_+|^2+|H_-|^2)\sin^2\theta_1\sin^2\theta_2 \\
&\quad +[\Re(H_-H_+^*)\cos 2\chi + \Im(H_-H_+^*)\sin 2\chi]2\sin^2\theta_1\sin^2\theta_2 \\
&\quad +[\Re(H_-H_0^* - H_+H_0^*)\cos\chi + \Im(H_-H_0^* - H_+H_0^*)\sin\chi]\sin 2\theta_1\sin 2\theta_2
\end{aligned}$$

where θ_1 (θ_2) is the decay polar angle of D (charged π) in the D^* (ρ) rest frame and χ is the azimuthal angle between the two decay planes in the B rest frame. Assuming CP symmetry in the decay, helicity amplitudes of B and $\bar{B}$ are related by $H_\lambda(\bar{B}) = H_{-\lambda}(B)$ $(\lambda = +,-,0)$ [40] which corresponds to flipping the sign of χ. The data samples of B and $\bar{B}$ are combined accordingly. By convention, H_0 is taken to be real and the normalization is $|H_+|^2 + |H_-|^2 + |H_0|^2 = 1$. Then, the expression is invariant under the exchange $H_\pm \leftrightarrow H_\mp^*$, which means that the data cannot distinguish which is H_+ and which is H_-. The $V-A$ nature of the interaction, however, indicates that $|H_+| < |H_-|$ for $\bar{B}$ and we thus define the larger of $|H_\pm|$ to be H_- (opposite for B). Table 3 shows the result of the fit. [41]

Table 3: Helicity amplitudes for $\bar{B}^0 \to D^{*+}\rho^-$ and $\bar{B}^- \to D^{*0}\rho^-$.

	$\bar{B}^0 \to D^{*+}\rho^-$		$\bar{B}^- \to D^{*0}\rho^-$	
	magnitude	phase	magnitude	phase
H_0	0.936	0	0.932	0
H_-	$0.317 \pm 0.052 \pm 0.013$	$0.19 \pm 0.23 \pm 0.14$	$0.283 \pm 0.068 \pm 0.039$	$1.13 \pm 0.27 \pm 0.17$
H_+	$0.152 \pm 0.058 \pm 0.037$	$1.47 \pm 0.37 \pm 0.32$	$0.228 \pm 0.069 \pm 0.036$	$0.95 \pm 0.31 \pm 0.19$

There are signs of non-zero phases. Systematics checks have been performed to confirm that it is not due to ρ width, non-resonant $\pi\pi$ component, nor due to backgrounds. The angular distribution expression above indicates that the non-trivial phases results in $\sin\chi$ or $\sin 2\chi$ distributions. Thus, in principle such terms can be seen directly in 1-dimensional distributions (with proper weighting as a function of $\theta_{1,2}$). The data is unfortunately unable to demonstrate such signatures beyond statistical errors. Alternatively, one could directly obtain the coefficient of each term of the angular distribution by the moment analysis. [42] The result shows that the coefficient $\Im(H_+H_-^*)$ is non-zero with 2.5σ significance. The non-zero phases of Table 3 are more significant, and it comes from the information contained in the relative values of the coefficients.

6 Conclusions

The ingredients required for the study of CP violation in B decays are steadily accumulating and we are closing in on actually measuring CP asymmetries. Many of the modes used in the measurements of CKM angles α and γ as well as the 'easy' β are already observed, and with a little more statistics we may even be able to see direct CP violation. There are strong indications that the necessary ingredients for such asymmetries - gluonic penguin and final state phases - indeed exist.

Acknowledgments

I thank Sandip Pakvasa and Tom Browder for useful conversations relating to the topics presented.

1. C. Caso *et.al.*, The European Physics Journal, **C3** 1 (1998).
2. Based on 2.02fb^{-1} of $\Upsilon 4S$ data. L. Gibbons *et.al.*(CLEO Collaboration), Phys. Rev.**D56**, 3783 (1997).
3. M. Neubert and C. T. Sachradja, Nucl. Phys. **B483**, 339 (1997).
4. Based on 3.2fb^{-1} of $\Upsilon 4S$ data. T. E. Coan *et.al.*(CLEO Collaboration), Phys. Rev. Lett. **80**, 1150 (1998).
5. Since the angular correlation is too weak for $P_D < 1.3$ GeV/c, there we use the separately-measured $Br(\bar{B} \to D, \bar{D})$ and $Br(\bar{B} \to \ell^- X)$, where $D, \bar{D}$ and the leptons are in the given momentum windows, to separate $(D\ell)_{opp.B}$ and $(D\ell)_{same\ B}$ assuming $(D\ell^+)_{same\ B} = 0$. See Ref 4 above.
6. G. Buchalla, I. Dunietz, and H. Yamamoto, Phys. Lett. **B364** 188 (1995).
7. M. Gronau and D. Wyler, Phys. Lett. **B265**, 172 (1991).
8. D. Atwood, I. Dunietz, and A. Soni, Phys. Rev. Lett. **78**, 3257 (1997).
9. M. Gronau, hep-ph/9802315.
10. Z.-Z. Xing, hep-ph/9804434.
11. J. Jang and P. Ko, hep-ph/9807496.
12. M. Gronau and J. Rosner, hep-ph/9807447.
13. Based on 3.1fb^{-1} of $\Upsilon 4S$ data. M. Athanas *et.al.*(CLEO Collaboration), Phys. Rev. Lett. **80**, 5493 (1998).
14. See for example, A. Buras and R. Fleischer, hep-ph/9810260 and references therein.

15. Based on 3.14fb^{-1} of $\Upsilon 4S$ data. R. Godang *et.al.*(CLEO Collaboration), Phys. Rev. Lett. **80**, 3456 (1998).
16. Based on 5.6fb^{-1} of $\Upsilon 4S$ data. CLEO Collaboration, the International Conference on High Energy Physics, Vancouver 1998, CLEO-CONF 98-20 and ICHEP98 858.
17. Based on 5.6fb^{-1} of $\Upsilon 4S$ data. CLEO Collaboration, the International Conference on High Energy Physics, Vancouver 1998, CLEO-CONF 98-9 and ICHEP98 860.
18. Based on 3.11fb^{-1} of $\Upsilon 4S$ data. CLEO Collaboration, the 1997 EPS conference, CLEO-CONF 97-22a and EPS97-333.
19. Based on 3.11fb^{-1} of $\Upsilon 4S$ data. CLEO Collaboration, the 1997 EPS conference, CLEO-CONF 97-23 and EPS97-335.
20. Based on 3.11fb^{-1} of $\Upsilon 4S$ data. T. Bergfeld *et.al.*(CLEO Collaboration), Phys. Rev. Lett.**81**, 272 (1998).
21. Based on 2.01fb^{-1} of $\Upsilon 4S$ data. M. S. Alam *et.al.*(CLEO Collaboration), Phys. Rev. Lett. **74**, 2885 (1995).
22. Based on 3.1fb^{-1} of $\Upsilon 4S$ data. CLEO Collaboration, the International Conference on High Energy Physics, Vancouver 1998, CLEO-CONF 98-2 and ICHEP98 857.
23. I. Halperin and A. Zhitnitsky, Phys. Rev. Lett, **80**, 438 (1998).
24. H. Y. Cheng and B. Tseng, Phys. Lett. **B415**, 263 (1997) and references therein.
25. W. S. Hou and B. Tseng, Phys. Rev. Lett. **80**, 434 (1998).
26. A. L. Kagan and A. A. Petrov, hep-ph/9707354.
27. A. Ali, J. Chay, C. Greub, P. Ko, Phys. Lett. **B424**, 161 (1998).
28. M. R. Ahmady, E. Kou, A. Sugamoto, Phys. Rev. **D58**, 14015 (1998).
29. D. Du, C. S. Kim, Y. Yang, Phys. Lett. **426**, 133 (1998).
30. H. Lipkin, Phys. Lett. **B254**, 247 (1991).
31. A. Datta, X. G. He, and S. Pakvasa, Phys. Lett. **B419**, 369 (1998).
32. D. Atwood and A. Soni, Phys. Lett. **B405**, 150 (1997).
33. M. Neubert *et.al.*, in Heavy Flavors, ed. A. J. Buras and M. Lindner, World Scientific, 1992; A. Deandrea *et.al.*, Phys. Lett. **B318**, 549 (1993).
34. Based on 5.59fb^{-1} of $\Upsilon 4S$ data. CLEO Collaboration, the International Conference on High Energy Physics, Vancouver 1998, CLEO-CONF 98-25 and ICHEP98 859.
35. I. Halperin and A. Zhitnitsky, Phys. Rev. **D56**, 7247 (1997).
36. F. Yuan and K. T. Chao, Phys. Rev. **D56**, 2495 (1997).
37. H. Fritzsch, Phys. Lett. **B415** 83 (1997).
38. Based on 3.1fb^{-1} of $\Upsilon 4S$ data. B. Nemati *et.al.*(CLEO Collaboration), Phys. Rev. **D57**, 5363 (1998).

39. Based on 3.1fb^{-1} of $\Upsilon 4S$ data. C. P. Jessop *et.al.*(CLEO Collaboration), Phys. Rev. Lett. **79**, 4533 (1997).
40. This is true to all orders under CP. Another relation found in the literature is $H_\lambda(\bar{B}) = H^*_{-\lambda}(B)$ which is true to the first order in perturbation under CPT. Since we are concerned with FSI which is a higher order process, this CPT relation does not apply.
41. Based on 3.1fb^{-1} of $\Upsilon 4S$ data. CLEO Collaboration, the International Conference on High Energy Physics, Vancouver 1998, CLEO-CONF 98-23 and ICHEP98 852.
42. A. Dighe, I. Dunietz, and R. Fleischer, hep-ph/9804253; H. Yamamoto, CLEO internal note, CBX 98-34.

B PHYSICS AT CDF

S. DONATI*
INFN - Sezione di Pisa, Via Vecchia Livornese 1291, I-56010 San Piero a Grado (Pisa), Italy
E-mail: donati@pi.infn.it

B physics results from the CDF Collaboration based on data collected during the 1992-1996 Tevatron run are presented. In particular, we report the discovery of the B_c meson in the semileptonic decay $B_c \to J/\psi l \nu X$, updates of b hadrons lifetime measurements, with a description of the B_s lifetime measurement, the $B_d^0 - \overline{B}_d^0$ mixing results and the limits set on rare B decay branching ratios. Current results are used to extrapolate B physics prospects for the future high luminosity run II.

1 Introduction

At the Fermilab Tevatron proton and antiproton beams are collided with a center of mass energy of 1.8 TeV. This is a very favorable environment to study B physics because the production cross section for b quarks is large and the resulting number of produced b hadrons is huge. An extrapolation from CDF cross section measurements indicates a total cross section of about 30 μb in the central region ($|\eta| < 1$), where CDF has most of its muon and tracking coverage. For typical Run I instantaneous luminosities (10^{31} cm^{-2}sec^{-1}) the $b\bar{b}$ production rate was 300 Hz, which was one order of magnitude larger than the CDF data logging rate. During run I the Tevatron delivered an integrated luminosity of 110 pb^{-1}: summing the neutral and charged species and including antiparticles, this corresponds to a total of about 10^9 B mesons produced in the central region with a transverse momentum greater than 6 GeV/c. Although B production is very copious, the problem of extracting a signal from background is particularly serious: the total inelastic $p\bar{p}$ cross section is 10^3 times larger than b production cross section and b production is peaked at low p_t, where most backgrounds are concentrated. This problem is faced in two steps: the trigger and the offline analysis. CDF based its run I trigger strategy on leptons:

- dilepton triggers: either two muons with $p_t >$ 2-3 GeV/c, or one muon with $p_t > 2$ GeV/c and one electron with $E_t > 5$ GeV;
- single lepton triggers: one muon with $p_t > 7.5$ GeV/c or one electron with $E_t > 8$ GeV;

*FOR THE CDF COLLABORATION

Dilepton triggers are used to select resonant states (J/ψ, ψ', Υ) and reconstruct exclusive b decays ($B \to J/\psi K$), or (away from resonances) for mixing studies and rare decays search. Single lepton triggers are the source of semi-exclusive b-hadron semileptonic decays used for lifetime measurements.

The CDF detector has been described in detail elsewhere[1]. The subdetectors which are relevant for B physics are the tracking, the muon systems and the electromagnetic calorimeter. The main components of the tracking system are the Silicon Vertex Detector[2] and the Central Tracking Chamber, which are contained in a 1.5 T magnetic field produced by a superconducting solenoid. Electrons are detected as showers in the electromagnetic calorimeter which surrounds the solenoid. Muons are detected by sets of drift chambers located in the back of the central calorimeter modules. The tracking system has excellent momentum and impact parameter resolution: $\sigma_{p_t}/p_t^2 \sim 10^{-3}$ and $\sigma_d = (40/p_t + 13)$ μm (p_t in GeV/c).

2 Observation of the B_c meson

The B_c meson is the bound state of a beauty and a charm quark. Nonrelativistic QCD potential models and the spectator model are expected to give a reliable description of its spectroscopy[3] and an estimate ot its lifetime[4]: the mass is expected between 6.2 and 6.3 GeV/c^2 and, depending on the adopted model, the lifetime is estimated between 0.4 and 1.4 ps. Although significantly shorter than other B mesons, the B_c lifetime is expected to be measurable with the CDF Silicon Vertex Detector. Fragmentation models predict that B_c production is suppressed by a factor of the order of 10^{-3} with respect to B_d and B_u mesons.

The most favorable and clear experimental signature is the decay into final states which include a J/ψ which have also a total branching ratio of the order of 20 %. We report here CDF search for the decay $B_c \to J/\psi l \nu$ where the lepton is either a muon or an electron[5 6]. The J/ψ is reconstructed through the decay $J/\psi \to \mu^+\mu^-$ which is selected at trigger level. The two muons and the third lepton are constrained to come from a common displaced vertex. Due to the presence of the neutrino in the final state, the only measurable quantities are the mass and the pseudoproper decay length of the trilepton system ($M(J/\psi l)$ and $ct^* = \frac{L_{xy} \times M(J/\psi l)}{p_t(J/\psi l)}$, where L_{xy} is the distance between the reconstructed decay vertex and the average beam position in the transverse plane). From Monte Carlo studies, which assume a B_c mass of 6.27 GeV/c^2, most signal events are expected to have $4 < M(J/\psi l) < 6$ GeV/c^2. To select possible decays of long lived particles a lower threshold on ct^* at 60 μm is set. The ct^* cut is chosen at about 1σ from zero (SVX provides a vertex position

Table 1. Summary of background sources and estimate of signal events for a $J/\psi l$ mass between 4 and 6 GeV/c^2.

	J/ψ e results	J/ψ μ results
False Electrons	$2.6 \pm 0.05 \pm 0.3$	
Conversions	$1.2 \pm 0.8 \pm 0.4$	
Total False Muons		$6.4 \pm 0.5 \pm 1.3$
Punch-Through		$0.88 \pm 0.13 \pm 0.33$
Decay-in-Flight		$5.5 \pm 0.5 \pm 1.3$
$B\overline{B}$ bck.	1.2 ± 0.5	0.7 ± 0.3
Total Background	5.0 ± 1.1	7.1 ± 1.5
Events observed in data	19	12
Net Signal	14.0	4.9
$P_{Counting}$(Null)	2.1×10^{-5}	0.084

resolution of 50 μm) in order to have higher efficiency on signal. This cut has beeen removed for the lifetime measurement.

Starting from a sample of 196,000 J/ψ reconstructed in SVX and after rejecting candidates compatible with being $B^+ \to J/\psi K^+$ decays and events where the electron was identified as coming from a photon conversion, a sample of 31 J/ψ-lepton candidates (19 $J/\psi e$ and 12 $J/\psi\mu$) is found. A correct understanding and estimate of backgrounds is essential to claim that B_c is found. The main sources of background are expected to be due to real J/ψ which form a good displaced vertex when paired to a hadron misidentified as third lepton, and to $b\bar{b}$ events with one b hadron decaying to a J/ψ and the other b hadron decaying semileptonically, with a topology of the event compatible with having the J/ψ and the lepton exiting from the same vertex. Hadrons misidentified as the third lepton are found to be the main source of background. For muons this is due to light hadrons (pions or kaons) which punch-through the calorimeter and are then detected in the muon chambers, or decay in flight and produce a muon with a kink small enough to be well linked to the track of the hadron. For electrons this happens when the shower of a hadron in the electromagnetic calorimeter is undistinguishable from that of an electron. The contribution of these sources is estimated in the following way: starting from the J/ψ-lepton sample, the parent J/ψ-track sample is obtained releasing the lepton identification criteria on the third track; this sample is then weighted with the probability, estimated from real data as a function of p_t, that a hadron is misidentified as lepton. With this method also the mass shape of background can be obtained. Real $J/\psi l$ background from $b\bar{b}$ events is estimated from Monte Carlo simulation. Other possible sources of background are electron production from photon conversions and events with

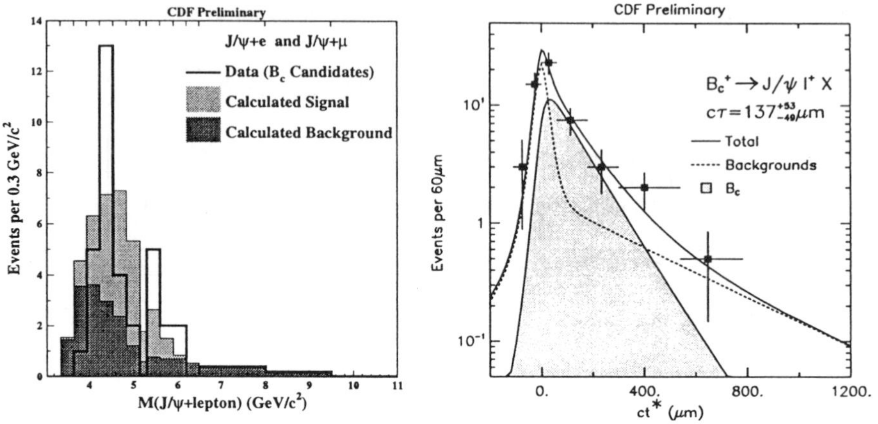

Figure 1. Left: Mass distribution of B_c candidates. The result of the fit for B_c signal (blue) and measured backgound (magenta) is superimposed. Events found to be compatible with being a $B^+ \to J/\psi K^+$ decay, when the kaon mass is assigned to the lepton, are rejected. These events would fall in the narrow $J/\psi l$ mass bin. Right: Lifetime distribution for data (crosses) with the result of the fit for signal (shaded histogram) and background (dashed line). The long lifetime component in the background distribution is due to B decays.

false J/ψ paired to a real lepton. A summary of all background sources and of the estimated signal in both channels is reported in Table 1.

The number of B_c mesons and the statistical significance of the excess is also estimated from a likelihood fit of the $J/\psi l$ mass distribution (fig. 1). The mass shape for signal and for background are respectively constrained to the results of signal simulation and of background measurement. The only free parameter returned by the fit is the number of B_c mesons, $N(B_c) = 20.4^{+6.2}_{-5.5}$. Signal significance is determined generating simulated experiments with no signal and the same level of background as data and counting how often the fit returns a value for $N(B_c)$ greater or equal than 20.4. The null hypothesis is rejected at the 4.8σ level.
To determine the B_c mass, the observed $J/\psi l$ mass distribution is fit for assumed B_c masses between 5.5 and 7.5 GeV/c^2. The dependence of the likelihood on the B_c mass shows a shallow minimum and provides the result:

$$M(B_c) \;=\; (6.40 \;\pm\; 0.39(stat.) \;\pm\; 0.13(syst.))\; GeV/c^2 \qquad (1)$$

This result confirms the initial value assumed for the mass (6.27 GeV/c^2). To determine the lifetime, the cut on the pseudoproper decay length (initially set at 60 μm) is dropped. The new ct^* distribution (fig. 1) is fitted assuming that the background is parametrized by a prompt contribution plus a negative and a positive exponentials, while the signal is parametrized by a sigle positive exponential on which a statistical correction for the missing neutrino p_t is applied. Both background and signal distributions are convoluted with the experimental resolution on the decay length. The fit returnes:

$$\tau(B_c) = (0.46^{+0.18}_{-0.16}(stat.) \pm 0.03(syst.))\ ps \qquad (2)$$

A measurement of the B_c production cross section times branching ratio is performed with respect to B^+ production times $B^+ \to J/\psi K^+$ decay rate. This is done in order to cancel several systematical uncertainties and we find:

$$\frac{\sigma(B_c^+) \cdot Br(B_c^+ \to J/\psi l^+ \nu)}{\sigma(B_u^+) \cdot Br(B_u^+ \to J/\psi K^+)} = 0.132^{+0.041}_{-0.037}(stat.) \pm 0.031(syst.)^{+0.032}_{-0.020}(lifet.) \qquad (3)$$

3 Lifetimes

B mesons lifetime measurements are very important to search for effects beyond the spectator decay model. The expected effects are much smaller than in the case of D mesons: a difference only of the order of 5-10 % is expected between B_u and B_d lifetimes. CDF has measured the lifetime of all the species of b hadrons. B_d and B_u lifetime has been measured both in fully reconstructed $B \to \Psi K$ exclusive decays[7] (Ψ is either a J/ψ or a $\psi(2S)$, K can be any of the following: $K^\pm$, $K^{*0} \to K\pi$, $K_s^0 \to \pi^+\pi^-$ or $K^{*+} \to K_s^0\pi^+$) and in partially reconstructed semileptonic decays[8] $B \to DlX$ (D can be any of the following: a) $D^0 \to K^-\pi^+$; b) $D^{*+} \to D^0\pi_s^+$, $D^0 \to K^-\pi^+$; c) $D^{*+} \to D^0\pi_s^+$, $D^0 \to K^-\pi^+X$). The average of the two measurements gives the following results:

$$\tau(B_u) = 1.66 \pm 0.05(stat \oplus syst)\ ps \qquad (4)$$

$$\tau(B_d) = 1.52 \pm 0.06\ (stat \oplus syst)\ ps \qquad (5)$$

$$\tau(B_u)/\tau(B_d) = 1.09 \pm 0.05\ (stat \oplus syst)\ ps \qquad (6)$$

These measuerements are very competitive with the results from other experiments and are approaching the precision necessary to test non spectator contributions. An update of a previous B_s lifetime measurement[9] in the semileptonic channels will be here presented. We anticipate the result:

$$\tau(B_s) = 1.36 \pm 0.09(stat.)^{+0.06}_{-0.05}(syst)\ ps \qquad (7)$$

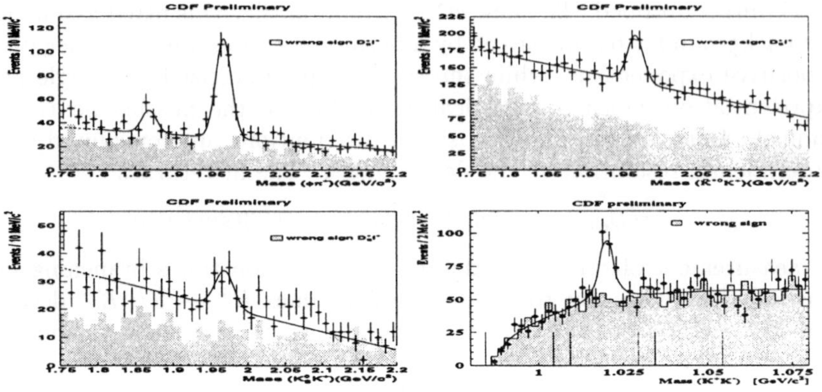

Figure 2. D_s invariant mass for the $B_s \to D_s^- l^+ \nu X$ events. Shaded histograms are the wrong sign $D_s l$ combinations which are used together with the D_s mass sidebands to determine the background shape in the lifetime fit. For the $D_s^- \to \phi\mu^-\nu$ channel the ϕ candidates mass is reported: in this case wrong sign combinations are $\mu^\pm\mu^\pm$ and $K^\pm K^\pm$ and the shaded histogram normalisation is rescaled to have the same number of signal and background events in the mass sidebands.

CDF has also measured Λ_b^0 lifetime[10] in the partially reconstructed decay $\Lambda_b^0 \to \Lambda_c^+ l^- X$, with the Λ_c^+ reconstructed in the resonant decay $\Lambda_c^+ \to K^- p \pi^+$. For completeness we also recall the result for B_c meson:

$$\tau(\Lambda_b^0) = 1.32 \pm 0.15(stat)\, 0.07(syst) ps \quad (8)$$

$$\tau(\Lambda_b^0)/\tau(B_d) = 0.85 \pm 0.10(stat) \pm 0.05(syst)\ ps \quad (9)$$

$$\tau(B_c) = 0.46^{+0.18}_{-0.16}(stat) \pm 0.03(syst)\ ps \quad (10)$$

3.1 Measurement of B_s Lifetime

We present a new result obtained using the partially reconstructed semi-leptonic decay $B_s \to D_s^- l^+ \nu X$, with the lepton being either an electron or a muon. The D_s is reconstructed in four modes: a) $D_s^- \to \phi\pi^-$,

b) $D_s^- \to K^{*0}K^-$, c) $D_s^- \to K_s^0 K^-$, d) $D_s^- \to \phi\mu^-\nu$. Signal events are events with a $D_s l$ invariant mass between 3 and 5 GeV/c^2. We find about 600 such events. Mass plots for the four decay channels are reported in fig.2. Control samples for combinatorial background are obtained from the D_s mass side-bands and from wrong sign lepton-D_s combinations. A source of background is due to $D_d^- \to K^{*0}\pi^-$ and $D_d^- \to K_s^0\pi^-$ events which can be reconstructed as D_s^- decays to $K^{*0}K^-$ and $K_s^0K^-$. This is caused by the poor CDF particle identification. Due to low signal statistics, no attempt of reducing this background is performed. Instead, we introduce the estimated fraction of D_d^- events in the signal samples as a parameter of the final lifetime fit. This fraction is estimated using the large mass difference between D_d^- and D_s^- and the shape of reflections in the mass distributions. Intersecting the D_s and the lepton trajectories in the transverse plane, the B_s decay length is determined with a resoution of about 100 μm. A correction for the missing transverse momentum of the neutrino is applied event by event to obtain the proper decay time. The combined result of the four modes has been anticipated in eq. (7). This is the most precise measurement from a single experiment and it is lower by about 2σ than the current world average (1.57 ± 0.08 ps[11]).

4 Mixing

The Standard Model allows $B^0 \to \overline{B}^0$ transitions via higher order weak interactions contributions. As a consequence of this, a B^0 produced at $t = 0$ has a certain probability to mix into a $\overline{B}^0$ and viceversa. To measure this effect as a function of proper time, it is necessary to determine the B flavor at decay and production time and to measure the time of decay. When this can be done the frequency of mixing can then be determined from the following asymmetry:

$$A(t) = \frac{dN/dt_{B^0\to B^0} - dN/dt_{B^0\to\overline{B}^0}}{dN/dt_{B^0\to B^0} + dN/dt_{B^0\to\overline{B}^0}} = \cos(\Delta m t) \qquad (11)$$

where Δm is the mass difference between the B^0 and $\overline{B}^0$ states. Since it is possible to assign wrong tags, a mistag probability (P_0) has to be introduced and the measured asymmetry becomes ($2P_0$ - 1)$\cdot\cos(\Delta m t) =$ $D_0 \cdot \cos(\Delta m t)$, with D_0 being the diluition of the tagging algorithm.

Experimentally the idea is to reconstruct one B in the $b \to cl\nu$ mode and tag the B flavor at decay time with the charge of the lepton. The time of decay is determined from the decay length in the same way as in the lifetime measurements. This is done using either exclusive D decay modes or inclusive secondary vertexing techniques, which take advantage of a higher statistics but pay the price of a lower resolution on the decay time. The B flavor at production time can be determined from the away side b[a] (using the lepton from the semileptonic decay, or a p_t-weighted charge average of the tracks in the b-jet), or from the tracks produced in the fragmentation of the same side b using the Same Side Tagging (SST) algorithm[12]. The SST method tags the B flavor using the charge of the pion produced with the minimum p_t^{rel} with respect to the system constitued by the B meson and the pion itself. CDF has recently measured B_d^0 mixing using this tagging technique[14] applied on a sample of partially reconstructed semileptonic decays $B^0 \rightarrow l^+ D^{(*)-} X$. The measurement determines simultaneously Δm_d and D_0:

$$\Delta m_d \; = \; 0.471^{+0.078}_{-0.068} \; \pm \; 0.034 \; ps^{-1} \tag{12}$$

$$D_0 \; = \; 0.18 \; \pm \; 0.03 \; \pm \; 0.02 \tag{13}$$

5 Rare decays

Evidence for physics beyond the Standard Model would result from an anomalous rate of rare decays like $B_d^0(B_s^0) \rightarrow \mu^+\mu^-$. The theoretical prediction for the branching ratio is at the level of $10^{-10}(10^{-9})$[15]. Experimentally this channel is particularly favorable for the presence of the two muons which can be requested at trigger level, taking thus advantage of the copious B production provided by the Tevatron. The analysis basically requires the muon pair to come from a secondary vertex ($c\tau > 100\ \mu$m) and to have an invariant mass within 75 MeV/c^2 (3σ) from the $B_d^0(B_s^0)$ mass. The request that B candidates are isolated, i.e. they carry most of the energy in the region surrounding the B candidates themselves, provides further background rejection[16]. A similar analysis is performed to set limits also on the branching ratio for the non resonant $B_u^+(B_d^0) \rightarrow \mu^+\mu^- K^+(K^{*0})$ decays, for which the theoretical prediction is $0.4(1.0) \cdot 10^{-6}$. The following 90 % c.l. upper limits are found:

$$Br(B_d^0(B_s^0) \rightarrow \mu^+\mu^-) \; < \; 6.8 \; \cdot \; 10^{-7} \; (2.0 \; \cdot \; 10^{-6}) \tag{14}$$

$$Br(B_u^+(B_d^0) \rightarrow \mu^+\mu^- K^+(K^{*0})) \; < \; 5.4 \; (4.1) \; \cdot \; 10^{-6} \tag{15}$$

which are still far away (three orders of magnitude) from theoretical predictions for the $B_d^0(B_s^0) \rightarrow \mu^+\mu^-$, while they are getting very close to constraining the theory in the $B_u^+(B_d^0) \rightarrow \mu^+\mu^- K^+(K^{*0})$ case.

---

[a] "Same side b" is the b reconstructed in the $b \rightarrow cl\nu$ mode. "Away side b" is the other b produced in the $p\bar{p}$ interaction: it can be used to tag the B flavor at production time (away side tagging methods).

CDF has also searched for the non Standard Model $B_d^0(B_s^0) \to e^\pm\mu^\mp$ decays. The simplest model which incorporates the lepton flavor non-conservation and admits the existence of a new force that mediates transitions between quark and leptons is the Pati-Salam model[17]. This model predicts heavy bosons called Pati-Salam leptoquarks (LQ) which would mediate the decays $B_d^0(B_s^0) \to e^\pm\mu^\mp$ [18, 19]. Limits on branching ratios are set and limits on LQ masses are derived (at 90 % c.l.):

$$Br(B_d^0(B_s^0) \to e^\pm\mu^\mp) \; < \; 3.5(6.1) \cdot 10^{-6} \tag{16}$$

$$M_{LQ}B_d^0(B_s^0) \; > \; 21.7(20.7)\; TeV/c^2 \tag{17}$$

6 Prospects for run II

CDF strategy for run II upgrades is to optimize quality of information in the central region and expand coverage. A new 3D Silicon Vertex detector and an Intermediate Silicon Layers detector will be installed in order to improve tracking robustness and vertex finding ability in the central region and extend tracking capability to the forward region. The Silicon Vertex Tracker[20] will allow online tracking of the Silicon Vertex[13]. B samples statistics will result significantly increased by the possibility to trigger on secondary vertices and otherwise unexplorable channels will be accessible. The increased coverage of lepton detectors will improve B tagging efficiency in the leptonic modes. CDF goals for B physics in run II are the observation of CP violation, of B_s mixing and of rare and radiative B decays. Although the sensitivity on the recent $\sin(2\beta)$ measurement[b] from run I data is low ($\sin(2\beta) = 1.8 \pm 1.1 \pm 0.3$)[21], this result establishes the feasibility of measuring CP symmetries in B meson decays at a hadron collider. With the expectation of 20 times more integrated luminosity in run II and the improved detector performance, CDF has very good prospects to observe CP violation in the $B_d^0 \to J/\psi K_s$ and maybe also in the $B_d^0 \to \pi^+\pi^-$ decays.

7 Conclusions

Recent B physics results from CDF experiment based on data collected during the 1992-1996 run have been reported: the B_c discovery, an updated B_s lifetime measurement, a new measurement of Δm_d obtained with the Same Side Tagging technique, and new limits set on rare B decays branching ratios.

[b]This result was not reported in these Proceedings because it was unpublished at the time of this Conference.

Extrapolations of current results give bright B physics prospects to the future high luminosity run II.

Acknowledgments

We thank the Fermilab staff and the technical staffs of the partecipating institutions for their vital contributions. This work was supported by the U.S. Department of Energy and National Science Foundation; the Italian Istituto Nazionale di Fisica Nucleare; the Ministry of Education, Science and Culture of Japan; The Natural Sciences and Engineering Research Council of Canada; the National Science Council of the Republic of China; the Swiss National Science Foundation; and the A.P. Sloan Foundation.

References

1. F. Abe et al., Nucl. Instr. Meth. **A 271**, 387 (1988).
2. D. Amidei et al., Nucl. Instr. Meth. **A 350**, 73 (1994).
3. E. Eichten et al., Phys. Rev. D **49**, 5845 (1994).
4. M. Benecke et al., Phys. Rev. D **53**, 4991 (1996).
5. F. Abe et al., FERMILAB-Pub-98/121-E, Submitted to Phys. Rev. D.
6. F. Abe et al., FERMILAB-Pub-98/157-E, Submitted to Phys. Rev. Lett.
7. F. Abe et al., The CDF Coll., Phys. Rev. D **57**, 5382 (1998).
8. F. Abe et al., FERMILAB-Pub-98/167-E, Submitted to Phys. Rev. Lett.
9. F. Abe et al., Phys. Rev. Lett. **77**, 1945 (1996).
10. F. Abe et al., Phys. Rev. Lett. **77**, 1439 (1996).
11. R. M. Barnett et al., Phys. Rev. D **54**, 1 (1996).
12. M. Gronau, A. Nippe and J. Rosner, Phys. Rev. D **47**, 1988 (1993).
13. The CDF II Coll., The CDF II Detector TDR,FERMILAB-Pub-96/390-E
14. F. Abe et al., Phys. Rev. Lett. **80**, 2057 (1998).
15. A. Ali, DESY 97-019, hep-ph/97022312.
16. F.Abe et al., Phys. Rev. D **57** Rapid Communications, R3811 (1998).
17. J. Pati and A. Salam, Phys. Rev. D **10**, 275 (1974).
18. G. Valencia, S. Willenbrock., Phys. Rev. D **50**, 6843 (1994).
19. A. V. Knznctsov and N. V. Mikhcev, Phys. Lett. B **329**, 295 (1994).
20. S. Belforte et al., IEEE Transactions on Nuclear Science, Vol. 42, no. 4, August 1995.
21. F. Abe et al., FERMILAB-PUB-98/189-E, Subm. to Phys. Rev. Lett.

CLOSING IN ON B-CP

David Atwood
Department of Physics and Astronomy, Iowa State University,
Ames, IA 50011
Email: atwood@iastate.edu

Amarjit Soni
Theory Group, Brookhaven National Laboratory,
Upton, NY 11973
E-mail: soni@penguin.phy.bnl.gov

Major new experimental efforts on detecting CP violation in B decays will very soon go on the air. Recent developments suggest that final state interaction phases in exclusive decays of B are unlikely to be small indicating the possibility of observable direct CP asymmetries in these channels. CLEO results on charmless hadronic modes suggest penguin amplitudes are rather big implying that the extraction of α from B^0-$\bar{B}^0$ alone will be difficult, thus necessitating also information from direct CP. Importance of $B(B_s)$ decays to two vectors for model independent tests of electroweak penguins and for extraction of $\alpha(\gamma)$ is emphasized. Inclusive $b \to sg^*$ and related modes, e.g. $B \to \eta' X_s$, are very good probes of CP-odd phase(s) due to beyond the standard model physics. On the other hand, $b \to dg^*$ and related modes, e.g. $B \to \eta' X_d$ are more suitable for CP violation due to the CKM phase. Two body b-quark decays: $b \to Mq_f$ leading to semi-inclusive, B decays, $B \to MX$ (with $2 \lesssim E_M \lesssim 2.8$ GeV), are very interesting and important. Their theory is relatively clean; partial rate asymmetries may be large in several cases (e.g. $M = K^*, K, \rho, \pi \ldots$) and a few cases provide very good probe of electroweak penguins.

1 Introduction and Summary[1]

Following is the outline of this talk:

I. Introduction and Summary

II. CLEO97: A robust penguin $\Rightarrow B^0$-$\bar{B}^0$ cannot do it alone.

III. In exclusive B-decays, final state rescattering phases are unlikely to be small.

IV. Repercussions of long distance rescattering for direct CP.

V. $B \to V_1 V_2$: Model independent tests for electroweak penguins and extraction of the angles of the unitarity triangle.

VI. Chasing LUCY with beauty: Search for a non-standard-phase via $b \to sg^*$, $B \to \eta' X_s$ and related modes.

VII. $b \to dg^*$, $B \to \eta' X_d$ and related modes for direct CP violation due SM.

VIII. Two body b decays: $b \to qM$ for direct CP studies and for looking for electroweak penguins.

The asymmetric B-factories at KEK and SLAC and the symmetric one at CORNELL will begin operation in about six months. There is great degree of expectation that these will soon provide us important new clues with regard to understanding an intriguing mystery of nature—CP violation.

Although the centerpiece of the asymmetric B-factories is the study of mixing induced CP via measurements of time dependent asymmetries in B^0-$\bar{B}^0$ decays, it is becoming increasingly clear that a precise determination of the angles of the unitarity triangle, essential for testing the SM, will require a broader understanding of CP violation i.e. involving extensive studies of direct CP violation dealing with charged as well as neutral B mesons.

The main point is that recent observation of the charmless, hadronic, exclusive modes by CLEO strongly indicate that the size of the penguin amplitudes is appreciable.[2,3,4] This renders the extraction of α or γ through B^0-$\bar{B}^0$ observations alone rather difficult. Subtraction of the penguin amplitudes will require study of CP asymmetries in charged B decays as well.

An essential ingredient that drives direct CP violation is the final state (FS) rescattering (CP-even) phase. Reliable calculation of this phase from theory still remains and outstanding and important challenge. Some phenomenological arguments as well as indirect indications suggest that these phases may not be small at the B mass. Thus sizeable direct CP violations in B-decays are very likely. Indeed, the FS rescattering phases in exclusive channels originate at least in part due to soft-physics (i.e. non-perturbative effects) and can lead to a different pattern of CP asymmetries[5] than hitherto envisioned to emerge from short distance operators. In exclusive modes (such as $K\pi$) appreciable asymmetries are plausible.[5]

Direct CP measurements in some of the exclusive channels are not entirely in vain insofar as determination of the unitarity angle goes. For example, measurements in the $K\pi$ can be used for quantitatively testing for the presence of electroweak penguins (EWP).[5] If EWP contributions to $K\pi$ are found to be negligible then B^0, $\bar{B}^0$, $B^\pm \to K\pi$ studies can be used to determine α.[7]

B, B_s decays to two vector mesons can be especially useful in this context.[9] Angular distribution of the decays of the vector mesons provide information on polarization of the vector mesons.[8] Model independent tests can be constructed to monitor the presence of EWP.[9] Modes that show negligible contribution from EWP can then be used to extract α and γ.[9]

Inclusive $b \to s$ transitions are a very good probe for non-standard CP violation phases.[10,11] Of particular interest is $b \to sg^*$ since it has $Br \sim 1\%$.[12] However, as such this mode is very difficult to detect. CLEO has recently reported a large signal for $B \to \eta' X_s$.[13] Experimental data, to date, supports the interpretation that an appreciable fraction of the signal comes from $g^* \to g\eta'$ through the anomalous coupling of the η' to the two gluons.[14] Consequently $B \to \eta' X_s$ becomes a good search for the non-standard-phase (NSP).[10,11] It may be useful to extend the search to other expected fragments (X_g) of $g^* \to X_g g$. Good examples of X_g (in addition to η') are $\eta(1440)$, $f_0(980)$, $f_2(1270)$, K^+K^-, $\pi^+\pi^-$...[10]

The penguin transition $b \to d$ on interference with the tree contribution $b \to u\bar{u}d$ is likely to be a good way to detect direct CP violation from the SM. The modes of interest here are completely analogous to those in the previous paragraph except for the replacement of $X_s \to X_d$.

The last topic to be emphasized in this talk is two body decays of the b quark: $b \to Mq_f$ where M is spin 0 or 1 meson.[15] These decays lead to semi-inclusive B decays. Theory for these should be cleaner than exclusive decays [such as $B \to K\pi$, $\rho\pi$...] and their experimental signatures are rather distinctive. Many of the modes could have large partial rate asymmetries (PRA). Two of the modes $B \to \rho^0 X_s$ and $\pi^0 X_s$ are also a good way to look for EWP as the latter dominate over other contributions.[15]

2 CLEO97: A Robust Penguin $\Rightarrow$ B^0-$\bar{B}^0$ Cannot do it Alone

After years of anticipation, in 1997, CLEO reported the first observation of some charmless, hadronic decay modes; both exclusive and inclusive.[2,3,4,13,16] For our purpose, the most interesting examples of these two categories are:[3,13]

$$Br(B \to K^+\pi^-) = (1.5 \pm .5 \pm .1) \times 10^{-5} \tag{1}$$

$$Br(B \to \eta' X_s) = (6.2 \pm 1.6(\text{stat}) \pm 1.3(\text{syst})^{+0.0}_{-1.5}(\text{bkg})) \times 10^{-4} \tag{2}$$

In addition, CLEO has reported lack of a statistically significant signal in the $\pi\pi$ mode:[2,16]

$$Br(B \to \pi^+\pi^-) \leq (1.5 \times 10^{-5}) \text{ @90\%CL} \tag{3}$$

A simple way to understand (1) and (3) is to assume that the $\pi\pi$ mode is dominated by tree. Due to the Cabibbo angle, the tree contribution to $K\pi$

must therefore be extremely small compared to (1). Thus $B \to K\pi$ with the stated Br is dominated by the penguin. Also the penguin contribution to $\pi\pi \sim |V_{td}|^2 * 1.5 \times 10^{-5}/|V_{ts}|^2 \sim 10^{-6}$, assuming $|V_{td}/V_{ts}| \sim \lambda \sim .22$, which is quite consistent with the existing constraints. Thus the amplitude ratio of tree vs penguin for $\pi\pi$ is about:

$$\frac{P_{\pi\pi}}{T_{\pi\pi}} \gtrsim .25 \tag{4}$$

Therefore, for a precision determination of the unitarity angle α, it does not appear safe to ignore the penguin contribution to $\pi\pi$. Assuming EWP contribution is negligible, extraction of α from $B \to \pi\pi$ is still possible but it requires observation of direct CP i.e. partial rate asymmetry (PRA) measurements in the $B^{\pm}$ will also be needed. In addition, the Br for $\pi^0\pi^0$, likely to be quite small, will also be required.[17]

In fact CLEO measurements (1) and (3) and the resultant $P_{\pi\pi}/T_{\pi\pi}$ also means that the penguin contribution for $B_s \to \rho K_s^0$ is likely to be very large. Indeed, since the tree is then color suppressed it seems the penguin over tree ratio for $B_s \to \rho K_s$:

$$\frac{P(B_s \to \rho K_s)}{T(B_s \to \rho K_s)} \sim 0(1) \tag{5}$$

Thus this "text-book example"[18] for extracting γ will also become very difficult to implement even when the study of B_s-$\bar{B}_s$ oscillations become an experimental reality.

Thus the first CLEO results on charmless B-decays are indicating that studies of B^0-$\bar{B}^0$ for the extraction of angles (α and γ) of the unitarity triangle are unlikely to suffice; information from direct CP in $B^{\pm}$ decays is also likely to play an important role. Interestingly there are good reasons to suspect that direct CP asymmetries in many exclusive modes may be appreciably large.

3 In Exclusive B-Decays Final State Rescattering Phases are Unlikely to be Small.

From the outset let us emphasize that a reliable methodology for calculating final state phases is not available. So, any conclusions about final state rescattering phases we arrive at will suffer from that drawback. Despite that reservation, it is still perhaps useful to point out that there are several reasons to suspect that final states rescattering phases in exclusive decays of B are unlikely to be negligibly small:

1. A long standing problem in B-decays is that the measured value of the semi-leptonic branching ratio is a bit (about 10%) less than the number predicted by theory. Most likely this difference between the two numbers originates from the error in the theoretical calculation of the hadronic B-width. It is quite possible that value of the total hadronic width calculated by using short distance perturbation theory is missing a contribution due to soft physics or non-perturbative effects.

2. The B-baryon lifetime is significantly different from that of B-mesons. Short-distance perturbation theory and/or heavy quark effective field theory apparatus is unable to account for such a difference. Most likely this originates from spectator quark interactions which are strongly intertwined with FS interactions.

3. There is a surprising reversal of trend as one goes from D to B decays. For D's one has[19]

$$\Gamma(D^0 \to K\pi)/\Gamma(D^+ \to K\pi) > 1. \tag{6}$$

In contrast, for the B's one finds

$$\Gamma(B^0 \to D\pi)/\Gamma(B^+ \to D\pi) < 1. \tag{7}$$

4. Donoghue *et al.*[20] have recently examined the dependence of the FS rescattering phases on the mass of the decaying b-quark; specifically they deal with the case of the $B \to K\pi$ mode. Assuming that the total cross-section for $K\pi$ obeys the same scaling laws as that for the other total hadronic cross sections[21] and using the optical theorem they show that the effects of elastic rescattering do not diminish as m_B gets larger.[20] In fact Donoghue *et al.*[20] claim that the effect of inelastic channels[22] is even much bigger than the elastic one leading them to the conclusion that FS rescattering phases are unlikely to disappear at m_B.

So, while a compelling theoretical argument is not available, it is quite plausible that final state rescattering phases in exclusive modes are appreciable; final word on the subject will have to come from experiment.

4 Repercussions of Long Distance Rescattering for Direct CP in Exclusive Modes

4.1 *Possibility of Large Direct CP in Exclusive Modes such as $K\pi$* [5]

FS phases due to long distance (LD) rescattering endow direct CP in exclusive modes a rich structure as can be understood by the use of the CPT theorem. Recall that the theorem requires that the total lifetime of B and $\bar{B}$ to be exactly equal whereas CP symmetry requires the partial widths of B and $\bar{B}$ into conjugate modes to be equal. As a result, when the CP violating partial width difference (PWD) for all the different modes are added up there has to be an exact cancellation to render the equality of the lifetimes. Furthermore, assuming isospin is exactly conserved by the strong interactions, the final states with different isospins cannot participate in the cancellation of PWD.

If multiparticle inelastic channels composed of light quarks make important contributions to the rescattering phases of FS such as $\pi\pi$, $K^*\pi$ etc. then the PWD in these FS must cancel against the PWD occurring amongst those multiparticle states responsible for the phase difference, as required by the CPT theorem. In other words, the cancellation of the PWD for FS such as $\pi\pi$, $K^*\pi$ etc. need not and will not occur with hadrons containing c, and $\bar{c}$ (i.e. $D^*\bar{D}$, $D\bar{D}_s$ etc.) in the FS. The latter has to be the case if rescattering phases were all of short distance (SD) origin coming from the penguin operators.[23,24]

For FS such as $K\pi$, $\pi\pi$... wherein the two mesons each have a non-vanishing isospin an interesting and important possibility for CP violation arises due to the rescattering effects caused by soft physics. For illustration, let us focus on $K\pi$. Then one way that CPT maintenance can occur in the presence of nonvanishing PRA's in the channel is for the PWD in the two possible modes of $B^-(B^0)$ to cancel against each other so that[5]

$$\begin{aligned}\delta(K^-\pi^0) &= -\delta(K^0\pi^-)\\ &= \delta(\bar{K}^0\pi^0) = -\delta(K^-\pi^+)\end{aligned} \tag{8}$$

where δ's denote the PWDs:

$$\delta(K^-\pi^0) \equiv \Gamma(B^- \to K^-\pi^0) - \Gamma(B^+ \to K^+\pi^0) \tag{9}$$

etc. A simple calculation then gives

$$\delta(K^-\pi^0) = 2\sqrt{2}|V_u|\,|V_c|\,|A|\,|D|\sin\gamma\sin\Phi \tag{10}$$

where $V_u = V_{ub}V^*_{us}$, $V_c = V_{cb}V^*_{cs}$, A is the $\Delta I = 1/2$ penguin amplitude for $b \to s$, D is the $\Delta I = 3/2$ tree amplitude, Φ is the strong rescattering phase, $\Phi = \arg(DA^*)$ and $\gamma \approx \arg(-V^*_{ub})$ is the CP-odd weak phase. The corresponding PRAs in these channels become:[5]

$$\begin{aligned}\text{PRA}(K^-\pi^0) &= \text{PRA}(\bar{K}^0\pi^0) &&= -2\text{PRA}(\bar{K}^0\pi^-)\\ & &&= -2\text{PRA}(K^-\pi^+)\\ &= \sqrt{2}r\sin\gamma\sin\Phi \end{aligned} \tag{11}$$

where $r = T_{K\pi}/P_{K\pi}$ for the tree versus penguin amplitudes for $K\pi$. Setting $\sin\gamma = \sin\Phi = 1$ for maximal phases:

$$\begin{aligned}\text{PRA}(K^-\pi^0)|_{\text{max}} &= \text{PRA}(\bar{K}^0\pi^0)|_{\text{max}} = -2\text{PRA}(\bar{K}^0\pi^-)|_{\text{max}}\\ &= -2\text{PRA}(K^-\pi^+)|_{\text{max}} = \sqrt{2}r\end{aligned} \tag{12}$$

Assuming $r \sim .3$, the PRA in $K^-\pi^0$ and $\bar{K}^0\pi^0$ modes can be $\sim 42\%$ whereas that in $\bar{K}^0\pi^-$ and $K^-\pi^+$ is $\sim 21\%$ (with the opposite sign).[5]

We thus see that FS rescattering phases can cause appreciable direct CP violating PRAs in exclusive modes such as $K\pi$.

4.2 Model Independent Test for EWP and/or New Physics in $K\pi$

Many studies of EWP exist by now.[25] However, in their numerical estimates these studies invoke many model dependent assumptions and approximations. Model independent assessment of EWP contributions are essential for possible application of these modes for extraction of the unitarity angles and/or for the search for new physics.

Assuming isospin conservation and making no additional assumptions or approximations allows us to arrive at a sum rule[5]

$$2|m_1|^2 - |m_2|^2 - |m_3|^2 + 2|m_4|^2 = 2|\bar{m}_1|^2 - \bar{m}_2|^2 - |\bar{m}_3|^2 + 2|\bar{m}_4|^2 \tag{13}$$

where m_1, m_2, m_3, m_4 are the amplitudes for B decays to $K^+\pi^0$, $K^0\pi^+$, $K^+\pi^-$ and $K^0\pi^0$ respectively.

A violation of this sum rule will be a model independent test for EWP. If, on the other hand, such a model independent test shows that EWP contribution is negligible, then the direct CP measurements in the $K\pi$ modes would also become very useful for the extraction of the angle α via a quadrangle construction[7] involving the four modes of $(B^+, \bar{B}^0)$ and (B^-, B^0) each. However, deducing the weak phase (α) through this construction also requires, in addition, time dependent oscillation studies of $(B^0, \bar{B}^0)$ to a self-conjugate final state, i.e. $K_s^0\pi^0$.[5,7]

4.3 Decays to Two Vectors: Extraction of α, γ and search for EWP[9]

The decays B, $B_s \to 2$ vector final states can be especially useful for determining the angles of the unitarity triangle, for model independent tests of EWP and for the search for new physics. For our method to work requires modes in which penguin and the tree are both contributing. Specifically, the final states need satisfy the following two conditions:

1. At least two decay amplitudes related by isospins must be involved.

2. The QCD penguin Hamiltonian must contributre only one isospin amplitude to the FS.

Some illustrative examples are $B \to K^*\omega$, $K^*\rho$, $\rho\omega$ and $\rho\phi$ and $B_s \to K^*\bar{K}^*$ and $K^*\rho$. The method provides a model independent test (see below) for EWP. Among the $B(B_s)$ modes mentioned above, those that are least affected by EWP can be used for the extraction of $\alpha(\gamma)$

Once conditions (1) and (2) above are satisfied B^-, $\bar{B}^0$ decays to a given FS can be used to write down a linear combination of the amplitudes that contains only the phase of the tree. As in the previous $K\pi$ case, to extract the weak phase α or γ, though, time dependent oscillation studies of B^0, $\bar{B}^0$ to the corresponding self conjugate FS are needed.

Another important ingredient in this study is the correlations between the decay distributions of the two vector mesons, or equivalently the correlations of their polarizations.

Extraction of the information about the CKM phases proceeds along the following key steps:

1. Following Gronau and London,[17] isolate a linear combination $(c_1 u_1^h + c_2 u_2^h)$ which contains only the weak phase of the tree. Here u_1^h and u_2^h are the corresponding amplitudes for B^- and $\bar{B}^0$ decays; the superscript $h = 1, 0, +1$ denotes the helicity of the vector particles. This combination of the amplitude and its CP conjugate will be related as follows:

$$(c_1 u_1^h + c_2 u_2^h)e^{-i\delta_T} = (c_1 \bar{u}_1^h + c_2 \bar{u}_2^h)e^{i\delta_T} \tag{14}$$

   where δ_T is the weak phase of the tree.

2. Angular distributions of the decay products of the two vector particles is used to obtain the magnitudes of the helicity amplitudes and the phases between the pairs of helicity amplitudes.

3. Since the phase differences between different helicities for the same FS can be obtained from the angular distributions, the set of three equations eqn. (14) (one for each helicity) can now be regarded as a set of three linear equations for the four remaining undetermined phases, ie. one phase for each of the two modes and two conjugate modes. If there is no EWP contamination, the equations will have a solution where the phases are unimodular as expected.[9] On the other hand, if such a solution cannot be found, it implies that eqn. (14) fails so EWP contamination or a new physics contribution is present. This test of EWP is completely model independent and is based only on isospin conservation.

4. Finally, as mentioned before, a time-dependent oscillation study of B^0-$\bar{B}^0$ to the corresponding self-conjugate FS needs to be done to fix the relative phase between the B^0 and the $\bar{B}^0$. This enables the phases between all pairs of amplitudes to become known so that the weak phase of the tree can thus be retrieved.

For $B \to K^*\omega$ the method should work especially well since the color allowed EWP contribution, (resulting in γ, $Z \to \omega$), carries the same weak phase as the strong penguin. Therefore, the extraction of α is rather clean, suffering only from the contamination of a color-suppressed EWP contribution which will most likely be sub-dominant. In any case, the tests for EWP mentioned above should be able to quantitatively monitor the extent of the contamination.[9]

For $B \to K^*\rho$, since $\bar{B}^0 \to \bar{K}^{*0}\rho^0$ is susceptible to EWP contamination a slightly more involved strategy which uses all four of the $K^*\rho$ modes but does allow a determination of α.[9]

Two other vector modes that may be useable for α extraction are $B \to \rho\omega$ and $B \to \rho\phi$. For $B \to \rho\omega$ to work, it will need to be demonstrated that EWP contamination is small as $B^0 \to \rho^0\omega$ receives color allowed EWP contribution from γ, $Z \to \rho^0$.[9] For $B \to \rho\phi$, EWP contribution is not expected to be a problem; however, its Br is likely to be $0(10^{-6})$ whereas for $K^*\omega$, $K^*\rho$ and $\rho\omega$ the Br's are likely to be $0(10^{-5})$.

The method can also be extended to determine γ via $B_s \to K^*\rho$ and $\bar{K}^*K^*$.

5 Chasing LUCY With Beauty: Search for a NSP Via $b \to sg^*$, $B \to \eta' X_s$ and Related Modes.

There are good reasons to think that in addition to the CKM phase of the SM, other CP violation phase(s) exist in nature. For one thing, it appears difficult to account for baryon asymmetry in the universe with the CKM phase. Besides, extensions of the SM with new scalars, fermions or gauge bosons almost invariably entail new phase(s). [LUCY is a generic name for a non-standard phase (NSP)]. With the advent of B-factories it is imperative to ask, how best can B-physics be used to search for a NSP.

For a variety of reasons the penguin transition, $b \to sg^*$ (where g^* is an on or off-shell gluon) is a highly suitable probe of LUCY. In inclusive processes the SM-CKM phase is likely to make only a negligible contribution, so non-standard CP violating effects may not get masked that easily. The fairly large $Br \sim$ 1–2% can be very helpful. However a completely inclusive search for $b \to sg^*$ is rather difficult.

CLEO experimentalists have recently provided an ingenious clue for harnessing an appreciable fraction of the $b \to sg^*$ signal. We are referring here to an observation of an unexpectedly large inclusive signal of η':[13]

$$Br(B \to \eta' X_s) = [6.2 \pm 1.6(\text{stat}) \pm 1.3(\text{syst})^{+0.0}_{-1.5}(\text{bkg})] \times 10^{-4} \qquad (15)$$

for $2.0 < P\eta' < 2.7$ GeV. The magnitude of this signal is about a factor of 5 bigger than estimates based on the SD Hamiltonian. Of course, since the experimental discovery was first reported a lot of theoretical effort has gone into understanding its origin.[14,26] An interesting suggestion is that an appreciable fraction of the η' signal originates from the coupling of the η' to two gluons via the QCD anomaly[14] (see Fig. 1). Recall that this anomalous coupling of the η' distinguishes the singlet η' from the flavored octet of pseudo-goldstone bosons and renders it heavier than the other members of the octet. The point is that the singlet axial current receives a contribution from the gauge sector:

$$\partial_\mu j_5^\mu = \frac{3\alpha_s}{4\pi} G_{\mu\nu}\tilde{G}^{\mu\nu} + 2i \sum_{q=uds} m_q \bar{q}\gamma^5 q \qquad (16)$$

Needless to say, quantitative estimates of this (or other) contributions to the signal for $\eta' X_s$ is very difficult. To begin with, the mechanism was found to be roughly able to account for the observed signal if one assumes that the $g^* - \eta' - g$ form factor stays approximately constant with q^2 over the relevant range, q^2 being the 4-momentum of g^*. Since then further study by CLEO[13] of the recoil mass distribution seems to give some additional support to the anomaly idea.[14]

It should be emphasized that the $g^* - \eta' - g$ form factor introduces appreciable uncertainty into this calculation; the off-shell gluon (g^*) carries a non-vanishing q^2, perhaps 0(a few GeV^2). There is no good reason to think

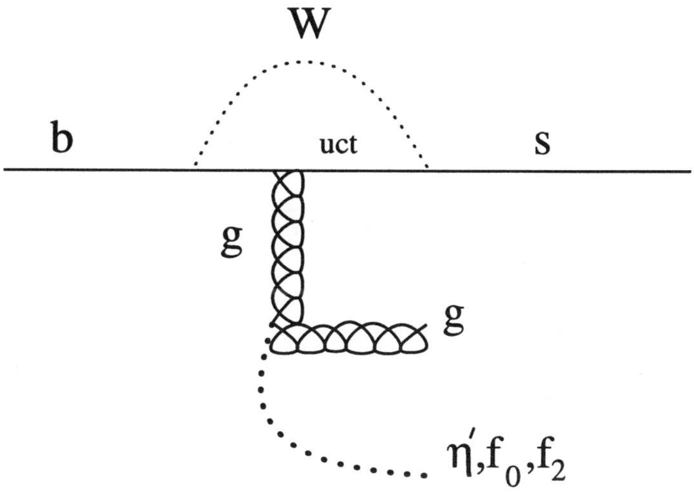

Figure 1: *The penguin diagram giving rise to $b \to sg^*$ followed by $g^* \to g + \eta'$, f_0, f_2 or other such states.*

that this form factor would be the same as the $\gamma^* - \pi^0 - \gamma$ one,[27] wherein it goes as $4\pi^2 f_\pi^2/q^2$, with $f_\pi = 130$ MeV. For the gluonic-rich η', the analogous form factor may involve a larger dimensionful parameter than f_π e.g. constituent quark mass and/or an effective gluon or glueball mass.

This mechanism (see Fig. 1) is especially significant for searches for CP violation[10,11] The key point is that the underlying fragmentation process, $g^* \to g\eta'$ renders the Feynman amplitude for $b \to s\eta' g$ for the inclusive $B \to \eta' X_s$ signal develop an imaginary part at the one loop. This (CP-even) absorptive part is essential in driving observable CP violating asymmetries. This is an interesting and perhaps an important difference between $b \to \eta' sg$ and the other large inclusive $b \to s$ process, namely $b \to s\gamma$ as, in perturbation theory, the latter cannot develop a strong phase at one-loop.

In the SM, though, as already mentioned the $b \to s\eta' g$ amplitude receives only a negligible CP-odd phase. Therefore detection of a largish CP asymmetry in this channel may be an indication for a non-standard source of CP violation.

For simplicity, we assume that the CKM phase in $b \to s$ penguin amplitude is negligible. The $b \to sg^*$ effective vertex can be parameterized as[10] (see. Fig. 2)

$$\Lambda_\mu^{bsg} = (v_t G_F/\sqrt{2})\bar{s}_i T_{ij}^a [-iF(q^2)(q^2 q_\mu - q_\mu \not{q})L$$

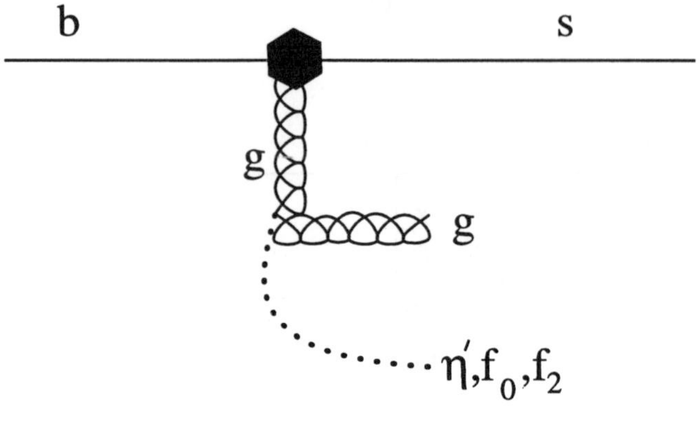

Figure 2: *The contributions of non-standard physics to* $b \to sg^*$ *followed by* $g^* \to g + \eta'$*,* f_0*,* f_2 *or other such states. The non-standard process is indicated by the hexagon.*

$$+(g_s/2\pi^2)m_b q_\mu \epsilon_\nu \sigma^{\mu\nu} G(q^2) R]b_j \tag{17}$$

Here

$$\begin{aligned} F(q^2) &= e^{i\delta_{st}} F_{SM} + e^{i\lambda_F} F_x \\ G(q^2) &= G_{SM} + e^{i\lambda_G} G_x \end{aligned} \tag{18}$$

where δ_{st} is the strong phase[23] generated by the absorptive part resulting form the $c\bar{c}$ cut for $q^2 > 4m_c^2$. To keep the discussion completely general we have introduced λ_F and λ_G as the CP-odd non-standard phases in the chromo-electric and chromo-magnetic form factors.

Although the inclusive η' signal is rather large, it is only about 5% of the total $b \to sg^*$. Besides, the η' detection efficiency is very low; at CLEO it is only about 5% for the inclusive signal.[13] It is, therefore, useful to ask, if the fragmentation reaction, $g^* \to g\eta'$ can lead to some other particle(s) in addition to the η', an appreciable fraction of the time.

Clearly the possible states that couple to two gluons have to have $J^{PC} = 0^{-+}, 0^{++}$ and 2^{++}. Some interesting examples are[10]

$$\begin{aligned} 0^{-+} &: \eta(958), \eta'(1440)\ldots \\ 0^{++} &: f_0(980), f_0(1370)\ldots \\ 2^{++} &: f_2(1270), f_2(1525)\ldots \\ \text{Continuum} &: \pi\pi, K\bar{K}, K\bar{K}\pi\ldots \end{aligned} \tag{19}$$

One can also arrive at these candidate states by looking at final states in radiative decays of the ψ.[10]

There are a couple of additional points worth keeping in mind:

1. At least in the case of ψ decays[10] there appears to be a close similarity between the states (0^{-+}) coupling to $G \cdot \tilde{G}$ versus those (0^{++}) coupling to $G \cdot G$:

$$\begin{aligned} Br(\psi \to \phi\eta'(958)) &= (3.3 \pm .4) \times 10^{-4} \\ Br(\psi \to \phi f_0(980)) &= (3.2 \pm .9) \times 10^{-4} \\ Br(\psi \to \omega\eta') &= (1.67 \pm .25) \times 10^{-4} \\ Br(\psi \to \omega f_0) &= (1.4 \pm .5) \times 10^{-4} \end{aligned} \tag{20}$$

2. Inspection of ψ decays[19] also seems to show that the signal for the $\eta(1440)$ may be comparable to that for the η'.

3. A particularly interesting continuum of states is $K\bar{K}$ as it leads to the overall signal of $B \to K\bar{K} + X_s$ (or $B \to K\bar{K} + X_d$) which may be quite distinctive. Once again, inspection of ψ decays suggests that the ratio of Br's: $(g^* \to K\bar{K} + g)/(g^* \to \eta' + g)$ could be around one or even somewhat bigger.

Calculation of the PRA proceeds along well known lines. Fig. 3(a) and 3(b) show the PRA as a function of the recoil mass (m_{rec}) for states of different J^{PC}. Fig. 3(a) is the case when beyond the SM (BSM) physics contributes only 10% to the production rate and Fig. 3(b) is assuming that 50% of the rate is due to new physics. In both figures we have set the CP-odd NSP to be maximal (i.e. λ_F, $\lambda_G = \pi/2$). These figures also show separately the cases when the chromoelectric form factor dominates (and the chromomagnetic is zero) and vice-versa. We see that PRA tend to be bigger (by 50–100%) for the case when chromoelectric dominates. Notice that the asymmetries are $\sim$ 12–34% in Fig. 3(b) and in Fig. 3(a), which is for the case when NSP is contributing only 10% to the rate, the asymmetries are still appreciable $\sim$ 8–17%. This is quite remarkable as it would be virtually impossible to detect the presence of the new physics, if BSM physics contributes only 10% to the rate, by comparison of the measured rate with theoretical expectation as the rate calculations are extremely unreliable. Search for CP violation can clearly be extremely worthwhile and decisive.

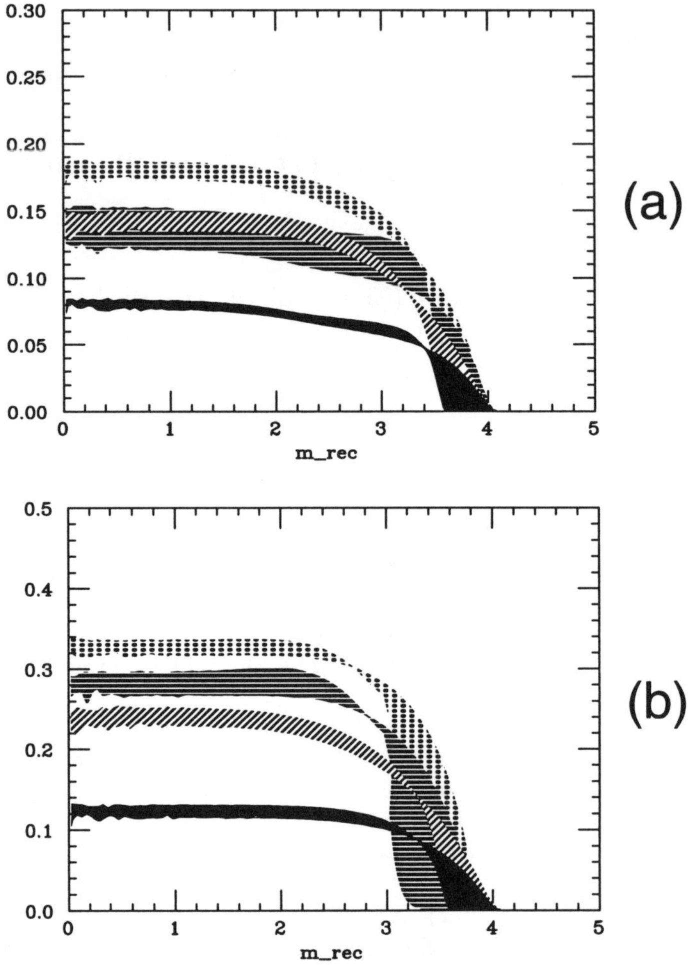

Figure 3: *(a)* |PRA| *versus* m_{rec} *assuming non-standard physics (NSP) contributes* $\sim 10\%$ *to the rate for each state. Also,* $\sin\lambda_F = \sin\lambda_G = 1$ *is used. The black shading shows the* |PRA| *for* $b \to s + g + 0^{-+}$ *assuming that* $F_x = 0$ *and taking* $m_{0^{-+}}$ *to vary from* $958MeV$ *to* $1725MeV$. *The horizontal striped region is the* |PRA| *for the same* 0^{-+} *states, now assuming that* $G_x = 0$. *Note that the region indicated by the diagonal stripes shows the* |PRA| *for* $b \to s + g + 0^{++}$ *assuming that* $F_x = 0$ *with* $m_{0^{++}}$ *ranging from* $980MeV$ *to* $1710MeV$*; the dotted region is the* |PRA| *for the* 0^{++} *assuming that* $G_x = 0$. *(b)* |PRA| *versus* m_{rec} *assuming NSP contributes* $\sim 50\%$ *to the rate for each state. The black shading (*$F_x = 0$*) and the horizontal striped (*$G_x = 0$*) are for* 0^{-+} *states as in Fig. 3(a). Diagonal striped (*$F_x = 0$*) and the dotted (*$G_x = 0$*) regions are now for* $b \to s + g + 2^{++}$ *with* $m_{2^{++}}$ *from* 1270 *to* $2300MeV$. *See also caption to Fig. 3(a).*

6 $b \to dg^*$, $B \to \eta' X_d$ and Related Modes for Direct CP violation due SM.[28]

The penguin $b \to d$ transition interfering with the tree $b \to u\bar{u}d$ can be a rich source of direct CP violation due to the SM-CKM phase. Specifically it is important to try to use $b \to dg^*$. Once again $g^* \to g\eta'$ and other fragments discussed in the preceding section can be used. While the Br for $B \to \eta' X_d$ is expected to be somewhat smaller ($\sim 7 \times 10^{-5}$) than $B \to \eta' X_s$, the PRA driven by the CKM phase are expected to be appreciably bigger for the $B \to \eta' X_d$ case compared to that for $B \to \eta' X_s$. For values of the CKM matrix which match current constraints, PRA $\sim O(12\%)$ for $B \to \eta' X_d$. The asymmetries for $B \to \eta(1440) + X_d$, $f_0(980) + X_d$, and $f_2(1270) + X_d$ are quite similar but somewhat bigger.[28]

7 Two Body Decays: $b \to qM$ for Direct CP Studies and for Looking for EWP.[15]

The importance of the two body decays of the b-quark is due to the fact that their theory is relatively simpler and many of them have relatively clean experimental signatures. To the extent that the spectator approximation works, these 2-body decays at the quark level end up materializing in quasi-two-body (QS2B) decays of the B-meson. Indeed, a famous recent example is $B \to \eta' X_s$ although from the point of view of the theory, the η' case is somewhat of a special one due to the unique relation of the η' to the QCD anomaly. As far as experimental detection goes, the inclusive $\eta\prime$ does serve an excellent example for the semi-inclusive decays $B \to MX$ that are under discussion in this section. As in the case for the η', the meson M originating from the 2-body decay $b \to Mq_f$ tends to obey two body kinematics and therefore has $E_M \sim 2.5 \pm .3$ GeV. The quark q_f leads to the debris of mesons with (1) relatively low average multiplicity ~ 3, (2) total energy $\sim 2 \pm .3$ GeV and (3) with negligible total transverse momentum (to $\vec{p}_M$ in the rest frame of B).[15]

Theoretical calculation for the QS2B decays are simpler than the traditional exclusive mode such as $B \to K\pi$, $\rho\pi \ldots$ A key difference is that the latter reactions require knowledge of exclusive form factors, e.g. $\langle B|J_1^h|\rho\rangle$ or $\langle B|J_2^h|K\rangle \ldots$ where J's are the appropriate quark level currents. The QS2B reactions bypass the need for these form factors as they represent a sum over the exclusive channels. Indeed, as a rule, the Br for the QS2B tends to be bigger compared to the exclusive two body modes, in some cases by factors of $\sim$ 5–10. In instances where the formation of M via the QS2B reactions is color suppressed whereas M can be made in a color allowed manner with the participation of the spectator for the corresponding exclusive two body mesons

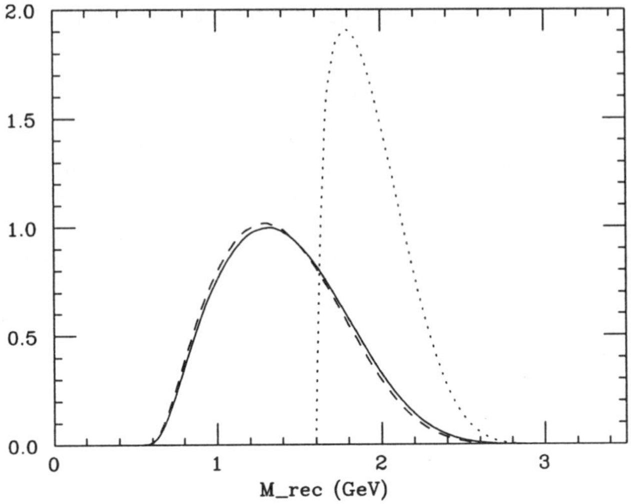

Figure 4: *The normalized recoil spectra for the quasi-twobody decays,* $b \to \pi^- u$ *(solid),* $K^{*-}u$ *(dashed) and* $D^- c$ *(dotted) are shown.*

reactions, then the latter can have a bigger Br compared to the QS2B. This may well, for example, be the case for $B^{\pm} \to \omega\rho^{\pm}$ versus $b \to \omega d$.

For these QS2B decays the CP-even FS interaction phase needed for generating the CP violating PRAs arises from the penguin Hamiltonian. Indeed since these are semi-inclusive reactions the calculations of the FS phase originating from the absorptive part of the penguin graph is close in spirit to the original quark level calculation.[23] It is quite likely that the FS rescattering effects due to LD or soft (non-perturbative) physics are much smaller in these inclusive processes than in the corresponding exclusive channels.[5]

A quick look at the Table shows that the PRA may be quite large in several channels,[15,29] e.g. $M = K^*, K, \rho, \pi \ldots$ Also given in the Table is $N_B^{3\sigma}$, the number of B mesons needed to obtain a $3-\sigma$ signal for the PRA, as well as an estimate of the EWP contribution. While it would be interesting and important to search for these PRA it would be very worthwhile even if the Br's of some of these channels are measured. The resulting input from experiment could be valuable in fine tuning the calculational parameters of this whole class of reactions. In this context, it is also worth noticing that the recoil spectrum (see Fig. 4) for these reactions can be calculated in much the same way as for the $B \to \gamma X_s$[30] (or $B \to \eta X_s$[14]). This means that the input parameters for the recoil spectrum can be refined by using data from all of these reactions.

Table 1: Some modes of interest; Br, PRA and $N_B^{3\sigma}$ along with EWP contributions (to color allowed channels only) are shown. Note $\gamma \equiv \arg(-V_{ub}^* V_{ud}/V_{cb}^* V_{cd})$. Note that the Br column does not include contributions from EWP.

Mode	Br	$\lvert\mathrm{PRA}\rvert/\sin\gamma(\%)$	$N_B^{3\sigma}\sin^2\gamma\epsilon_{eff}/10^6$	Br due to EWP
$\pi^- u$	1.3×10^{-4}	8	12	
$\rho^- u$	3.5×10^{-4}	8	4	
$\pi^0 d$	2.4×10^{-6}	36	28	4.7×10^{-8}
$\rho^0 d$	5.9×10^{-6}	38	10	1.3×10^{-7}
ωd	5.8×10^{-6}	39	10	7.0×10^{-9}
ϕd	2.3×10^{-7}	0		7.0×10^{-9}
$K^0 s$	2.5×10^{-6}	5	1200	
$K^{0^*} s$	2.9×10^{-6}	16	120	
$D^- c$	1.7×10^{-3}	2	17	
$D^{*-} c$	2.2×10^{-3}	2	13	
$K^- u$	2.9×10^{-5}	33	3	
$K^{*-} u$	5.1×10^{-5}	51	1	
$\bar{K}^0 d$	2.0×10^{-5}	1	3000	
$\bar{K}^{0^*} d$	2.6×10^{-5}	3	540	
$\pi^0 s$	9.8×10^{-8}	0		1.6×10^{-6}
$\rho^0 s$	2.5×10^{-7}	0		4.3×10^{-6}
ωs	1.3×10^{-6}	0		4.7×10^{-7}
ϕs	6.3×10^{-5}	0		4.7×10^{-7}
$D_s^- c$	4.2×10^{-2}	0.1	300	
$D_s^{*-} c$	5.3×10^{-2}	0	300	

Finally, we draw attention to the fact that $B^0 \to \rho^0 X_s$ and $B^0 \to \pi^0 X_s$ appear highly suited for searching for EWP.[15] This can be qualitatively understood as the EWP contributions are color allowed whereas the others are color suppressed. The contribution from the tree graph is also CKM suppressed. The EWP appears to be the dominant contributor[15] with $Br(B^0 \to \rho^0 X_s) \sim 4\times10^{-6}$ and $Br(B^0 \to \pi^0 X_s) \sim 2\times10^{-6}$. The same comments may also be true for the related reactions, $B^+ \to \rho^0 X_s$, $B^+ \to \pi^0 X_s$, except that in this case there is a possible background from the tree process $\bar{b} \to u\bar{u}\bar{s}$ wherein the $\bar{u}$ can combine with the spectator to form the ρ^0 or π^0 in a color allowed manner. Although most of the background from the tree process would be expected to give rise to lower momentum ρ^0 or π^0, one can estimate that the high momentum component of the ρ^0 or π^0 energy spectrum can produce a signal comparable to the EWP contribution.

Acknowledgements

We are grateful to the organizers of the International Workshop, and especially to George Hou, for the very warm and gracious hospitality. This research was supported in part by US DOE Contract Nos. DE-FG01-94ER40817 (ISU) and DE-AC02-98CH10886 (BNL).

References

1. This paper is based on invited talks given at the Fourth International Workshop on Particle Physics Phenomenology, Taiwan (June, 1998), BABAR Workshop, SLAC (March, 1998) and BELLE Workshop (June, 1998).
2. See, e.g., K. Lingel, T. Skwarnicki and J.G. Smith, hep-ex/980415, submitted to Ann. Rev. Nucl. & Part. Sci.
3. R. Godang *et al.* (CLEO Collaboration), *Phys. Rev. Lett.* **80**, 3456 (1998).
4. T. Bergfeld *et al.* (CLEO Collaboration), *Phys. Rev. Lett.* **81**, 272 (1998).
5. D. Atwood and A. Soni, *Phys. Rev.***D58**, 036005 (1998). For final state rescattering effects see also [6].
6. J.-M. Gérard and J. Weyers, hep-ph/9711469; M. Neubert, *Phys. Lett.* **B424**, 152 (1998); A. Falk, A.L. Kagan, Y. Nir and A. Petrov, *Phys. Rev.***D57**, 4290 (1998) and M. Gronau and J. Rosner, *Phys. Rev.***D57**, 2752 (1998).
7. Y. Nir and H. Quinn, *Phys. Rev. Lett.* **67**, 541 (1991); M. Gronau, *Phys. Lett.* **B265**, 389 (1991).
8. For analysis of two vector final states see, e.g.: G. Valencia, *Phys. Rev.***D39**, 3339 (1989); I. Dunietz, H.R. Quinn, A. Snyder, W. Toki and H. Lipkin, *Phys. Rev.***D43**, 2193 (1991); R. Aleksan, I. Dunietz and B. Kayser, Z. Phys. **C54**, 653 (1992) and G. Kramer, W.F. Palmer and H. Simma, *Nucl. Phys.* **B428**, 77 (1994).
9. D. Atwood and A. Soni, hep-ph/9805212 .
10. D. Atwood and A. Soni, *Phys. Rev. Lett.* **79**, 5206 (1997).
11. W.-S. Hou and B. Tseng, *Phys. Rev. Lett.* **80**, 434 (1998).
12. W.-S. Hou, A. Soni and H. Steger, *Phys. Rev. Lett.* **59**, 1521 (1987).
13. T.E. Browder *et al.* (CLEO Collaboration), hep-ex/9804018
14. D. Atwood and A. Soni, *Phys. Lett.* **B405**, 150 (1997).
15. D. Atwood and A. Soni, hep-ph/9804393.

16. See the talks by J. Roy and J. Alexander, ICHEP98, Vancouver (July 1998).
17. M. Gronau and D. London, *Phys. Rev. Lett.* **65**, 3381 (1990).
18. See, e.g., A.I. Sanda in "BNL Summer Study on CP Violation", Eds. S. Dawson and A. Soni, p. 115 (World Scientific).
19. See R.M. Barnett *et al.*, *Phys. Rev.***D54**, 1 (1996).
20. J.F. Donoghue, E. Golowich, A.A. Petrov and J.M. Soares, *Phys. Rev. Lett.* **77**, 2178 (1996)
21. See also, A. Donnachie and P.V. Landshoff, *Phys. Lett.* **B296**, 227 (1992); H. Lipkin, *Phys. Lett.* **B335**, 500 (1994).
22. See also, L. Wolfenstein, *Phys. Rev.***D52**, 537 (1995).
23. M. Bander, D. Silverman and A. Soni, *Phys. Rev. Lett.* **43**, 242 (1979).
24. J.-M. Gérard and W.-S. Hou, *Phys. Rev.***D43**, 2909 (1991).
25. See, e.g., R. Fleischer, Z. Phys. **C62**, 81 (1994); N.G. Deshpande and X.-G. He, *Phys. Rev. Lett.* **74**, 26 (1995); M. Gronau, O.F. Hernandez, D. London and J.L. Rosner, *Phys. Rev.***D52**, 6374 (1995) and A. J. Buras and R. Fleischer, *Phys. Lett.* **B365**, 390 (1996).
26. A. Ali and C. Greub, *Phys. Rev.***D57**, 2996 (1998); *Phys. Lett.* **B424**, 161 (1998); A. Datta, X.-G. He and S. Pakvasa, *Phys. Lett.* **B419**, 369 (1998); H.Y. Cheng and B. Tseng, *Phys. Lett.* **B415**, 263 (1997); A. Kagan and A. Petrov, hep-ph/9707354; N. Deshpande, B. Dutta and S. Oh, *Phys. Rev.***D57**, 5723 (1998).
27. S.J. Brodsky and G.P. Lepage, *Phys. Rev.* **D22**, 2157(1980); A. Anselm, A. Johansen, E. Leader and L. Lukaszuk, Z. Phys. **A359**, 457(1997); J.-M. Gérard and T. Lahna, *Phys. Lett.* **B356**, 381(1995).
28. D. Atwood and A. Soni (to be published).
29. For the K, K^* case see also, A. Datta, X.-G. He and S. Pakvasa, *Phys. Rev.***D57**, 6829 (1998).
30. A. Ali and C. Greub, *Phys. Lett.* **B259**, 182 (1991).

HADRONIC CHARMLESS B DECAYS AND NONFACTORIZABLE EFFECTS

HAI-YANG CHENG
Institute of Physics, Academia Sinica, Taipei, Taiwan 115, ROC

Hadronic charmless B decays and their nonfactorizable effects are reviewed.

1 Introduction

A remarkable progress in the study of exclusive charmless B decays has been made recently. Experimentally, CLEO [1] has discovered many new two-body decay modes

$$B \to \eta' K^{\pm},\ \eta' K_S^0,\ \pi^{\pm} K_S^0,\ \pi^{\pm} K^{\mp},\ \omega K^{\pm}, \tag{1}$$

and a possible evidence for $B \to \phi K^*$. Moreover, CLEO has provided new improved upper limits for many other decay modes. Some of the CLEO data are surprising from the theoretical point of view: The measured branching ratios for $B^{\pm} \to \eta' K^{\pm}$ and $B^{\pm} \to \omega K^{\pm}$ are about several times larger than the naive theoretical estimate.

2 Difficulties with naive factorization

The relevant effective weak Hamiltonian for hadronic weak B decay is of the form

$$\mathcal{H}_{\rm eff} = \frac{G_F}{\sqrt{2}} \left[\lambda_u (c_1 O_1^u + c_2 O_2^u) + \lambda_c (c_1 O_1^c + c_2 O_2^c) - \lambda_t \sum_{i=3}^{10} c_i O_i \right], \tag{2}$$

where $\lambda_i = V_{ib} V_{iq}^*$, O_{3-6} are the QCD penguin operators and O_{7-10} the electroweak penguin operators. The Wilson coefficients $c_i(\mu)$ in Eq. (2) have been evaluated to the next-to-leading order (NLO) and they depend on the choice of the renormalization scheme. The mesonic matrix elements are customarily evaluated using the factorization hypothesis. Under this assumption, the 3-body hadronic matrix element $\langle M_1 M_2 | O | B \rangle$ is approximated as the product of two matrix elements $\langle M_1 | J_{1\mu} | 0 \rangle$ and $\langle M_2 | J_2^{\mu} | B \rangle$. Although this approach for matrix elements is very simple, it encounters two major difficulties. First, the hadronic matrix element under factorization is renormalization scale μ

independent as the vector or axial-vector current is partially conserved. Consequently, the amplitude $c_i(\mu)\langle O\rangle_{\rm fact}$ is not truly physical as the scale dependence of Wilson coefficients does not get compensated from the matrix elements. Second, in the naive factorization approach, the relevant Wilson coefficient functions for color-allowed external W-emission (or so-called "class-I") and color-suppressed (class-II) internal W-emission amplitudes are given by $a_1 = c_1 + c_2/N_c$, $a_2 = c_2 + c_1/N_c$, respectively, with N_c the number of colors. However, naive factorization fails to describe class-II decay modes. For example, the ratio $R = \Gamma(D^0 \to \overline{K}^0\pi^0)/\Gamma(D^0 \to K^-\pi^+)$ is predicted to be $\sim \frac{1}{50}$, while experimentally[2] $R = 0.51 \pm 0.07$. This implies that it is necessary to take into account nonfactorizable contributions to the decay amplitude in order to render the color suppression of internal W-emission ineffective.

2.1 Scale and scheme independence of physical amplitudes

Under the factorization hypothesis, we would like to know if it is possible to obtain physical amplitudes independent of the choice of the renormalization scale and scheme. The answer is yes. The scale and scheme dependence of the hadronic matrix elements can be calculated in perturbation theory at the one-loop level[3,4,5]. Schematically,

$$\langle O(\mu)\rangle = g(\mu)\langle O\rangle_{\rm tree}, \qquad \langle \mathcal{H}_{\rm eff}\rangle = c^{\rm eff}\langle O\rangle_{\rm tree}, \tag{3}$$

with $g(\mu)$ being the perturbative corrections to the four-quark operators renormalized at the scale μ. Formally, one can show that $c^{\rm eff} = g(\mu)c(\mu)$ is μ and renormalization scheme independent. It is at this stage that the factorization approximation is applied to the hadronic matrix elements of the operator O at tree level. The physical amplitude obtained in this manner is guaranteed to be renormalization scaheme and scale independent. [a]

The penguin-type corrections to $g(\mu)$ depend on k^2, the gluon's momentum squared, so are the effective Wilson coefficient functions. To NLO, we obtain[7]

$$\begin{aligned}
c_1^{\rm eff} &= 1.149, & c_2^{\rm eff} &= -0.325,\\
c_3^{\rm eff} &= 0.0211 + i0.0045, & c_4^{\rm eff} &= -0.0450 - i0.0136,\\
c_5^{\rm eff} &= 0.0134 + i0.0045, & c_6^{\rm eff} &= -0.0560 - i0.0136,\\
c_7^{\rm eff} &= -(0.0276 + i0.0369)\alpha, & c_8^{\rm eff} &= 0.054\,\alpha,\\
c_9^{\rm eff} &= -(1.318 + i0.0369)\alpha, & c_{10}^{\rm eff} &= 0.263\,\alpha,
\end{aligned} \tag{4}$$

at $k^2 = m_b^2/2$.

[a] This formulation is different from the one advocated in[6] in which the μ dependence of the Wilson coefficients $c_i(\mu)$ are canceled out by that of the nonfactorization parameters $\varepsilon_8(\mu)$ and $\varepsilon_1(\mu)$ so that the effective parameters $a_i^{\rm eff}$ are μ independent.

2.2 Generalized factorization

Because there is only one single form factor (or Lorentz scalar) involved in the class-I or class II decay amplitude of $B \to PP,\ PV$ decays, the effects of nonfactorization can be lumped into the effective parameters a_1 and a_2 [8]:

$$a_1^{\rm eff} = c_1^{\rm eff} + c_2^{\rm eff}\left(\frac{1}{N_c} + \chi_1\right), \qquad a_2^{\rm eff} = c_2^{\rm eff} + c_1^{\rm eff}\left(\frac{1}{N_c} + \chi_2\right), \tag{5}$$

where χ_i are nonfactorizable terms and receive main contributions from the color-octet current operators. Since $|c_1/c_2| \gg 1$, it is evident from Eq. (5) that even a small amount of nonfactorizable contributions will have a significant effect on the color-suppressed class-II amplitude. If $\chi_{1,2}$ are universal (i.e. process independent) in charm or bottom decays, then we still have a generalized factorization scheme in which the decay amplitude is expressed in terms of factorizable contributions multiplied by the universal effective parameters $a_{1,2}^{\rm eff}$. For $B \to VV$ decays, this new factorization implies that nonfactorizable terms contribute in equal weight to all partial wave amplitudes so that $a_{1,2}^{\rm eff}$ *can* be defined. It should be stressed that, contrary to the naive one, the improved factorization does incorporate nonfactorizable effects in a process independent form. Phenomenological analyses of two-body decay data of D and B mesons indicate that while the generalized factorization hypothesis in general works reasonably well, the effective parameters $a_{1,2}^{\rm eff}$ do show some variation from channel to channel, especially for the weak decays of charmed mesons [8,9,10]. An eminent feature emerged from the data analysis is that $a_2^{\rm eff}$ is negative in charm decay, whereas it becomes positive in the two-body decays of the B meson [8,12,6]:

$$a_2^{\rm eff}(D \to \overline{K}\pi) \sim -0.50\,, \qquad a_2^{\rm eff}(B \to D\pi) \sim 0.20 - 0.28\,. \tag{6}$$

It follows that

$$\chi_2(D \to \overline{K}\pi) \sim -0.36\,, \qquad \chi_2(B \to D\pi) \sim 0.12 - 0.19\,. \tag{7}$$

The observation $|\chi_2(B)| \ll |\chi_2(D)|$ is consistent with the intuitive picture that nonperturbative soft gluon effects become stronger when the final-state particles move slower, allowing more time for significant final-state interactions after hadronization [8]. Phenomenologically, it is often to treat the number of colors N_c as a free parameter to model the nonfactorizable contribution to hadronic matrix elements and its value can be extracted from the data of two-body nonleptonic decays. Theoretically, this amounts to defining an effective number of colors $N_c^{\rm eff}$, called $1/\xi$ in [11], by $1/N_c^{\rm eff} \equiv (1/N_c) + \chi$. It is clear from

(7) that

$$N_c^{\rm eff}(D \to \overline{K}\pi) \gg 3, \qquad N_c^{\rm eff}(B \to D\pi) = 1.8 - 2.2 \approx 2\,. \tag{8}$$

3 Nonfactorizable effects in charmless B decays

We next study the nonfactorizabel effects in charmless rare B decays. We note that the effective Wilson coefficients appear in the factorizable decay amplitudes in the combinations $a_{2i} = c_{2i}^{\rm eff} + \frac{1}{N_c} c_{2i-1}^{\rm eff}$ and $a_{2i-1} = c_{2i-1}^{\rm eff} + \frac{1}{N_c} c_{2i}^{\rm eff}$ $(i = 1, \cdots, 5)$. As discussed in Sec. 2.2, nonfactorizable effects in the decay amplitudes of $B \to PP$, VP can be absorbed into the parameters $a_i^{\rm eff}$. This amounts to replacing N_c in $a_i^{\rm eff}$ by $(N_c^{\rm eff})_i$. Explicitly,

$$a_{2i}^{\rm eff} = c_{2i}^{\rm eff} + \frac{1}{(N_c^{\rm eff})_{2i}} c_{2i-1}^{\rm eff}, \qquad a_{2i-1}^{\rm eff} = c_{2i-1}^{\rm eff} + \frac{1}{(N_c^{\rm eff})_{2i-1}} c_{2i}^{\rm eff}\,. \tag{9}$$

It is customary to assume in the literature that $(N_c^{\rm eff})_1 \approx (N_c^{\rm eff})_2 \cdots \approx (N_c^{\rm eff})_{10}$; that is, the nonfactorizable term is usually assumed to behavor in the same way in penguin and non-penguin decay amplitudes. A closer investigation shows that this is not the case. We have argued in [7] that nonfactorizable effects in the matrix elements of $(V-A)(V+A)$ operators are *a priori* different from that of $(V-A)(V-A)$ operators. One reason is that the Fierz transformation of the $(V-A)(V+A)$ operators $O_{5,6,7,8}$ is quite different from that of $(V-A)(V-A)$ operators $O_{1,2,3,4}$ and $O_{9,10}$. As a result, contrary to the common assumption, $N_c^{\rm eff}(LR)$ induced by the $(V-A)(V+A)$ operators are theoretically different from $N_c^{\rm eff}(LL)$ generated by the $(V-A)(V-A)$ operators [7]. Hence, it is plausible to assume that

$$\begin{aligned} N_c^{\rm eff}(LL) &\equiv \left(N_c^{\rm eff}\right)_1 \approx \left(N_c^{\rm eff}\right)_2 \approx \left(N_c^{\rm eff}\right)_3 \approx \left(N_c^{\rm eff}\right)_4 \approx \left(N_c^{\rm eff}\right)_9 \approx \left(N_c^{\rm eff}\right)_{10}, \\ N_c^{\rm eff}(LR) &\equiv \left(N_c^{\rm eff}\right)_5 \approx \left(N_c^{\rm eff}\right)_6 \approx \left(N_c^{\rm eff}\right)_7 \approx \left(N_c^{\rm eff}\right)_8, \end{aligned} \tag{10}$$

and that $N_c^{\rm eff}(LR) \neq N_c^{\rm eff}(LL)$. In principle, $N_c^{\rm eff}$ can vary from channel to channel, as in the case of charm decay. However, in the energetic two-body B decays, $N_c^{\rm eff}$ is expected to be process insensitive as supported by data [6].

3.1 Classification of charmless B decays

By studying the $N_c^{\rm eff}$-dependence of the effective parameters a_i's (for simplicity, we will drop the superscript "eff" henceforth), we learn that (i) the dominant coefficients are a_1, a_2 for current-current amplitudes, a_4 and a_6 for QCD penguin-induced amplitudes, and a_9 for electroweak penguin-induced amplitudes, and (ii) a_1, a_4, a_6 and a_9 are $N_c^{\rm eff}$-stable, while the other coefficients depend strongly on $N_c^{\rm eff}$. Therefore, for those charmless B decays whose decay amplitudes depend dominantly on $N_c^{\rm eff}$-stable coefficients, their decay rates can be reliably predicted within the factorization approach even in the absence of information of nonfactorizable effects. By contrast, the decay modes involving the coefficients a_2, a_3 and a_5 are sensitive to $N_c^{\rm eff}$ and hence the nonfactorizable effects.

In order to study hadronic charmless B decays, it is useful to classify the decay modes into several different categories. Besides the widely used three classes I, II, III for tree-dominated decay modes, the penguin-dominated charmless rare B decays also can be classified into three classes:

- Class-IV for those decays whose amplitudes are governed by the parameters a_4 and a_6 in the combination $a_4 + Ra_6$, where the coefficient R arises from the $(S-P)(S+P)$ part of the operator O_6. In general, $R = 2m_{P_b}^2/[(m_1+m_2)(m_b-m_3)]$ for $B \to P_aP_b$ with the meson P_b being factored out in the factorizable approximation, $R = -2m_{P_b}^2/[(m_1+m_2)(m_b+m_3)]$ for $B \to V_aP_b$, and $R = 0$ for $B \to P_aV_b$ and $B \to V_aV_b$ with V_b being factorizable. Note that a_4 is always accompanied by a_{10}, and a_6 by a_8. Examples are $\overline{B}_d \to K^-\pi^+, \overline{K}^0\pi^0$, $B^- \to K^-\pi^0$, $\overline{B}_s \to K^+K^-, K^0\overline{K}^0, \cdots$.

- Class-V modes for those decays whose amplitudes are governed by the effective coefficients a_3, a_5, a_7 and a_9 in the combinations $a_3 \pm a_5$ and/or $a_7 \pm a_9$. Examples are $\overline{B}_d \to \phi\pi^0$, $B^- \to \phi\pi^-$, $\overline{B}_s \to \phi\pi^0$.

- Class-VI involving the interference of class-IV and class-V decays, e.g. $B \to K\eta'$, $K\omega$, $K\phi$ $(B = B_u, B_d, B_s)$.

Sometimes the tree and penguin contributions are comparable. For example, decays $\overline{B}_s \to K^0\omega, K^{*0}\omega$ fall into the classes of II and VI.

3.2 Some general features for penguin-dominated proceses

For penguin-dominated decay modes, some observations can be made:

- For class-IV modes, the decay rates obey the pattern:

$$\Gamma(B \to P_aP_b) > \Gamma(B \to P_aV_b) \sim \Gamma(B \to V_aV_b) > \Gamma(B \to V_aP_b), \quad (11)$$

  where $M = P_b$ or V_b is factorizable under the factorization assumption. For example, the branching ratios for $\overline{B}^0 \to K^-\pi^+, K^{*-}\pi^+, K^{*-}\rho^+, K^-\rho^+$ are predicted to be $\sim 1.5 \times 10^{-5}$, 0.7×10^{-5}, 0.6×10^{-5} and 0.5×10^{-6}

respectively. This hierarchy follows from various interference between the penguin terms characterized by the effective coefficients a_4 and a_6. On the contrary, in general $\Gamma(B \to P_a V_b) > \Gamma(B \to P_a P_b)$ for tree-dominated decays because the vector meson has three different polarization states.

- Among the two-body charmless B decays, the class-III decay modes $B^- \to \eta' K^-$, $\overline{B}_d \to \eta' K^0$ and $\overline{B}_s \to \eta\eta', \eta'\eta'$ have the largest branching ratios. Theoretically, $\mathcal{B}(B^- \to \eta' K^-) \approx \mathcal{B}(\overline{B}_d \to \eta' K^0) \sim 4 \times 10^{-5}$ and $\mathcal{B}(\overline{B}_s \to \eta\eta', \eta'\eta') \sim 2 \times 10^{-5}$. These decay modes receive two different sets of penguin terms proportional to $a_4 + Ra_6$ with $R > 0$. By contrast, VP, VV modes in charm decays or bottm decays involving charmed mesons usually have larger branching ratios than the PP mode.

- The decay amplitudes of $B \to M\pi^0, M\rho^0$ with $\pi^0(\rho^0)$ being factored out contain the electroweak penguin contributions proportional to $\frac{3}{2}(-a_7 + a_9)X_u^{(BM,\pi^0)}$ and $\frac{3}{2}(a_7 + a_9)X_u^{(BM,\rho^0)}$, respectively,[13] with

$$X_u^{(BM,\pi^0)} = \langle \pi^0 | (\bar{u}u)_{V-A} | 0 \rangle \langle M | (\bar{q}b)_{V-A} | \overline{B} \rangle. \tag{12}$$

For $\overline{B}_s \to \eta'\pi, \eta'\rho, \phi\pi, \phi\rho$, QCD penguin contributions are canceled out so that these decays are dominated by electroweak penguins. Hence, a measurement of them can be used to determine the effective coefficient a_9. [b]

3.3 Nonfactorizable effects in spectator amplitudes

We focus on class-III decay modes dominated by the spectator diagrams induced by the current-current operators O_1 and O_2 and are sensitive to the interference between external and internal W-emission amplitudes. Good examples are the class-III modes: $B^\pm \to \omega\pi^\pm$, $\pi^0\pi^\pm$, $\eta\pi^\pm$, $\pi^0\rho^\pm, \cdots$, etc. Considering $B^\pm \to \omega\pi^\pm$, we find that the branching ratio is sensitive to $1/N_c^{\rm eff}$ and has the lowest value of order 2×10^{-6} at $N_c^{\rm eff} = \infty$ and then increases with $1/N_c^{\rm eff}$. The 1997 CLEO measurement yields[15]

$$\mathcal{B}(B^\pm \to \omega\pi^\pm) = (1.1^{+0.6}_{-0.5} \pm 0.2) \times 10^{-5}. \tag{13}$$

---

[b] It has been suggetsed in [14] that $\overline{B}_d \to \overline{K}^0\rho^0$ can be utilized to extract a_9. However, this method relies on the cancellation between the QCD penguin terms characterized by a_4 and a_6. In general, this cancellation is not complete and this makes this decay mode less clean than $\overline{B}_s \to (\eta', \phi)(\pi, \rho)$ for determining a_9.

Consequently, $1/N_c^{\rm eff} > 0.35$ is preferred by the data [7]. Because this decay is dominated by tree amplitudes, this in turn implies that $N_c^{\rm eff}(V-A) < 2.9$. If the value of $N_c^{\rm eff}(V-A)$ is fixed to be 2, the branching ratio for positive ρ, which is preferred by the current analysis [16], will be of order $(0.9-1.0)\times 10^{-5}$, which is very close to the central value of the measured one. Unfortunately, the significance of $B^\pm \to \omega\pi^\pm$ is reduced in the recent CLEO analysis and only an upper limit is quoted [17]: $\mathcal{B}(B^\pm \to \pi^\pm\omega) < 2.3\times 10^{-5}$. Since $\mathcal{B}(B^\pm \to K^\pm\omega) = (1.5^{+0.7}_{-0.6}\pm 0.2)\times 10^{-5}$ and $\mathcal{B}(B^\pm \to h^\pm\omega) = (2.5^{+0.8}_{-0.7}\pm 0.3)\times 10^{-5}$ with $h=\pi,\ K$, the central value of $\mathcal{B}(B^\pm \to \pi^\pm\omega)$ remains about the same as (11). The fact that $N_c^{\rm eff}(LL)\sim 2$ is preferred in charmless two-body decays of the B meson is consistent with the nonfactorizable term extracted from $B\to (D,D^*)\pi,\ D\rho$ decays, namely $N_c^{\rm eff}(B\to D\pi)\approx 2$. Since the energy release in the energetic two-body decays $B\to\omega\pi$, $B\to D\pi$ is of the same order of magnitude, it is thus expected that $N_c^{\rm eff}(LL)|_{B\to\omega\pi}\approx 2$.

Just like the decay $B^-\to\pi^-\omega$, the branching ratio of $B^-\to\pi^-\pi^0$ also increases with $1/N_c^{\rm eff}$. The CLEO measurement is [18]

$$\mathcal{B}(B^\pm\to\pi^\pm\pi^0) = \left(0.9^{+0.6}_{-0.5}\right)\times 10^{-5} \quad < 2.0\times 10^{-5}. \tag{14}$$

However, the errors are so large that it is meaningless to put a sensible constraint on $N_c^{\rm eff}(LL)$. Nevertheless, we see that in the range [5] $0\le 1/N_c^{\rm eff}\le 0.5$, $N_c^{\rm eff}(LL)\approx 2$ is most favored.

In analogue to the decays $B\to D^{(*)}\pi(\rho)$, the interference effect of spectator amplitudes in class-III charmless B decay can be tested by measuring the ratios:

$$R_1 \equiv 2\,\frac{\mathcal{B}(B^-\to\pi^-\pi^0)}{\mathcal{B}(\bar B^0\to\pi^-\pi^+)},\quad R_2 \equiv 2\,\frac{\mathcal{B}(B^-\to\rho^-\pi^0)}{\mathcal{B}(\bar B^0\to\rho^-\pi^+)},\quad R_3 \equiv 2\,\frac{\mathcal{B}(B^-\to\pi^-\rho^0)}{\mathcal{B}(\bar B^0\to\pi^-\rho^+)}. \tag{15}$$

Evidently, the ratios R_i are greater (less) than unity when the interference is constructive (destructive). Numerically we find

$$R_1 = \begin{cases}1.74,\\ 0.58,\end{cases}\qquad R_2 = \begin{cases}1.40,\\ 0.80,\end{cases}\qquad R_3 = \begin{cases}2.50 & \text{for } N_c^{\rm eff}=2,\\ 0.26 & \text{for } N_c^{\rm eff}=\infty.\end{cases} \tag{16}$$

Hence, a measurement of R_i (in particular R_3), which has the advantage of being independent of the Wolfenstein parameters ρ and η, will constitute a very useful test on the effective number of colors $N_c^{\rm eff}(LL)$. The present experimental information on $\overline{B}^0\to\pi^+\pi^-$ is [18]

$$\mathcal{B}(\overline{B}^0\to\pi^\pm\pi^\mp) = (0.7\pm 0.4)\times 10^{-5} \quad < 1.5\times 10^{-5}. \tag{17}$$

As far as the experimental central value of R_1 is concerned, it appears that $1/N_c^{\rm eff}\sim 0.5$ is more favored than any other small values of $1/N_c^{\rm eff}$.

3.4 *Nonfactorizable effects in penguin amplitudes*

The penguin amplitude of the class-VI mode $B \to \phi K$ is proportional to $(a_3 + a_4 + a_5)$ and hence sensitive to the variation of $N_c^{\rm eff}$. Neglecting W-annihilation and space-like penguin diagrams, we find [7] that $N_c^{\rm eff}(LR) = 2$ is evidently excluded from the present CLEO upper limit [17]

$$\mathcal{B}(B^\pm \to \phi K^\pm) < 0.5 \times 10^{-5}, \tag{18}$$

and that $1/N_c^{\rm eff}(LR) < 0.23$ or $N_c^{\rm eff}(LR) > 4.3$. A similar observation was also made in [19]. The branching ratio of $B \to \phi K^*$, the average of ϕK^{*-} and ϕK^{*0} modes, is also measured recently by CLEO with the result [17]

$$\mathcal{B}(B \to \phi K^*) = \left(1.1^{+0.6}_{-0.5} \pm 0.2\right) \times 10^{-5}. \tag{19}$$

We find that the allowed region for $N_c^{\rm eff}(LR)$ is $4 \gtrsim N_c^{\rm eff}(LR) \gtrsim 1.4$. This is in contradiction to the constraint $N_c^{\rm eff}(LR) > 4.3$ derived from $B^\pm \to \phi K^\pm$. In fact, the factorization approach predicts that $\Gamma(B \to \phi K^*) \approx \Gamma(B \to \phi K)$ when the W-annihilation type of contributions is neglected. The current CLEO measurements (18) and (19) are obviously not consistent with the prediction based on factorization. One possibility is that generalized factorization is not applicable to $B \to VV$. Therefore, the discrepancy between $\mathcal{B}(B \to \phi K)$ and $\mathcal{B}(B \to \phi K^*)$ will measure the degree of deviation from the generalized factorization that has been applied to $B \to \phi K^*$. It is also possible that the absence of $B \to \phi K$ events is a downward fluctuation of the experimental signal. At any rate, in order to clarify this issue and to pin down the effective number of colors $N_c^{\rm eff}(LR)$, we urgently need measurements of $B \to \phi K$ and $B \to \phi K^*$, especially the neutral modes, with sufficient accuracy.

The decay mode $B \to \eta' K$ also provides another useful information on $N_c^{\rm eff}(LR)$. The discrepancy between the experimental measurements

$$\begin{aligned} \mathcal{B}(B^\pm \to \eta' K^\pm) &= \left(6.5^{+1.5}_{-1.4} \pm 0.9\right) \times 10^{-5}, \\ \mathcal{B}(B^0 \to \eta' K^0) &= \left(4.7^{+2.7}_{-2.0} \pm 0.9\right) \times 10^{-5} \end{aligned} \tag{20}$$

and the theoretical estimates [20,4,21] of order 1×10^{-5} seems to call for some new mechanisms unique to the η' production or even some new physics beyond the Standard Model.

In the conventional way of treating $N_c^{\rm eff}(LR)$ and $N_c^{\rm eff}(LL)$ in the same manner, the branching ratio of $B^\pm \to \eta' K^\pm$ can be enhanced to the order of $(2-3) \times 10^{-5}$ due to the small running strange quark mass at $\mu = m_b$ and $SU(3)$ breaking in the decay constants f_8 and f_0 (corresponding to the dashed

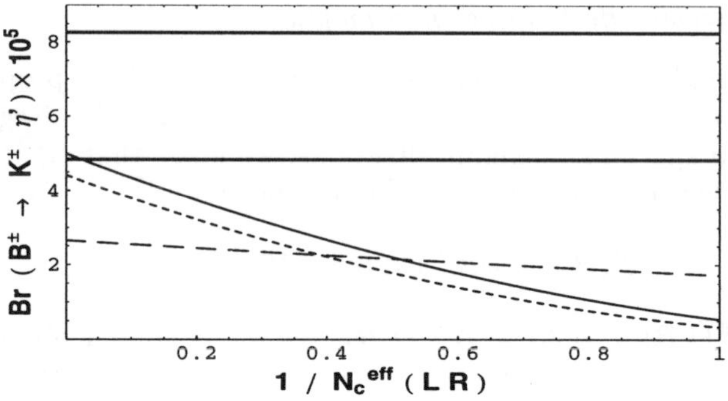

Figure 1: The branching ratio of $B^{\pm} \to \eta' K^{\pm}$ as a function of $1/N_c^{\rm eff}(LR)$ with $N_c^{\rm eff}(LL)$ being fixed at the value of 2 and $\eta = 0.34$ and $\rho = 0.16$. The charm content of the η' with $f_{\eta'}^c = -6\,{\rm MeV}$ contributes to the solid curve, but not to the dotted curve. The anomaly contribution to $\langle \eta' | \bar{s}\gamma_5 s | 0 \rangle$ is included. For comparison, the prediction for the case that $N_c^{\rm eff}(LL) = N_c^{\rm eff}(LR)$ as depicted by the dashed curve is also shown. The solid thick lines are the CLEO measurements with one sigma errors.

curve in Fig. 1). It should be emphasized that this prediction has taken into account the anomaly effect in the matrix element $\langle \eta' | \bar{s}\gamma_5 s | 0 \rangle$. Specifically,

$$\langle \eta' | \bar{s}\gamma_5 s | 0 \rangle = -i \frac{m_{\eta'}^2}{2m_s} \left(f_{\eta'}^s - f_{\eta'}^u \right), \tag{21}$$

where the QCD anomaly effect is manifested by the decay constant $f_{\eta'}^u$. Since $f_{\eta'}^u \sim \frac{1}{2} f_{\eta'}^s$ and the decay amplitude is dominated by $(S-P)(S+P)$ matrix elements, it is obvious that the decay rate of $B \to \eta' K$ would be (wrongly) enhanced considerably in the absence of the anomaly term in $\langle \eta' | \bar{s}\gamma_5 s | 0 \rangle$.

It has been advocated that the new internal W-emission contribution coming from the Cabibbo-allowed process $b \to c\bar{c}s$ followed by a conversion of the $c\bar{c}$ pair into the η' via two gluon exchanges may play an important role since its mixing angle $V_{cb}V_{cs}^*$ is as large as that of the penguin amplitude and yet its Wilson coefficient a_2 is larger than that of penguin operators. The decay constant $f_{\eta'}^c$, defined by $\langle 0 | \bar{c}\gamma_\mu\gamma_5 c | \eta' \rangle = i f_{\eta'}^c q_\mu$, has been estimated to be $f_{\eta'}^c = (50 - 180)$ MeV, based on the OPE, large-N_c approach and QCD low energy theorems [22]. Recent refined estimates [23,24] give $f_{\eta'}^c = -(2 \sim 15)$ MeV, which is in strong contradiction in magnitude and sign to the estimate of [22]. It turns out that if $N_c^{\rm eff}(LL)$ is treated to be the same as $N_c^{\rm eff}(LR)$, this new mechanism is not welcome for explaining $\mathcal{B}(B \to \eta' K)$ at small $1/N_c^{\rm eff}$ due to the fact that its contribution is proportional to a_2, which is negative at small $1/N_c^{\rm eff}$.

We have shown in [7] that if $N_c^{\rm eff}(LL) \sim 2$ and $N_c^{\rm eff}(LR) > N_c^{\rm eff}(LL)$, $\mathcal{B}(B^\pm \to \eta' K^\pm)$ at $1/N_c^{\rm eff}(LR) \le 0.2$ will be enhanced considerably from $(2.5-3)\times 10^{-5}$ to $(3.7-5)\times 10^{-5}$ (see Fig. 1). First, the η' charm content contribution now contributes in the right direction to the decay rate irrespective of the value of $N_c^{\rm eff}(LR)$ as a_2 now is always positive. Second, the interference between the spectator amplitudes of $B^\pm \to \eta' K^\pm$ is constructive. Third, the term proportional to $2(a_3 - a_5)X_u^{(BK,\eta')} + (a_3 + a_4 - a_5)X_s^{(BK,\eta')}$ is enhanced when $(N_c^{\rm eff})_3 = (N_c^{\rm eff})_4 = 2$. The agreement with experiment for $B^\pm \to \eta' K^\pm$ thus provides another strong support for $N_c^{\rm eff}(LL) \sim 2$ and for the relation $N_c^{\rm eff}(LR) > N_c^{\rm eff}(LL)$.

4 Final-state interactions and $B \to \omega K$

The CLEO observation [17] of a large branching ratio for $B^\pm \to \omega K^\pm$

$$\mathcal{B}(B^\pm \to \omega K^\pm) = \left(1.5^{+0.7}_{-0.6} \pm 0.2\right) \times 10^{-5}, \tag{22}$$

is difficult to explain at first sight. Its factorizable amplitude is of the form

$$A(B^- \to \omega K^-) \propto (a_4 + Ra_6)X^{(B\omega,K)} + (2a_3 + 2a_5 + \frac{1}{2}a_9)X^{(BK,\omega)} + \cdots, \tag{23}$$

with $R = -2m_K^2/(m_b m_s)$, where ellipses represent for contributions from W-annihilation and space-like penguin diagrams. It is instructive to compare this decay mode closely with $B^- \to \rho K^-$

$$A(B^- \to \rho^0 K^-) \propto (a_4 + Ra_6)X^{(B\rho,K)} + \frac{3}{2}a_9 X^{(BK,\rho)} + \cdots. \tag{24}$$

Due to the destructive interference between a_4 and a_6 penguin terms, the branching ratio of $B^\pm \to \rho^0 K^\pm$ is estimated to be of order 5×10^{-7}. The question is then why is the observed rate of the ωK mode much larger than the ρK mode ? By comparing (23) with (24), it is natural to contemplate that the penguin contribution proportional to $(2a_3+2a_5+\frac{1}{2}a_9)$ accounts for the large enhancement of $B^\pm \to \omega K^\pm$. However, this is not the case: The coefficients a_3 and a_5, whose magnitudes are smaller than a_4 and a_6, are not large enough to accommodate the data unless $N_c^{\rm eff}(LR) < 1.1$ or $N_c^{\rm eff}(LR) > 20$ (see Fig. 9 of [7]).

So far we have neglected three effects in the consideration of $B^\pm \to \omega K^\pm$: W-annihilation, space-like penguin diagrams and final-state interactions (FSI). It turns out that FSI may play the dominant role for $B^\pm \to \omega K^\pm$. The weak decays $B^- \to K^{*-}\pi^0$ via the penguin process $b \to su\bar{u}$ and $B^- \to K^{*0}\pi^-$ via $b \to sd\bar{d}$ followed by the quark rescattering reactions $\{K^{*-}\pi^0,\ K^{*0}\pi^-\} \to \omega K^-$ contribute constructively to $B^- \to \omega K^-$, but destructively to $B^- \to \rho K^-$. Since the branching ratios for $B^- \to K^{*-}\pi^0$ and $K^{*0}\pi^-$ are large, of order $(0.5-0.8)\times 10^{-5}$, it is conceivable that a large branching ratio for $B^\pm \to \omega K^\pm$ can be achieved from FSI via inelastic scattering. Moreover, if FSI dominate, it is expected that $\mathcal{B}(B^\pm \to \omega K^\pm) \approx (1+\sqrt{2})^2\mathcal{B}(B^0 \to \omega K^0)$.

5 Conclusions

To summarize, the CLEO data of $B^\pm \to \omega\pi^\pm$ available last year clearly indicate that $N_c^{\rm eff}(LL)$ is favored to be small, $N_c^{\rm eff}(LL) < 2.9$. This is consistent with the observation that $N_c^{\rm eff}(LL) \approx 2$ in $B \to D\pi$ decays. Unfortunately, the significance of $B^\pm \to \omega\pi^\pm$ is reduced in the recent CLEO analysis and only an upper limit is quoted. Therefore, a measurement of its branching ratio is urgently needed. In analogue to the class-III $B \to D\pi$ decays, the interference effect of spectator amplitudes in charged B decays $B^- \to \pi^-\pi^0,\ \rho^-\pi^0,\ \pi^-\rho^0$ is sensitive to $N_c^{\rm eff}(LL)$; measurements of them [see (15)] will be very useful to pin down the value of $N_c^{\rm eff}(LL)$.

As for $N_c^{\rm eff}(LR)$, we found that the constraints on $N_c^{\rm eff}(LR)$ derived from $B^\pm \to \phi K^\pm$ and $B \to \phi K^*$ are no consistent. Under the factorization hypothesis, the decays $B \to \phi K$ and $B \to \phi K^*$ should have almost the same branching ratios, a prediction not borne out by current data. Therefore, it is crucial to measure the charged and neutral decay modes of $B \to \phi(K, K^*)$ in order to see if the generalized factorization approach is applicbale to $B \to \phi K^*$ decay. Nevertheless, our analysis of $B \to \eta' K$ indicates that $N_c^{\rm eff}(LL) \approx 2$ is favored and $N_c^{\rm eff}(LR)$ is preferred to be larger. Since the energy release in the energetic two-body charmless B decays is not less than that in $B \to D\pi$ decays, it is thus expected that

$$|\chi(2-\text{body rare B decay})| \lesssim |\chi(B \to D\pi)|. \tag{25}$$

It follows from Eq. (7) that $N_c^{\rm eff}(LL) \approx N_c^{\rm eff}(B \to D\pi) \sim 2$ and $N_c^{\rm eff}(LR) \sim 2-5$, depending on the sign of χ. Therefore, we conjecture that $N_c^{\rm eff}(LR) \sim 5 > N_c^{\rm eff}(LL) \sim 2$.

Acknowledgments

I would like to thank the organizer Guey-Lin Lin for this well run and stimulating workshop. This work was supported in part by the National Science Council of the Republic of China under Grant NSC87-2112-M006-018.

1. For a review of CLEO measurements on charmless B decays, see K. Lingel, T. Skwarnicki, and J.G. Smith, hep-ex/9804015.
2. C. Caso *et al.* (Particle Data Group), Eur. Phys. J. **C3**, 1 (1998).
3. A.J. Buras, M. Jamin, M.E. Lautenbacher, and P.H. Weisz, *Nucl. Phys.* B**370**, 69 (1992); R. Fleischer, *Z. Phys.* C**58**, 483 (1993);
4. G. Kramer, W.F. Palmer, and H. Simma, *Z. Phys.* C**66**, 429 (1995); *Nucl. Phys.* B**428**, 77 (1994).
5. A. Ali and C. Greub, *Phys. Rev.* D**57**, 2996 (1998).
6. M. Neubert and B. Stech, CERN-TH/97-99 [hep-ph/9705292], to appear in *Heavy Flavours*, edited by A.J. Buras and M. Lindner, 2nd ed. (World Scientific, Singapore).
7. H.Y. Cheng and B. Tseng, hep-ph/9803457, to appear in Phys. Rev. D.
8. H.Y. Cheng, *Phys. Lett.* B**395**, 345 (1994); in *Particle Theory and Phenomenology,* XVII International Karimierz Meeting on Particle Physics, Iowa State University, May 1995, edited by K.E. Lassila *et al.* (World Scientific, Singapore, 1996), p.122 [hep-ph/9506340].
9. A.N. Kamal, A.B. Santra, T. Uppal, and R.C. Verma, *Phys. Rev.* D**53**, 2506 (1996).
10. H.Y. Cheng, *Z. Phys.* C**69**, 647 (1996).
11. M. Bauer, B. Stech, and M. Wirbel, *Z. Phys.* C**34**, 103 (1987).
12. H.Y. Cheng and B. Tseng, *Phys. Rev.* D**51**, 6295 (1995).
13. Y.H. Chen, H.Y. Cheng, and B. Tseng, in preparation.
14. A. Ali, G. Kramer and C.D. Lü, hep-ph/9804363.
15. CLEO Collaboration, M.S. Alam *et al.,* CLEO CONF 97-23 (1997).
16. F. Parodi, P. Roudeau, and A. Stocchi, hep-ph/9802289.
17. CLEO Collaboration, T. Bergfeld *et al.,* CLNS 97/1537 [hep-ex/9803018].
18. CLEO Collaboration, R. Godang *et al., Phys. Rev. Lett.* **80**, 3456 (1998).
19. N.G. Deshpande, B. Dutta, and S. Oh, OITS-644 [hep-ph/9712445].
20. L.L. Chau, H.Y. Cheng, W.K. Sze, H. Yao, and B. Tseng, *Phys. Rev.* D**43**, 2176 (1991).
21. D.S. Du and L. Guo, *Z. Phys.* C**75**, 9 (1997); D.S. Du, M.Z. Yang, and D.Z. Zhang, *Phys. Rev.* D**53**, 249 (1996).
22. I. Halperin and A. Zhitnitsky, *Phys. Rev.* D**56**, 7247 (1997).
23. A. Ali, J. Chay, C. Greub, and P. Ko, *Phys. Lett.* B**424**, 161 (1998).
24. A.A. Petrov, JHU-TIPAC-97016 [hep-ph/9712497]; T. Feldmann, P. Kroll, and B. Stech, hep-ph/9802409; F. Araki, M. Musakhanov, and H. Toki, hep-ph/9803356.

PERTURBATIVE QCD ANALYSIS OF EXCLUSIVE B MESON DECAYS

HSIANG-NAN LI

Department of Physics, National Cheng-Kung University, Tainan, Taiwan 701, Republic of China

E-mail: hnli@mail.ncku.edu.tw

We review the perturbative QCD formalism for exclusive heavy meson decays, concentrating on the three-scale factorization theorem for nonleptonic processes. The formalism is then extended to the radiative decay $B \to K^*\gamma$, which occurs through penguin diagrams. It is observed that the contributions from the operators other than the penguin one $b \to s\gamma$ are not negligible. From the best fit to the experimental data of the branching ratio $\mathcal{B}(B \to K^*\gamma)$, we extract the B meson wave function, which possesses a sharp peak in the region with a small momentum fraction.

1 Introduction

Recently, perturbative QCD (PQCD) has been proposed to be an alternative approach to the study of heavy hadron decays [1,2,3], which complements the heavy quark effective theory [4] and the Bauer-Stech-Wirbel (BSW) method [5]. In this talk we review the PQCD analysis of exclusive B meson decays, concentrating on the three-scale factorization theorem for nonleptonic processes [6]. The PQCD formalism is then applied to the radiative decay $B \to K^*\gamma$, which involves the contributions from penguin diagrams. According to the factorization theorem, a heavy meson decay rate is expressed as the convolution of a hard subamplitude with meson wave functions. The former is calculable in usual perturbation theory, while the latter must be extracted from experimental data or derived by using nonperturbative methods, such as QCD sum rules. Since the K^* meson wave function has been known from the sum rule analyses [7], the $B \to K^*\gamma$ decay is an ideal process, from which the unknown B meson wave function can be determined. With the B meson wave function obtained here, we are able to make predictions for other B meson decay modes.

The operator for exclusive semileptonic heavy meson decays, such as $B \to D^{(*)}l\nu$, is

$$\frac{G_F}{\sqrt{2}} V_{cb}\bar{\nu}\gamma_\mu(1-\gamma_5)l\bar{c}\gamma^\mu b\,, \tag{1}$$

where G_F is the Fermi coupling constant and V_{cb} the Cabibbo-Kabayashi-Maskawa (CKM) matrix element. The formalism for semileptonic decays is simpler, and part of that for nonleptonic decays. Hence, we shall explain only the factorization of nonleptonic heavy meson decays below, which occur through the effective Hamiltonian,

$$H_{\text{eff}} = \frac{G_F}{\sqrt{2}} V_{ij} V_{kl}^* [c_1(\mu) O_1(\mu) + c_2(\mu) O_2(\mu)] \,, \tag{2}$$

with the $V - A$ current $(\bar{q}q) = \bar{q}\gamma_\mu(1-\gamma_5)q$, the four-fermion operators $O_1 = (\bar{q}_l q_k)(\bar{q}_j q_i)$ and $O_2 = (\bar{q}_j q_k)(\bar{q}_l q_i)$. The Wilson coefficients $c_{1,2}$, organizing the large logarithms from the hard gluon corrections to all orders, describe the evolution from the W boson mass M_W to a lower scale μ with the matching conditions $c_1(M_W) = 1$ and $c_2(M_W) = 0$.

The most widely adopted approach to exclusive nonleptonic heavy meson decays is the BSW model[5], in which the factorization hypothesis on the matrix elements of the operators $O_{1,2}$ is assumed. In this model decay rates are expressed in terms of various hadronic transition form factors. Employing the Fierz transformation, the coefficient of the form factors corresponding to external W boson emissions is $a_1 = c_1 + c_2/N$, and that corresponding to internal W boson emissions is $a_2 = c_2 + c_1/N$, N being the number of colors. The form factors may be related to each other by heavy quark symmetry, and parametrized by different ansatz. Nonfactorizable contributions, which can not be expressed in terms of hadronic transition form factors, and nonspectator contributions from W boson exchanges are neglected.

Though the BSW model is simple and gives predictions in fair agreement with experimental data, it encounters several difficulties. It has been known that the large N limit of $a_{1,2}$, *i.e.*, the choice $a_1 = c_1(M_c) \approx 1.26$ and $a_2 = c_2(M_c) \approx -0.52$, with M_c the c quark mass, explains the data of charm decays[5]. However, the same large N limit of $a_1 = c_1(M_b) \approx 1.12$ and $a_2 = c_2(M_b) \approx -0.26$, M_b being the b quark mass, does not apply to the bottom case. That is, the different mechanism between charm and bottom decays can not be understood in the BSW approach. Moreover, it has been difficult to explain the two ratios associated with the $B \to J/\psi K^{(*)}$ decays[8], $R = \mathcal{B}(B \to J/\psi K^*)/\mathcal{B}(B \to J/\psi K)$ and $R_L = \mathcal{B}(B \to J/\psi K_L^*)/\mathcal{B}(B \to J/\psi K^*)$, simultaneously using the BSW model, where $\mathcal{B}(B \to J/\psi K)$ is the branching ratio of the decay $B \to J/\psi K$. This controversy was found to be attributed to the neglect of the nonfactorizable internal W-emission amplitudes, which are in fact of the same order as the factorizable ones[9,10].

The above difficulties have been overcome by a modified PQCD formalism, the three-scale factorization theorem[6]. In the next section we shall summarize the basic ideas of this theorem.

2 Three-scale factorization theorem

Nonleptonic heavy meson decays involve three scales: the W boson mass M_W, at which the matching conditions of the effective Hamiltonian to the original Hamiltonian are defined, the typical scale t of a hard subamplitude, which reflects the dynamics of heavy meson decays, and the factorization scale $1/b$, with b the conjugate variable of parton transverse momenta. The dynamics below $1/b$ is regarded as being completely nonperturbative, and parametrized into a meson wave fucntion $\phi(x)$, x being the momentum fraction. PQCD for the scale above $1/b$ is reliable, and radiative corrections produce two types of large logarithms $\ln(M_W/t)$ and $\ln(tb)$. The former are summed to give the evolution from M_W down to t described by the Wilson coefficients $c(t)$. While the latter are summed to give the evolution from t to $1/b$.

There exist also double logarithms $\ln^2(Pb)$ from the overlap of collinear and soft divergences, P being the dominant light-cone component of a meson momentum. The resummation of these double logarithms leads to a Sudakov form factor $\exp[-s(P,b)]$, which suppresses the long-distance contributions in the large b region, and improves the applicability of PQCD around the energy scale of few GeV. The b quark mass scale is located in the range of applicability. This is the motivation we develop the PQCD formalism for heavy hadron decays. For the detailed derivation of the Sudakov factor for heavy meson decays, refer to [1,2]. With all the large logarithms organized, the remaining finite contributions are absorbed into a hard b quark decay subamplitude $H(t)$. Because of Sudakov suppression, the perturbative expansion of H in the coupling constant α_s makes sense.

Therefore, a three-scale factorization formula possesses the typical expression,

$$c(t) \otimes H(t) \otimes \phi(x) \otimes \exp\left[-s(P,b) - 2\int_{1/b}^{t} \frac{d\bar{\mu}}{\bar{\mu}} \gamma_q(\alpha_s(\bar{\mu}))\right], \qquad (3)$$

where the exponential containing the quark anomalous dimension γ_q describes the evolution from t to $1/b$ mentioned above. As the hard scale t runs to below M_b and M_c, the constructive and destructive interferences involved in bottom and charm decays, respectively, appear naturally. Furthermore, not only factorizable, but nonfactorizable and nonspectator contributions can be evaluated in a systematic way. With the inclusion of nonfactorizable contributions, we find that $a_{1,2}$ restore their original role of the Wilson coefficients, instead of being treated as the BSW free parameters. The branching ratios of the various decay modes $B \to D^{(*)}\pi(\rho)$ and the ratios R and R_L associated with the $B \to J/\psi K^{(*)}$ decays can all be well explained by the above formalism [11].

We review the evaluation of the factorizable, nonfactorizable and nonspectator contributions to exclusive nonleptonic decays $B \to D^{(*)}\pi(\rho)$ in the three-scale PQCD factorization theorem. The factorizable external W-emission amplitudes define the $B \to D^{(*)}$ transition form factors ξ_i, $i = +, -, V, A_1, A_2$, and A_3, through the hadronic matrix elements,

$$
\begin{aligned}
\langle D(P_2)|V^\mu|B(P_1)\rangle &= \sqrt{M_B M_D}[\xi_+(\eta)(v_1+v_2)^\mu + \xi_-(\eta)(v_1-v_2)^\mu] \,, \\
\langle D^*(P_2)|V^\mu|B(P_1)\rangle &= i\sqrt{M_B M_{D^*}}\xi_V(\eta)\epsilon^{\mu\nu\alpha\beta}\epsilon^*_\nu v_{2\alpha} v_{1\beta} \,, \\
\langle D^*(P_2)|A^\mu|B(P_1)\rangle &= \sqrt{M_B M_{D^*}}[\xi_{A_1}(\eta)(\eta+1)\epsilon^{*\mu} - \xi_{A_2}(\eta)\epsilon^*\cdot v_1 v_1^\mu \\
&\quad -\xi_{A_3}(\eta)\epsilon^*\cdot v_1 v_2^\mu] \,. \qquad (4)
\end{aligned}
$$

P_1 (P_2), M_B ($M_{D^{(*)}}$) and v_1 (v_2) are the momentum, the mass, and the velocity of the B ($D^{(*)}$) meson, satisfying the relation $P_1 = M_B v_1$ ($P_2 = M_{D^{(*)}} v_2$). ϵ^* is the polarization vector of the D^* meson. The velocity transfer $v_1 \cdot v_2$ in two-body nonleptonic decays takes the maximal value $\eta = (1+r^2)/(2r)$ with $r = M_{D^{(*)}}/M_B$. In the infinite mass limit of M_B and $M_{D^{(*)}}$ the form factors ξ_i obey the relations $\xi_+ = \xi_V = \xi_{A_1} = \xi_{A_3} = \xi$ and $\xi_- = \xi_{A_2} = 0$, where ξ is the so-called Isgur-Wise (IW) function [12].

Equation (4) is the standard definition of the form factors ξ_i. In our approach, however, the factorization formulas for ξ_i will contain the Wilson coefficients as the extra convoluiton factors. ξ_i include the contributions from the hadronic matrix element of O_1 and from the color-suppressed matrix element of O_2. Therefore, their factorization formulas involve the Wilson coefficient $a_1 = c_1 + c_2/N$. The form factors for the internal W-emission diagrams include the factorizable contributions from the matrix elements of O_2 and from the color-suppressed matrix element of O_1, which then contain the Wilson coefficient $a_2 = c_2 + c_1/N$. The form factors for the W-exchange diagrams include the factorizable contributions from the matrix elements of O_2 and from the color-suppressed matrix element of O_1. Hence, they also contain the Wilson coefficient a_2.

The nonfactorizable contributions to the $B \to D^{(*)}\pi$ decays are color-suppressed, which can not be expressed in terms of hadronic form factors. A nonfactorizable diagram associated with external W emissions involves a hard exchanged gluon, which, for example, attaches the spectator quark of the B meson and a valence quark of the pion. The resultant amplitude depends on the Wilson coefficient c_2/N. A nonfactorizable diagram associated with internal W emissions contains a hard gluon, which attaches the spectator quark of the B meson and a valence quark of the D meson. The resultant amplitude depends on the Wilson coefficient c_1/N. Similarly, a nonfactorizable amplitude associated with W exchanges contains the Wilson coefficient c_1/N. For the explicit expressions of the above form factors and amplitudes, refer to[11].

3 The $B \to K^*\gamma$ decay

We now apply the three-scale factorization theorem to the radiative decay $B \to K^*\gamma$[13], whose effective Hamiltonian is

$$H_{\rm eff} = -\frac{G_F}{\sqrt{2}} V_{ts}^* V_{tb} \sum_{i=1}^{12} c_i(\mu) O_i(\mu) \; . \tag{5}$$

It will be found that only the operators $O_2 = (\bar{s}c)(\bar{c}b)$, $O_7 = \frac{e}{8\pi^2} M_b \bar{s}\sigma^{\mu\nu}(1+\gamma_5)bF_{\mu\nu}$, and $O_8 = -\frac{g}{8\pi^2} M_b \bar{s}\sigma^{\mu\nu}(1+\gamma_5)T^a b G^a_{\mu\nu}$, with $\sigma^{\mu\nu} = (i/2)[\gamma^\mu, \gamma^\nu]$, are important. The μ dependence of the corresponding Wilson coefficients $c_2(\mu)$, $c_7(\mu)$ and $c_8(\mu)$ can be quoted from[14].

We assign $P_1 = (M_B/\sqrt{2})(1, 1, \mathbf{0}_T)$ and $P_2 = (M_B/\sqrt{2})(1, r^2, \mathbf{0}_T)$ with $r = M_{K^*}/M_B$, M_{K^*} being the K^* meson mass, as the B and K^* meson momenta, respectively, in the rest frame of the B meson. Assume that the light valence quark in the B (K^*) meson carries the momentum k_1 (k_2). k_1 has a minus component k_1^-, giving the momentum fraction $x_B = k_1^-/P_1^-$, and small transverse components $\mathbf{k}_{BT}$. k_2 has a large plus component k_2^+, giving $x_K = k_2^+/P_2^+$, and small $\mathbf{k}_{KT}$. The photon momentum is then $P_3 = P_1 - P_2$, whose nonvanishing component is only P_3^-.

The decay amplitude is expressed as

$$M = \epsilon_\gamma \cdot \epsilon_{K^*}^* M_1 + i\epsilon_{\mu\rho+-}\epsilon_\gamma^\mu \epsilon_{K^*}^{*\rho} M_2 \; , \tag{6}$$

with ϵ_γ and ϵ_{K^*} the polarization vectors of the photon and of the K^* meson, respectively, and $M_{1(2)} = M_{1(2)2} + M_{1(2)7} + M_{1(2)8}$, where the terms on the right-hand side come from the operators O_2, O_7, and O_8, respectively. From Eq. (6), it is obvious that only the K^* mesons with transverse polarizations are produced in the decay. The factorization formulas for the amplitudes M_{ij} are, in terms of the overall factor $\Gamma^{(0)} = (G_F/\sqrt{2})(e/\pi)V_{ts}^* V_{tb} C_F M_B^5$, listed below. The $B \to K^*\gamma$ decay rate is then given by

$$\Gamma = \frac{1-r^2}{32\pi M_B}(|M_1|^2 + |M_2|^2) \; . \tag{7}$$

The amplitude from O_2 is written as

$$M_{12} = \Gamma^{(0)} \frac{4}{3} \int_0^1 dx \int_0^{1-x} dy \int_0^1 dx_B dx_K \int_0^{1/\Lambda} b db \phi_B(x_B)\phi_{K^*}(x_K)$$
$$\times \alpha_s(t_2) c_2(t_2) \exp[-S(x_B, x_K, t_2, b, b)]$$

$$\times[(1-r^2+2rx_K+2x_B)y-(rx_K+3x_B)(1-x)]$$
$$\times\frac{(1-r)(1-r^2)x_Kx}{xy(1-r^2)x_KM_B^2-M_c^2}H_2(Ab,\sqrt{|B_2^2|b})\ , \tag{8}$$

with

$$A^2=x_Kx_BM_B^2\ ,\quad t_2=\max(A,\sqrt{|B_2^2|},1/b)\ ,$$
$$B_2^2=x_Kx_BM_B^2-\frac{y}{1-x}(1-r^2)x_KM_B^2+\frac{M_c^2}{x(1-x)}\ , \tag{9}$$

where the integrations over x and y are associated with the charm loop, from which the photon and the hard exchanged gluon are radiated. The variable b_B (b_K), conjugate to the parton transverse momentum k_{BT} (k_{KT}), represents the transverse extent of the B (K^*) meson. t_2 is the characteristic scale of the hard subamplitude,

$$H_2(Ab,\sqrt{|B_2^2|b})=K_0(Ab)-K_0(\sqrt{|B_2^2|b})\ ,\quad B_2^2>0\ ,$$
$$=K_0(Ab)-i\frac{\pi}{2}H_0^{(1)}(\sqrt{|B_2^2|b})\ ,\quad B_2^2<0\ , \tag{10}$$

as mentioned before.

There are two diagrams associated with the operator O_7, where the hard gluon attaches the two quarks in the B meson and in the K^* meson separately. The corresponding amplitude is

$$M_{17}=\Gamma^{(0)}2\int_0^1 dx_Bdx_K\int_0^{1/\Lambda}b_Bdb_Bb_Kdb_K\phi_B(x_B)\phi_{K^*}(x_K)(1-r^2)$$
$$\times\Big[rH_7^{(a)}(Ab_K,\sqrt{|B_7^2|b_B},\sqrt{|B_7^2|b_K})F_7(t_{7a})$$
$$+[1+r+(1-2r)x_K]H_7^{(b)}(Ab_B,C_7b_B,C_7b_K)F_7(t_{7b})\Big]\ , \tag{11}$$

with

$$B_7^2=(x_B-r^2)M_B^2\ ,\quad C_7^2=x_KM_B^2$$
$$t_{7a}=\max(A,\sqrt{|B_7^2|},1/b_B,1/b_K)\ ,\quad t_{7b}=\max(A,C_7,1/b_B,1/b_K)\ . \tag{12}$$

The function F_7 denotes the product

$$F_7(t)=\alpha_s(t)c_7(t)\exp[-S(x_B,x_K,t,b_B,b_K)]\ . \tag{13}$$

The hard functions

$$H_7^{(a)}(Ab_K, \sqrt{|B_7^2|}b_B, \sqrt{|B_7^2|}b_K) =$$
$$K_0(Ab_k)h(\sqrt{|B_7^2|}b_B, \sqrt{|B_7^2|}b_K)\,, \quad B_7^2 > 0\,,$$
$$K_0(Ab_k)h'(\sqrt{|B_7^2|}b_B, \sqrt{|B_7^2|}b_K)\,, \quad B_7^2 < 0\,, \tag{14}$$

with

$$h = \theta(b_B - b_K)K_0(\sqrt{|B_7^2|}b_B)I_0(\sqrt{|B_7^2|}b_K) + (b_B \leftrightarrow b_K)\,,$$
$$h' = i\frac{\pi}{2}\left[\theta(b_B - b_K)H_0^{(1)}(\sqrt{|B_7^2|}b_B)J_0(\sqrt{|B_7^2|}b_K) + (b_B \leftrightarrow b_K)\right]\,, \tag{15}$$

and

$$H_7^{(b)}(Ab_B, C_7 b_B, C_7 b_K) = K_0(Ab_B)h(C_7 b_B, C_7 b_K)\,, \tag{16}$$

come from the two diagrams.

Four diagrams are associated with the operator O_8, where the photon is radiated by each quark in the B and K^* mesons. The corresponding amplitude is

$$\begin{aligned} M_{18} = & -\Gamma^{(0)}\frac{1}{3}\int_0^1 dx_B dx_K \int_0^{1/\Lambda} b_B db_B b_K db_K \phi_B(x_B)\phi_{K^*}(x_K) \\ & \times\{(1 - r^2 + x_B)(rx_K + x_B)H_8^{(a)}(Ab_K, B_8 b_B, B_8 b_K)F_8(t_{8a}) \\ & +[(2 - 3r)x_K - x_B + r(1 - x_K)(rx_K - 2rx_B + 3x_B)] \\ & \times H_8^{(b)}(Ab_B, C_8 b_B, C_8 b_K)F(t_{8b}) \\ & +(1 + r)(1 - r^2)[(1 + r)x_B - rx_K] \\ & \times H_8^{(c)}(\sqrt{|A'^2|}b_K, D_8 b_B, D_8 b_K)F(t_{8c}) \\ & -[(1 - r^2)((1 - r^2)(2 - x_K) + (1 + 3r)(2x_K - x_B)) \\ & +2r^2 x_K(x_K - x_B)] \\ & \times H_8^{(d)}(\sqrt{|A'^2|}b_B, E_8 b_B, E_8 b_K)F_8(t_{8d})\} \end{aligned} \tag{17}$$

with

$$\begin{aligned} & A'^2 = (1 - r^2)(x_B - x_K)M_B^2\,, \quad B_8^2 = (1 - r^2 + x_B)M_B^2\,, \\ & C_8^2 = (1 - x_K)M_B^2\,, \quad D_8^2 = (1 - r^2)x_B M_B^2\,, \quad E_8^2 = (1 - r^2)x_K M_B^2\,, \\ & t_{8a} = \max(A, B_8, 1/b_B, 1/b_K)\,, \quad t_{8b} = \max(A, C_8, 1/b_B, 1/b_K)\,, \\ & t_{8c} = \max(\sqrt{|A'^2|}, D_8, 1/b_B, 1/b_K)\,, \quad t_{8d} = \max(\sqrt{|A'^2|}, E_8, 1/b_B, 1/b_K)\,. \end{aligned} \tag{18}$$

The function F_8 denotes the product

$$F_8(t) = \alpha_s(t)c_8(t)\exp[-S(x_B, x_K, t, b_B, b_K)] \,. \tag{19}$$

The hard functions

$$\begin{aligned}
&H_8^{(a)}(Ab_K, B_8b_B, B_8b_K) = H_7^{(b)}(Ab_K, B_8b_B, B_8b_K) \,, \\
&H_8^{(b)}(Ab_B, C_8b_B, C_8b_K) = K_0(Ab_B)h'(C_8b_B, C_8b_K) \\
&H_8^{(c)}(\sqrt{|A'^2|}b_K, D_8b_B, D_8b_K) = \\
&\qquad K_0(\sqrt{|A'^2|}b_K)h(D_8b_B, D_8b_K) \,, \quad A'^2 \geq 0 \,, \\
&\qquad i\frac{\pi}{2}H_0^{(1)}(\sqrt{|A'^2|}b_K)h(D_8b_B, D_8b_K) \,, \quad A'^2 < 0 \,, \\
&H_8^{(d)}(\sqrt{|A'^2|}b_B, E_8b_B, E_8b_K) = \\
&\qquad K_0(\sqrt{|A'^2|}b_B)h'(E_8b_B, E_8b_K) \,, \quad A'^2 \geq 0 \,, \\
&\qquad i\frac{\pi}{2}H_0^{(1)}(\sqrt{|A'^2|}b_B)h'(E_8b_B, E_8b_K) \,, \quad A'^2 < 0
\end{aligned} \tag{20}$$

are derived from the four diagrams.

The amplitudes with the structure $i\epsilon_{\mu\rho+-}\epsilon_\gamma^\mu\epsilon_{K^*}^{*\rho}$ possess similar expressions, and are not presented here. The exponentials $\exp(-S)$ appearing in M_{ij} are the complete Sudakov factor, with the exponent

$$S = s(x_BP_1^-, b_B) + 2\int_{1/b_B}^t \frac{d\mu}{\mu}\gamma_q(\alpha_s(\mu))$$
$$+s(x_KP_2^+, b_K) + s((1-x_K)P_2^+, b_K) + 2\int_{1/b_K}^t \frac{d\mu}{\mu}\gamma_q(\alpha_s(\mu)). \tag{21}$$

The wave functions ϕ_B and ϕ_{K^*} satisfy the normalization, $\int_0^1 \phi_i dx = f_i/(2\sqrt{6})$, with the decay constant f_i, $i = B$ and K^*. The wave funciton for the K^* meson with transverse polarizations has been derived using QCD sum rules [7], given by

$$\phi_{K^*} = \sqrt{6}f_{K^*}x(1-x)[0.7-(1-2x)^2] \,. \tag{22}$$

As to the B meson wave functions, we investigate two models [15,16],

$$\phi_B^{(I)}(x) = \frac{N_Bx(1-x)^2}{M_B^2 + C_B(1-x)} \,, \tag{23}$$

$$\phi_B^{(II)}(x) = N'_B\sqrt{x(1-x)}\exp\left[-\frac{1}{2}\left(\frac{xM_B}{\omega}\right)^2\right] \,, \tag{24}$$

with the normalization constants N_B and N'_B, and the shape parameters C_B and ω.

Table 1: Amplitudes M_{ij} in the unit of 10^{-6} GeV^{-2}.

M_{12}/Γ_0	M_{17}/Γ_0	M_{18}/Γ_0
$-6.20 - 20.38i$	$11.10 - 198.08i$	$-1.94 - 0.29i$
M_{22}/Γ_0	M_{27}/Γ_0	M_{28}/Γ_0
$-4.84 - 16.17i$	$-11.10 + 198.08i$	$2.13 + 0.36i$

4 Results

In the evaluation of the various form factors and amplitudes, we adopt $G_F = 1.16639 \times 10^{-5}$ GeV^{-2}, the decay constants $f_B = 200$ MeV and $f_{K^*} = 220$ MeV, the CKM matrix elements $|V_{ts}^* V_{tb}| = 0.04$, the masses $M_c = 1.5$ GeV, $M_B = 5.28$ GeV and $M_{K^*} = 0.892$ GeV, and the $\bar{B}^0$ meson lifetime $\tau_{B^0} = 1.53$ ps [11]. It turns that no matter what value of the shape parameter C_B is chosen, the model $\phi_B^{(I)}$ in Eq. (23) with a flat profile leads to results much smaller than the CLEO data of the branching ratio $\mathcal{B}(B \to K^*\gamma) = (4.2 \pm 0.8 \pm 0.6) \times 10^{-5}$ [17]: the maximal prediction from $\phi_B^{(I)}$, corresponding to the shape parameter $C_B = -M_B^2$, is about 1.1×10^{-5}. The model $\phi_B^{(II)}$ in Eq. (24) with a sharp peak at small x then serves a possible candidate. It is indeed found that as $\omega = 0.425$ GeV, a prediction 4.19×10^{-5} for the branching ratio is reached, which is equal to the central value of the experimental data. If varying the shape parameter to $\omega = 0.42$ GeV and to $\omega = 0.43$ GeV, we obtain the branching ratios 4.36×10^{-5} and 4.08×10^{-5}, respectively. It indicates that the allowed range of ω is wide due to the large uncertainties of the data.

The detailed contribution from each amplitude M_{ij} is listed in the following table in the unit 10^{-6} GeV^{-2}. It is observed that the contributions from the operator O_2 are comparable with those from the penguin operator O_7. The amplitudes associated with O_8 are less important. The other operators give contributions smaller than those from O_8 by an order of magnitude, which are thus not included here. Furthermore, the imaginary parts of $M_{1(2)7}$ from the timelike interal quarks play an essential role for the explanation of the data.

In this work we have determined the B meson wave function,

$$\phi_B(x) = 1.81359\sqrt{x(1-x)} \exp\left[-\frac{1}{2}\left(\frac{xM_B}{0.425 \text{ GeV}}\right)^2\right], \tag{25}$$

from the best fit to the experimental data of $\mathcal{B}(B \to K^*\gamma)$. We stress that, however, Eq. (25) is not conclusive because of the large allowed range of the shape parameter ω. A more precise B meson wave function can be obtained by considering a global fit to the data of various decay modes, including $B \to$

$D^{(*)}\pi(\rho)$ [18]. Once the B meson wave fucntion is fixed, we shall employ it in the evaluation of the nonleptonic charmless decays.

Acknowledgments

This work was supported by the National Science Council of Republic of China under the Grant No. NSC-88-2112-M006-013.

References

1. H-n. Li and H.L. Yu, *Phys. Rev. Lett.* **74**, 4388 (1995); *Phys. Lett.* B **353**, 301 (1995);*Phys. Rev.* D **53**, 2480 (1996).
2. H-n. Li, *Phys. Rev.* D **52**, 3958 (1995).
3. H-n. Li and H.L. Yu, *Phys. Rev.* D **53**, 4970 (1996).
4. H. Georgi, *Phys. Lett.* B **240**, 447 (1990).
5. M. Bauer, B. Stech and M. Wirbel, *Z. Phys.* C **29**, 637 (1985);*Z. Phys.* C **34**, 103 (1987).
6. C.H. Chang and H-n. Li, *Phys. Rev.* D **55**, 5577 (1997).
7. V.L. Chernyak and A.R. Zhitnitsky, *Phys. Rep.* 112, 173 (1984).
8. M. Gourdin, A.N. Kamal, and X.Y. Pham, *Phys. Rev. Lett.* **73**, 3355 (1994).
9. C.E. Carlson and J. Milana, *Phys. Rev.* D **49**, 5908 (1994).
10. C.Y. Wu, T.W. Yeh, and H-n. Li, *Phys. Rev.* D **55**, 237 (1997).
11. T.W. Yeh and H-n. Li, *Phys. Rev.* D **56**, 1615 (1997).
12. N. Isgur and M.B. Wise, *Phys. Lett.* B **232**, 113 (1989);*Phys. Lett.* B **237**, 527 (1990).
13. H-n. Li and G.L. Lin, in preparation.
14. A.J. Buras *et al.*, *Nucl. Phys.* B **424**, 374 (1994).
15. R. Akhoury, G. Sterman and Y.-P Yao, *Phys. Rev.* D **50**, 358 (1994).
16. M. Bauer and M. Wirbel, *Z. Phys.* C **42**, 671 (1989).
17. S. Stone, to appear in Proceedings of the Fourth Workshop on Particle Physics Phenomenology, Kaohsiung, Taiwan, 1998.
18. H-n. Li and B. Melic, in preparation.

Contribution of $b \to sgg$ through the QCD anomaly in exclusive decays $B \to (\eta', \eta)(K, K^*)$

P. Ko
Department of Physics
Korea Advance Institute of Science and Technology
Taejon 305-701, KOREA
E-mail: pko@charm.kaist.ac.kr

The charm content in η' is estimated using the QCD $U(1)$ axial anomaly applied to $b \to sgg \to \eta' s$. We find $f_{\eta'}^{(c)} \approx -3.1$ MeV for $m_c = 1.3$ GeV. Our estimate agrees with other independent methods by Feldmann and Kroll, and Araki, Musakhanov and Toki. The resulting branching ratios for $B \to \eta' K$ in the generalized factorization model are marginally consistent with the CLEO data, although they are lower than the data by ~ 2.

1 Introduction

In the year of 1997, the CLEO collaboration reported measurements in a number of exclusive two-body non-leptonic decays of the type $B \to h_1 h_2$, where h_1 and h_2 are light mesons and the inclusive decay $B^{\pm} \to \eta' X_s$. [1-4] In particular, large branching ratios into the final states including η' are reported : [1]

$$BR(B^{\pm} \to \eta' + X_s) = (6.2 \pm 1.6 \pm 1.3) \times 10^{-4} \quad (1)$$
$$(\text{for } 2.0 \text{ GeV} \le p_{\eta'} \le 2.7 \text{ GeV}),$$
$$BR(B^{\pm} \to \eta' + K^{\pm}) = (7.1^{+2.5}_{-2.1} \pm 1.0) \times 10^{-5}, \quad (2)$$
$$BR(B^0 \to \eta' K^0) = (5.3^{+2.8}_{-2.2} \pm 1.2) \times 10^{-5} \; . \quad (3)$$

They did not observe any decay involving the (ηK) or $(\eta, \eta')K^*$ modes. These measurements have stimulated a lot of theoretical activity, [5-17] both in the inclusive and exclusive decays involving η' mesons.

Such unexpectedly large branching ratios of B decays into the final states with an η' meson led to an idea that the charm content in η' meson might be very large. The amplitude for $b \to s(\bar{c}c) \to s(\eta', \eta)$ can be parametrized as $\langle \eta' | \bar{c} \gamma_\mu \gamma_5 c | 0 \rangle = -i f_{\eta'}^c q_\mu$, and similarly for the η case. The relevant quantity $f_{\eta^{(\prime)}}^c$ is often referred to as the charm content of the $\eta^{(\prime)}$. This quantity is *a priori* not known, but it was claimed that the data (1)-(3) can be accomodated if $f_{\eta'}^c$ is as large as $(50 - 180)$ MeV. [7,8] However, this is uncomfortably large compared to the typical meson decay constants, *e.g.*, $f_\pi = 132$ MeV. Later it turned out that they could be determined in a number of different ways, also including the B-decays combined with the QCD $U_{\rm A}(1)$ anomaly [18] being discussed here.

In this talk, I'd like to present another method for computing the contribution of the amplitudes $b \to s(gg) \to s(\eta', \eta)$. This method is based on calculating the amplitude for the chromomagnetic penguin process $b \to sgg$, followed by the transitions $gg \to (\eta', \eta)$ which are calculated using the QCD anomaly, determining both the sign and magnitude of these contributions. The branching ratios for $B^\pm \to (\eta', \eta)(K^\pm, K^{*\pm})$ and $B^0 \to (\eta', \eta)(K^0, K^{*0})$ based on the QCD-anomaly method are calculated in this letter and compared with the present CLEO measurements and with the ones in Ref. [5]. We find that the theoretical branching ratios for $B^\pm \to \eta' K^\pm$ and $B^0 \to \eta' K^0$ are almost equal and both are in the range $(2-4) \times 10^{-5}$, in agreement with the estimates in Ref. [5].

2 Estimate of b $\to (\eta, \eta')$s via QCD anomaly

Let us consider the charm-anticharm pair into two gluons, followed by the transition $gg \to \eta^{(')}$ (see Fig. 1). The first part of this two-step process, i.e. $b \to s(\bar{c}c \to g(k_1)g(k_2))$ which amounts to calculating the charm-quark-loop from which two gluons are emitted, has been worked out by Simma and Wyler [19] in the context of a calculation in the full theory. Their result is readily translated to our effective theory approach and can be compactly written as a new (induced) effective Hamiltonian $H_{\rm eff}^{gg}$,

$$\begin{aligned} H_{\rm eff}^{gg} &= -\frac{\alpha_s}{2\pi}\left(C_2^{\rm eff} + \frac{C_1^{\rm eff}}{N_c}\right)\frac{G_F}{\sqrt{2}} V_{cb} V_{cs}^* \Delta i_5\left(\frac{q^2}{m_c^2}\right)\frac{1}{k_1 \cdot k_2} \\ &\times G_a^{\alpha\beta}(D_\beta \tilde{G}_{\alpha\mu})_a \, \bar{s}\gamma^\mu(1-\gamma_5)b \quad , \end{aligned} \tag{4}$$

with $\tilde{G}_{\mu\nu} = \frac{1}{2}\epsilon_{\mu\nu\alpha\beta}G^{\alpha\beta}$ ($\epsilon_{0123} = +1$). The $C_{1,2}^{\rm eff}$ are modified Wilson coefficients for the current-current operators as described in Ref. [5]. In this formula, which holds for on-shell gluons ($q^2 = (k_1 + k_2)^2 = 2k_1 \cdot k_2$), the sum over color indices is understood. The function $\Delta i_5(q^2/m_c^2)$ is defined as

$$\Delta i_5(z) = -1 + \frac{1}{z}\left[\pi - 2\arctan\left(\frac{4}{z} - 1\right)^{1/2}\right]^2, \text{ for } 0 < z < 4 \quad . \tag{5}$$

Note that the $b \to sgg$ calculation brings in an explicit factor of α_s. However, this explicit α_s factor gets absorbed into the matrix element of the operator resulting from the anomaly. So, to the order that we are working, we use the coefficients $C_1^{\rm eff}$ and $C_2^{\rm eff}$ in Eq. (4). The $u\bar{u}$ contribution in Fig. 1 is suppressed due to the unfavourable CKM factors. The $t\bar{t}$ contribution is included in the effective Hamiltonian via the $bsgg$ piece present in the operator O_g. However, in the factorization framework, the $bsgg$ term in O_g does not contribute to the decays discussed. So, the $c\bar{c}$ contribution in Fig. 1 is the only one that survives.

Working out the hadronic matrix element of Eq. (4) using factorization, we now need to evaluate the matrix elements :

$$\langle \eta^{(\prime)} | G_a^{\alpha\beta} (D_\beta \tilde{G}_{\alpha\mu,a}) | 0 \rangle , \tag{6}$$

which can be written as

$$G_a^{\alpha\beta} (D_\beta \tilde{G}_{\alpha\mu})_a = \partial_\beta (G_a^{\alpha\beta} \tilde{G}_{\alpha\mu,a}) - (D_\beta G^{\alpha\beta})_a \tilde{G}_{\alpha\mu,a} . \tag{7}$$

Now we can discard the second term since it is suppressed by an additional power of g_s which follows on using the equation of motion, and furthermore, the first term is enhanced by N_c in the large N_c limit. The matrix elements of $\partial_\beta (G_a^{\alpha\beta} \tilde{G}_{\alpha\mu,a})$ are related to those of $G\tilde{G}$:

$$\partial_\beta \langle \eta^{(\prime)} | G_a^{\alpha\beta} \tilde{G}_{\alpha\mu,a} | 0 \rangle = \frac{iq_\mu}{4} \langle \eta^{(\prime)} | G_a^{\alpha\beta} \tilde{G}_{\alpha\beta,a} | 0 \rangle \quad . \tag{8}$$

The conversion of the gluons into η and η' is described by an amplitude which is fixed by the $SU(3)$ symmetry and the axial $U(1)$ current triangle anomaly. The matrix elements for $G\tilde{G}$ can be written as [20]

$$\langle \eta^{(\prime)} | \frac{\alpha_s}{4\pi} G_a^{\alpha\beta} \tilde{G}_{\alpha\beta,a} | 0 \rangle = m^2_{\eta^{(\prime)}} f^u_{\eta^{(\prime)}} \quad . \tag{9}$$

In Eq. (9) the decay constants $f^u_{\eta'}$ and f^u_η read

$$f^u_\eta = \frac{f_8}{\sqrt{6}} \cos\theta_8 - \frac{f_0}{\sqrt{3}} \sin\theta_0 \quad , \quad f^u_{\eta'} = \frac{f_8}{\sqrt{6}} \sin\theta_8 + \frac{f_0}{\sqrt{3}} \cos\theta_0 \quad , \tag{10}$$

where the coupling constants f_8, f_0 and the mixing angles θ_8 and θ_0 have been introduced earlier. We follow here the two-angle (η, η') mixing formalism of Ref. [21], where the mass eigenstates $|\eta\rangle$ and $|\eta'\rangle$ have the following decompositions:

$$\begin{aligned} |\eta\rangle &= \cos\theta_8 |\eta_8\rangle - \sin\theta_0 |\eta_0\rangle , \\ |\eta'\rangle &= \sin\theta_8 |\eta_8\rangle + \cos\theta_0 |\eta_0\rangle . \end{aligned} \tag{11}$$

Collecting the individual steps, the matrix elements in Eq. (6) can be written as

$$\langle \eta^{(\prime)}(q) | \frac{\alpha_s}{4\pi} G_a^{\alpha\beta} (D_\beta \tilde{G}_{\alpha\mu a}) | 0 \rangle = iq_\mu \frac{m^2_{\eta^{(\prime)}}}{4} f^u_{\eta^{(\prime)}} \quad . \tag{12}$$

One would have naively expected that the gluonic matrix elements are small since they contain an extra factor of α_s. However, as shown by Eq. (12), this is obviously not the case and the gluon operator with α_s as a whole is responsible for the invariant mass of the $\eta^{(\prime)}$ mesons. Also, the combination entering in Eq. (12) involving the product of α_s and the gluon field operators is independent of the renormalization scale.

Before closing this section, let me comment on the alternative method to estimate $f^c_{\eta'}$ using the $\eta_c - \eta'$ mixing (Fig. 2 (a)) :

$$J/\psi \to \eta_c^* + \gamma \tag{13}$$

For example, in Ref. [5], these quantities were determined from the decays $J/\psi \to (\eta, \eta', \eta_c)\gamma$, extending the usual (η, η')-mixing formalism [22] to the (η_c, η', η) system. Using the measured decay widths for the decays $J/\psi \to (\eta, \eta', \eta_c)\gamma$ and $(\eta_c, \eta', \eta) \to \gamma\gamma$ yields $|f^c_{\eta'}| \simeq 5.8$ MeV and $|f^c_{\eta}| \simeq 2.3$ MeV. [5] However, in such a scheme, the intermediate η_c is far off-shell, and one has to know the form factor for $J/\psi - \eta_c - \gamma$ vertex. Moreover, this picture is not consistent with electromagnetic gauge invariance, since at the quark level, the above transition occurs through $c\bar{c} \to \gamma + gg$ followed by $gg \to \eta'$ (Fig. 2 (b)). Note that there is a diagram in which the photon is emitted in the middle of two gluons. Such diagram cannot be represented by the simple $\eta_c - \eta$ mixing picture. One cannot simply neglect this diagram either, since it would violate electromagnetic gauge invariance. Therefore, the usual picture of $\eta_c - \eta'$ mixing for $J/\psi \to \eta'\gamma$ may not be a good one, although it is widely used in the literature.

3 Estimate of the decay rate for $\mathbf{B} \to (\eta', \eta)(\mathbf{K}, \mathbf{K}^*)$

In order to compute the complete amplitude for the exclusive decays, one has to combine the contribution from the decay $b \to s(c\bar{c}) \to s(gg) \to s\eta^{(\prime)}$ discussed in the previous section with all the others arising from the four-quark and chromomagnetic operators, as detailed in Ref. [5]. The resulting amplitudes in the factorization approximation are listed in Ref. [18]. By comparing two expressions in Refs. [5] and [18], we can make the following identification :

$$-\Delta i_5(m^2_{\eta'}/m^2_c) f^u_{\eta'} \to f^c_{\eta'}, \qquad -\Delta i_5(m^2_{\eta}/m^2_c) f^u_{\eta} \to f^c_{\eta}. \tag{14}$$

Therefore, we have a simple relation between the decay constants $f^c_{\eta'}$, f^c_{η}, introduced in the intrinsic charm content method, and the form factor Δi_5 entering via the operator in Eq. (4). The idea of intrinsic charm quark content of η' and η and the contribution of the operator in Eq. (4) are related since this operator comes from the charm quark loop. Using the best-fit values of the (η,η')-mixing parameters from Ref. [23], yielding $\theta_8 = -22.2^\circ, \theta_0 = -9.1^\circ, f_8 = 168$ MeV, $f_0 = 157$ MeV, which in turn yields $f^u_{\eta'} = 63.6$ MeV and $f^u_\eta = 77.8$ MeV, the relations in Eq. (14) give $f^c_{\eta'} \sim -3.1$ MeV (-2.3 MeV) and $f^c_\eta \sim -1.2$ MeV (-0.9 MeV), with m_c having the value 1.3 GeV (1.5 GeV). In our approach the relative signs of the contributions from O^c_1 and O^c_2 to the other contributions are determined; we obtain the negative-$f^c_{\eta'}$ (and f^c_η) solution of the two possible ones. Our estimate of $f_{\eta'}$ is consistent with the results obtained by Feldmann, Kroll and Stech, [24]

$$f_{\eta'} = -(6.3 \pm 0.6)\ \text{MeV}, \quad f_\eta = -(2.4 \pm 0.2)\ \text{MeV}, \tag{15}$$

which is based on a new scheme for the flavor mixing and the axial anomaly. Also there appeared Ref. [25] by Araki *et al.*, who did an independent calculation in the same way as Haperin and Zhitnitsky and got

$$f^{(c)}_{\eta'} = -(12.3 - 18.4)\ \text{MeV}, \tag{16}$$

after correcting some mistakes in the orginal calculations in Ref. [7]. By now it is amusing that three independent methods for estimating $f^{(c)}_{\eta'}$ agree on its sign and the size within a factor of $2-3$. So one can say that the issue of the charm content in η' is now settled with reasonable confidence.

We plot the resulting branching ratios $BR(B^\pm \to \eta' K^{0\pm})$ and $BR(B^0 \to \eta' K^0)$ as functions of the parameter ξ in Figs. 3 and 4 for three different sets of CKM elements ρ and η in the Wolfenstein parametrization : [26]

$$(\rho,\eta) = (0.05, 0.36),\ (0.30, 0.42),\ \text{and}\ (0, 0.22) \tag{17}$$

The branching ratios for the neutral B-meson decays are averages with the corresponding charged conjugated decays in the figures. The numerical results for the branching ratios depend on various input parameters, the details of which can be found in Ref. [18]. Let us discuss the sensitivity on various parameters in brief. The branching ratios show a mild dependence (of order 10%) on the CKM parameters. However the branching ratios depend on the s-quark mass very sensitively, because the amplitude for $B \to \eta' K$ contains terms with the $1/m_s$ factor. In this work, we used $\overline{m}_s(2.5\ \text{GeV}) = 122$ MeV. However, if the present high values of the branching ratios for $B^{0(\pm)} \to \eta' K^{0(\pm)}$ continue to persist, one might have to consider smaller values of $\overline{m_s}$. We also expect progress in calculating quark masses on the lattice, sharpening the theoretical estimates presented here. Dependence on ξ amounts to between 20% and 35% depending on the other parameters if one varies ξ in the range $0 \le \xi \le 0.5$.

In all cases, the branching ratios are larger for $\xi = 0$. Taking into account the parametric dependences just discussed, we note that the theoretical branching ratio $BR(B^{0(\pm)} \to \eta' K^{0(\pm)})$ are uncertain by a factor 2. For the ratio of the branching ratios $BR(B^\pm \to \eta' K^\pm)/BR(B^0 \to \eta' K^0)$, which is useful quantity since it is practically independent of the form factors and most input parameters, the residual uncertainty is due to the CKM-parameter dependence of this ratio. It is estimated as about 10%. We get (for $0 \leq \xi \leq 0.5$)

$$\frac{BR(B^\pm \to \eta' K^\pm)}{BR(B^0 \to \eta' K^0)} = 0.9 - 1.02 \; . \tag{18}$$

The present experimental value of this ratio as calculated by adding the experimental errors in the numerator and denominator in quadrature is 1.34 ± 0.85. One can also study other modes involving K^* instead of K. Summarizing the results of Ref. [18], the branching ratios for the decay modes $B^\pm \to \eta K^\pm$ and $B^0 \to \eta K^0$ are smaller compared to their η'-counterparts by at least an order of magnitude, namely, $BR(B^\pm \to \eta K^\pm) = (1-2) \times 10^{-6}$ and a similar value for the neutral B decay mode. On the other hand, the branching ratios for the decay modes $B^\pm \to \eta(K^\pm, K^{*\pm})$ and $B^0 \to \eta(K^0, K^{*0})$ are all comparable to each other somewhere in the range $(1-3) \times 10^{-6}$.

4 Concluding Remarks

In this talk, we have presented an independent estimate of the charm content of $\eta^{(\prime)}$ meson using the process $b \to s(c\bar{c}) \to s(gg) \to s(\eta', \eta)$ and QCD anomaly. We could fix its sign as well as the size. Our results are in good agreement with other results based on independent methods, and thus we can be confident by now that the charm content in $\eta^{(\prime)}$ meson is indeed small. One could further study the branching ratios in $B \to (\eta', \eta)(K, K^*)$ in the generalized factorization approximation. Our result is marginally consistent with the CLEO data, considering various theoretical uncertainties in this kind of game (Figs. 3 and 4). Also our result is drastically different from the ones which follow in other scenarios. Hence, ongoing and future experiments will be able to test the predictions of the present approach as well as of the competing ones, such as models based on the dominance of the intrinsic charm contributions in η', as suggested in Refs. [7,8], or models in which dominant role is attributed to the soft-gluon-fusion process to form an η or η'. [16,17]

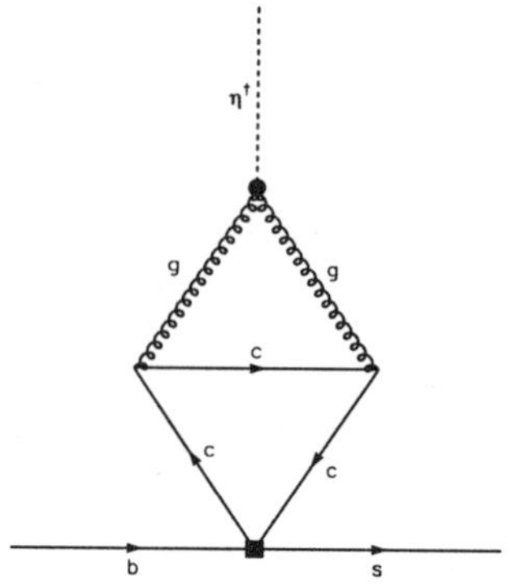

Figure 1: Feynman diagram contributing to the processes $b \to s(c\bar{c}) \to s(gg) \to s\eta^{(')}$ in the full and effective theory. The lower vertex in the diagram on the right is calculated with the insertion of the operators O_1^c and O_2^c in the effective Hamiltonian approach; the upper vertex in both the full and effective theory is determined by the QCD triangle anomaly.

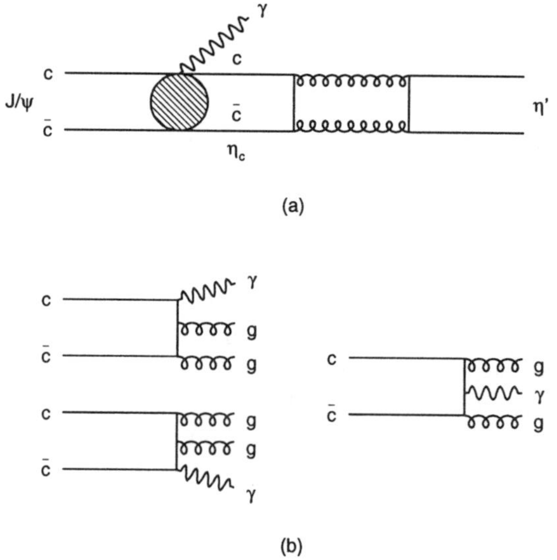

Figure 2: Feynman diagram contributing to the processes $J/\psi \to \gamma\eta^{'}$ (a) in the $\eta_c - eta^{'}$ mixing picture, and (b) in the parton picture.

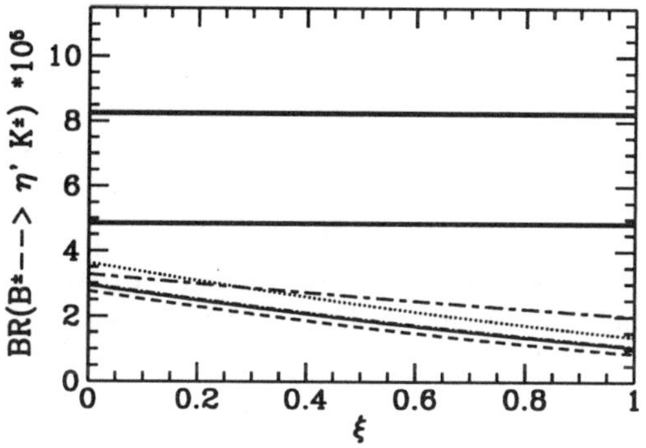

Figure 3: The branching ratio $BR(B^\pm \to \eta' K^\pm)$ plotted against the parameter ξ. The lower three curves correspond to the value $\overline{m_s}(2.5 \text{ GeV}) = 122$ MeV and the three choices of the CKM parameters: $\rho = 0.05, \eta = 0.36$ (solid curve); $\rho = 0.30, \eta = 0.42$ (dashed curve); $\rho = 0, \eta = 0.22$ (dashed-dotted curve). The upper two curves correspond to the value $\overline{m_s}(2.5 \text{ GeV}) = 100$ MeV, $\rho = 0.05, \eta = 0.36$ and $f^c_{\eta'} = -2.3$ MeV from the QCD-anomaly method (dotted curve) and $f^c_{\eta'} = -5.8$ MeV from [5] (long-short dashed curve). The horizontal thick solid lines represent the present CLEO measurements (with $\pm 1\sigma$ errors).

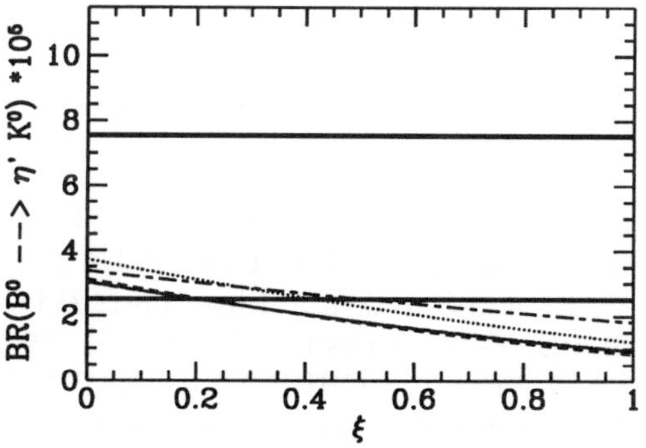

Figure 4: The branching ratio $BR(B^0 \to \eta' K^0)$ plotted against the parameter ξ. The legends are the same as in Fig. 3, and the horizontal thick solid lines represent the present CLEO measurements (with $\pm 1\sigma$ errors).

Acknowledgments

The author thanks the organizing committee of this workshop, H.Y. Cheng, W.-S. Hou, H.-N. Li and G.-L. Lin for the inivitation. He also thanks A. Ali, J. Chay and C. Greub for enjoyable collaboration on the subject presented in this talk. This work is supported in part by the Korea Science and Engineering Foundation (KOSEF) Contract No. 971-0201-002-2, the Ministry of Education grant BSRI 98-2418, the KOSEF through the SRC program of SNU-CTP, and the Distinguished Scholar Exchange Program of Korea Research Foundation.

References

1. J. Smith (CLEO Collaboration), talk presented at the 1997 ASPEN winter conference on Particle Physics, Aspen, Colorado, 1997 ; S. Anderson et al. (CLEO Collaboration), CLEO CONF 97-22a and EPS 97-333 (1997).
2. R. Godang et al. (CLEO Collaboration), preprint CLNS 97-1522, CLEO 97-27 (1997).
3. M.S. Alam et al. (CLEO Collaboration), CLEO CONF 97-23 and EPS 97-335 (1997).
4. J. Roy (CLEO Collaboration), invited talk at the Heavy Flavour Workshop, Rostock, September 1997.
5. A. Ali and C. Greub, Phys. Rev.**D 57**, 2996 (1998).
6. D. Atwood and A. Soni, Phys. Lett. **B405** (1997) 150 and Phys. Rev. Lett. **79**, 5206 (1997).
7. I. Halperin and A. Zhitnitsky, Phys. Rev. **D56** (1997) 7247; Phys. Rev. Lett. **80**, 438 (1998).
8. E. Shuryak and A.R. Zhitnitsky, Phys. Rev. **D 57**, 2001 (1998).
9. W.S. Hou and B. Tseng, Phys. Rev. Lett. **80**, 434 (1998).
10. F. Yuan and K.T. Chao, Phys. Rev. **D56** (1997) 2495.
11. A. Datta, X.-G. He and S. Pakvasa, Phys. Lett. **B419**, 369 (1998).
12. H.-Y. Cheng and B. Tseng, Phys.Lett.**B415**, 263 (1997).
13. A. Kagan and A.A. Petrov , UCHEP-27, UMHEP-443, hep-ph/9707354.
14. A.S. Dighe, M. Gronau and J. Rosner, Phys. Rev. Lett. **79** (1997) 4333.
15. N.G. Deshpande, B. Dutta and S. Oh, Phys. Rev. **D 57**, 5723 (1998).
16. M.R. Ahmady, E. Kou and A. Sugomoto, Phys. Rev. **D 58**, 014015 (1998).
17. D. Du, C.S. Kim and Y. Yang, Phys. Lett. **B426**, 133 (1998).
18. A. Ali, J. Chay, C. Greub and P. Ko, Phys. Lett. **B 424**, 161 (1998).
19. H. Simma and D. Wyler, Nucl. Phys. **B344**, 283 (1990).
20. M. Voloshin and V. Zakharov, Phys. Rev. Lett. **45** (1980) 688;

P. Ball, J.-M.. Frère and M. Tytgat, Phys. Lett. **B365** (1996) 367.

21. H. Leutwyler, Nucl. Phys. Proc. Suppl.**64** 223 (1998) ; P. Herrera-Sikoldy, J.I. Latorre, P. Pascual and J. Taron, Phys. Lett. **B 419**, 326 (1998).
22. F.J. Gilman and R. Kaufman, Phys. Rev. **D36** (1987) 2761.
23. T. Feldmann and P. Kroll, preprint WUB 97-28, hep-ph/9711231.
24. T. Feldmann, P. Kroll and B. Stech, WU-B-98-2, hep-ph/9802409.
25. F. Araki, M. Musakhanov, H. Toki, Oaksa University Preprint, hep-ph/9803356.
26. L. Wolfenstein, Phys. Rev. Lett. **51** (1983) 1945.

SENSITIVITY OF HEAVY MESON SEMILEPTONIC DECAYS TO FORM FACTOR MODELS

DAE SUNG HWANG
Department of Physics, Sejong University, Seoul 143–747, Korea
E-mail: dshwang@kunja.sejong.ac.kr

DO-WON KIM
Department of Physics, Kangnung National University, Kangnung 210-702, Korea
E-mail: dwkim@phys1.kangnung.ac.kr

We study the $d\Gamma/dq^2$ spectra and the branching fractions of the heavy meson exclusive semileptonic decays with the lepton mass effects into consideration. In D meson decays we study the sensitivity to form factors models. We find that for the decays to pseudoscalar meson the results are sensitive to the property of the form factor $F_1(q^2)$, and for the decays to vector meson the results are sensitive to the form factor $A_1(q^2)$ and not on the other ones ($A_0(q^2)$, $A_2(q^2)$ and $V(q^2)$). In B meson decays we study $\bar{B}^0 \to D^{(*)+}l^-\bar{\nu}$ using experimentally measured form factors, and find the $d\Gamma/dq^2$ spectra and branching fractions for $l = e$, μ and τ.

1 Introduction

The CP-violation phenomenon is discovered only in the $K_L \to \pi\pi$ decay and the charge asymmetry in the decay $K_L \to \pi^{\pm}l^{\mp}\nu$ for more than 30 years. The B-factories at KEK and SLAC are under construction for the discovery of CP-violation in the B meson system. The mechanism of CP-violation through the complex phase of the CKM three family mixing matrix[1] is presently considered standard for the CP-violation. In order to measure the CKM matrix elements accurately, it is important to know the hadronic form factors of the transition matrix elements well. For the heavy to heavy transitions the heavy quark effective theory provides good informations for the form factors. However, for the heavy to light transitions the understanding of the form factors is still limited and this fact hinders the extractions of the CKM matrix elements from experimental results significantly. At the same time, we will be able to have important clues for the internal structures of hadrons by knowing these form factors well. Semileptonic decay processes are good sources for the knowledge of the form factors both experimentally and theoretically, and the lepton mass effects in heavy meson exclusive semileptonic decays were studied by Körner and Schuler[2].

We derive the formulas for $d\Gamma/dq^2$ with non-zero lepton mass in the forms which are efficient to study the form factor dependences. This formula for the pseudoscalar to pseudoscalar transition was also given by Khodjamirian et al. [3]. By using these formulas we study the $d\Gamma/dq^2$ spectra and branching fractions of the exclusive semileptonic D meson decays: $D^0 \to K^{(*)-}e^+\nu$, $D^0 \to K^{(*)-}\mu^+\nu$, $D^0 \to \pi^-(\rho^-)e^+\nu$ and $D^0 \to \pi^-(\rho^-)\mu^+\nu$. In this anslysis we employ three models of form factors and show how the results are influenced by the difference of form factors. For the decays to a pseudoscalar meson and a lepton pair, the results are sensitive to the property of the form factor $F_1(q^2)$. For the decays to a vector meson and a lepton pair, the results are sensitive to the form factor $A_1(q^2)$, and not on the other ones ($A_0(q^2)$, $A_2(q^2)$ and $V(q^2)$). We also study the B meson decays $\bar{B}^0 \to D^{(*)+}l^-\bar{\nu}$ by using experimentally measured form factors, and find the $d\Gamma/dq^2$ spectra and branching fractions for $l = e$, μ and τ.

2 Semileptonic Decays of Heavy Mesons

From Lorentz invariance one finds the decomposition of the hadronic matrix element for pseudoscalar to pseudoscalar meson transition in terms of hadronic form factors:

$$\begin{aligned} &< P(p)|J_\mu|P(P) > \\ &= (P+p)_\mu f_+(q^2) + (P-p)_\mu f_-(q^2) \\ &= \left((P+p)_\mu - \frac{M^2-m^2}{q^2} q_\mu\right) F_1(q^2) + \frac{M^2-m^2}{q^2}\, q_\mu\, F_0(q^2), \end{aligned} \tag{1}$$

where $J_\mu = \bar{q'}\gamma_\mu(1-\gamma_5)q$. We use the following notations: M represents initial meson mass, m final meson mass, m_l lepton mass, P initial meson momentum, p final meson momentum, and $q_\mu = (P-p)_\mu$. The form factors $F_1(q^2)$ and $F_0(q^2)$ correspond to 1^- and 0^+ exchanges, respectively. At $q^2 = 0$ we have the constraint $F_1(0) = F_0(0)$, since the hadronic matrix element in (1) is nonsingular at this kinematic point.

The q^2 distribution of the semileptonic decay $D^0 \to K^-l^+\nu$ is given in terms of the hadronic form factors $F_1(q^2)$ and $F_0(q^2)$ as:

$$\frac{d\Gamma(D^0 \to K^-l^+\nu)}{dq^2} = \frac{G_F^2}{24\pi^3}\, |V_{cs}|^2\, K(q^2)\, (1 - \frac{m_l^2}{q^2})^2 \times \tag{2}$$

$$[\,(K(q^2))^2\,(1 + \frac{1}{2}\frac{m_l^2}{q^2})\,|F_1(q^2)|^2 \,+\, M^2\,(1 - \frac{m^2}{M^2})^2\,\frac{3}{8}\,\frac{m_l^2}{q^2}|F_0(q^2)|^2\,],$$

where $K(q^2)$, momentum of the final meson in the D meson rest frame, is given by

$$K(q^2) = \frac{1}{2M}\left((M^2+m^2-q^2)^2 - 4M^2m^2\right)^{\frac{1}{2}}, \tag{3}$$

and the physically allowed range of q^2 is given by

$$m_l^2 \le q^2 \le (M-m)^2. \tag{4}$$

For $m_l = 0$, (2) is reduced to the commonly used well-known formula:

$$\frac{d\Gamma(D^0 \to K^- l^+ \nu)}{dq^2} = \frac{G_F^2}{24\pi^3} |V_{cs}|^2 (K(q^2))^3 |F_1(q^2)|^2, \tag{5}$$

and $0 \le q^2 \le (M-m)^2$. We note in (5) that only $F_1(q^2)$ contributes for $m_l = 0$, however, for $m_l \ne 0$ $F_0(q^2)$ also contributes as we can see in (2).

From Lorentz invariance one finds the decomposition of the hadronic matrix element for pseudoscalar to vector meson transition in terms of hadronic form factors:

$$\begin{aligned} &< V(p)|J_\mu|P(P) > \\ &= \varepsilon^{*\nu}(p)\Big((M+m)g_{\mu\nu}A_1(q^2) - 2\frac{P_\mu P_\nu}{M+m}A_2(q^2) + \frac{q_\mu P_\nu}{M+m}A_3(q^2) \\ &+ i\varepsilon_{\mu\nu\rho\sigma}\frac{P^\rho p^\sigma}{M+m}V(q^2)\Big), \end{aligned} \tag{6}$$

where $\varepsilon_{0123} = 1$ and

$$2mA_0(q^2) = (M+m)A_1(q^2) - \frac{M^2-m^2+q^2}{M+m}A_2(q^2) + \frac{q^2}{M+m}A_3(q^2). \tag{7}$$

The form factors $V(q^2)$, $A_1(q^2)$, $A_2(q^2)$ and $A_0(q^2)$ correspond to 1^-, 1^+, 1^+ and 0^- exchanges, respectively. At $q^2 = 0$ we have the constraint $2mA_0(0) = (M+m)A_1(0) - (M-m)A_2(0)$, since the hadronic matrix element in (6) is nonsingular at this kinematic point.

After a rather lengthy calculation, the q^2 distribution of the semileptonic decay $D^0 \to K^{*-} l^+ \nu$ is given in terms of the hadronic form factors $A_1(q^2)$, $A_2(q^2)$, $A_3(q^2)$ and $V(q^2)$ as:

$$\begin{aligned} \frac{d\Gamma(D^0 \to K^{*-} l^+ \nu)}{dq^2} &= \frac{G_F^2}{32\pi^3} |V_{cs}|^2 \frac{1}{M^2} K(q^2) (1 - \frac{m_l^2}{q^2})^2 \times \\ \{|A_1(q^2)|^2 &\frac{(M+m)^2}{m^2}[\frac{1}{3}(MK)^2(1-\frac{m_l^2}{q^2}) + q^2m^2 + (MK)^2\frac{m_l^2}{q^2} + \frac{1}{2}m^2m_l^2] \\ +\mathrm{Re}(A_1(q^2)A_2^*(q^2))&[-\frac{M^2-m^2-q^2}{m^2}[\frac{2}{3}(MK)^2(1-\frac{m_l^2}{q^2}) + 2(MK)^2\frac{m_l^2}{q^2} \\ &+\frac{1}{2}(M^2+m^2-q^2)m_l^2] + (M^2-m^2+q^2)m_l^2] \end{aligned} \tag{8}$$

Table 1: The values of the power of pole for the three models used in this work.

	n_{F_1}	n_{F_0}	n_V	n_{A_1}	n_{A_2}	n_{A_0}
WSB(pole/pole)	1	1	1	1	1	1
Model I(pole/const.)	1	0	1	0	1	1
Model II(dipole/pole)	2	1	2	1	2	2

$$
\begin{aligned}
&+|A_2(q^2)|^2\frac{1}{(M+m)^2m^2}(MK)^2[\frac{4}{3}(MK)^2(1-\frac{m_l^2}{q^2})+4(MK)^2\frac{m_l^2}{q^2}+2M^2m_l^2] \\
&+|V(q^2)|^2\frac{q^2}{(M+m)^2}[\frac{8}{3}(MK)^2(1-\frac{m_l^2}{q^2})+4(MK)^2\frac{m_l^2}{q^2}] \\
&+|A_3(q^2)|^2\frac{q^2}{(M+m)^2m^2}\frac{1}{2}(MK)^2m_l^2 \\
&-\mathrm{Re}(A_3(q^2)A_2^*(q^2))\frac{1}{(M+m)^2m^2}(M^2-m^2+q^2)(MK)^2m_l^2 \\
&+\mathrm{Re}(A_3(q^2)A_1^*(q^2))\frac{1}{m^2}(MK)^2m_l^2\}.
\end{aligned}
$$

In the case of the B to D meson (heavy to heavy) transition, the heavy quark effective theory (HQET) gives the useful relations between the relevant form factors[4]:

$$
\begin{aligned}
F_1(q^2)=V(q^2) &= A_0(q^2) = A_2(q^2) = \frac{M+m}{2\sqrt{Mm}}\,\mathcal{F}(y), \qquad (9)\\
F_0(q^2)=A_1(q^2) &= \frac{2\sqrt{Mm}}{M+m}\,\frac{y+1}{2}\,\mathcal{F}(y),
\end{aligned}
$$

where $y=(M^2+m^2-q^2)/(2Mm)=E_{D^{(*)}}/m$ ($E_{D^{(*)}}$ is the energy of $D^{(*)}$ meson in the B meson rest frame), and $\mathcal{F}(y)$ is a form factor which becomes the Isgur-Wise function in the infinite heavy quark mass limit. When we use the relations (9), for $m_l=0$ the formula (8) becomes the well-known formula for the B to D^* transition:

$$
\begin{aligned}
\frac{d\Gamma(\bar{B}^0\to D^{*+}l^-\bar{\nu})}{dq^2} &= \frac{G_F^2}{48\pi^3}\,|V_{cb}|^2\,m^3\,(M-m)^2\,\sqrt{y^2-1}\,(y+1)^2\times \\
&\quad \{1+\frac{4y}{y+1}\,\frac{1-2yr+r^2}{(1-r)^2}\}\,(\mathcal{F}_{D^*}(y))^2, \qquad (10)
\end{aligned}
$$

where $r=m/M$.

Table 2: The values of pole masses (GeV) used in numerical calculations.

Current Relevant Form Factors	$m(0^-)$ A_0	$m(1^-)$ F_1 , V	$m(0^+)$ F_0	$m(1^+)$ A_1 , A_2
$\bar{s}c$	1.97	2.11	2.60	2.53
$\bar{d}c$	1.87	2.01	2.47	2.42

Table 3: The values of the form factors at $q^2 = 0$ used in numerical calculations.

	$F_1(0) = F_0(0)$	$V(0)$	$A_1(0)$	$A_2(0)$	$A_0(0)$
$D \to K^{(*)}$	0.762	1.226	0.880	1.147	0.733
$D \to \pi(\rho)$	0.692	1.225	0.775	0.923	0.669

3 $D^0 \to K^{(*)-} l^+ \nu$

For the form factors concerned with the exclusive semileptonic decays of D meson, we can not use the relations (9) of the HQET. Therefore, in the study of D meson decays we use models for form factors. The pole-dominance idea suggests the following q^2 dependence of the form factors [5]:

$$f_i(q^2) = f_i(0) \frac{1}{(1 - \frac{q^2}{m_{f_i}^2})^{n_{f_i}}}, \tag{11}$$

where n_{f_i} and m_{f_i} are corresponding power and pole mass of the form factors $f_i(q^2)$, respectively. The WSB model [5] adopts $n_{f_i} = 1$. However, the exact values of n_{f_i} are not known. The relations (9) of the HQET gives the following approximate relation among the powers of the form factors:

$$n_{F_1} = n_V = n_{A_0} = n_{A_2} = n_{F_0} + 1 = n_{A_1} + 1. \tag{12}$$

The above relation can be suggested for the form factors of the heavy to light transition, even though it is not clear whether this relation for the heavy to heavy transition still holds in this case. The lattice calculations show clearly that the form factors F_1, V and A_0 are more rapidly increasing functions of q^2 than the form factors F_0 and A_1 [6], which implies the relation (12). Non-perturbative analysis of QCD also suggests this relation [7]. Therefore, we will adopt two other models incorporating the relation (12), as well as the WBS model [5], in the present study of the exclusive semileptonic decays of D meson. We organize in Table 1 the values of the powers n_{f_i} of the three models which we use in this work. For the values of the pole masses and those of the form factors at $q^2 = 0$, we use the values organized in Table 2 and 3, which were

Table 4: The obtained branching fractions and their ratios for $D^0 \to K^- l^+ \nu$.

	$\mathcal{B}(D^0 \to K^- e^+ \nu)$	$\mathcal{B}(D^0 \to K^- \mu^+ \nu)$	$\mathcal{B}(e^+) : \mathcal{B}(\mu^+)$
WSB	3.49×10^{-2}	3.41×10^{-2}	1 : 0.98
Model I	3.49×10^{-2}	3.38×10^{-2}	1 : 0.97
Model II	4.78×10^{-2}	4.67×10^{-2}	1 : 0.98

Table 5: The obtained branching fractions and their ratios for $D^0 \to K^{*-} l^+ \nu$.

	$\mathcal{B}(D^0 \to K^{*-} e^+ \nu)$	$\mathcal{B}(D^0 \to K^{*-} \mu^+ \nu)$	$\mathcal{B}(e^+) : \mathcal{B}(\mu^+)$
WSB	3.88×10^{-2}	3.68×10^{-2}	1 : 0.95
Model I	3.28×10^{-2}	3.10×10^{-2}	1 : 0.95
Model II	3.85×10^{-2}	3.66×10^{-2}	1 : 0.95

given by Wirbel, Stech and Bauer [5]. Their precise values are not known and they should be different for each of the three models. However, their exact values are not crucial in this work, since our main purpose of the present work is to obtain the characters of the $d\Gamma/dq^2$ spectra and the branching fractions in each model and clarify their model dependences.

For $D^0 \to K^- l^+ \nu$, we use the formula (2) with non-zero lepton mass, instead of the commonly used formula (5) which is true for zero lepton mass. The obtained $d\Gamma(D^0 \to K^- e^+ \nu)/dq^2$ spectrum and branching fractions are presented in Figure 1 and Table 4, for each of the three models we adopt in this work: WSB, Model I and Model II explained in Table 1. We find that the shape of spectrum of Model II is different from those of WSB and Model I in Figure 1. That is, the value of n_{F_1} determines the shape of spectrum. The experimental result of the E687 Collaboration for this spectrum was given in Fig. 3 (a) of Ref. 8, and the shape of their spectrum favors Model II better than WSB and Model I. In the experimental extraction of the value of $F_1^{DK}(0)$, the simple pole of the form factors has been commonly assumed [9]. Under this assumption, $F_1^{DK}(0) = 0.75 \pm 0.02 \pm 0.02$ was extracted from the experimentally measured branching fraction $\mathcal{B}(D^0 \to K^- e^+ \nu) = (3.68 \pm 0.21) \times 10^{-2}$. However, if we assume in this extraction Model II which has the dipole form factor for $F_1(q^2)$, we would get

$$F_1^{DK}(0) = 0.75 \times \sqrt{\frac{3.49 \times 10^{-2}}{4.78 \times 10^{-2}}} = 0.75 \times 0.85 = 0.64 \qquad (13)$$

for the mean value. In (13) we used our results in Table 4 of the branching fraction $\mathcal{B}(D^0 \to K^- e^+ \nu)$ for WSB ($n_{F_1} = 1$) and Model II ($n_{F_1} = 2$), which

Table 6: The obtained branching fractions and their ratios for $D^0 \to \pi^- l^+ \nu$.

	$\mathcal{B}(D^0 \to \pi e^+ \nu)$	$\mathcal{B}(D^0 \to \pi \mu^+ \nu)$	$\mathcal{B}(e^+) : \mathcal{B}(\mu^+)$
WSB	2.92×10^{-3}	2.88×10^{-3}	1 : 0.99
Model I	2.92×10^{-3}	2.86×10^{-3}	1 : 0.98
Model II	5.94×10^{-3}	5.86×10^{-3}	1 : 0.99

Table 7: The obtained branching fractions and their ratios for $D^0 \to \rho^- l^+ \nu$.

	$\mathcal{B}(D^0 \to \rho e^+ \nu)$	$\mathcal{B}(D^0 \to \rho \mu^+ \nu)$	$\mathcal{B}(e^+) : \mathcal{B}(\mu^+)$
WSB	2.77×10^{-3}	2.66×10^{-3}	1 : 0.96
Model I	2.19×10^{-3}	2.10×10^{-3}	1 : 0.96
Model II	2.77×10^{-3}	2.66×10^{-3}	1 : 0.96

were obtained by using the same value of $F_1^{DK}(0)$. That is, the experimentally extracted value of $F_1^{DK}(0)$ is much dependent on the form factor models used in the analysis. In reality, it is not yet established which type of form factors is the right one, and the spectrum of the E687 Collaboration [8] prefers the dipole type for $F_1(q^2)$, as we mentioned in the above.

For $D^0 \to K^{*-} l^+ \nu$, we use the formula (8) with non-zero lepton mass. The obtained spectra and branching fractions are presented in Figure 2 and Table 5. We find that the results of WSB and Model II are almost the same, and they are significantly different from the results of Model I. This fact implies that the value of n_{A_1} determines the spectra and branching fractions of $D^0 \to K^{*-} l^+ \nu$.

4 $D^0 \to \pi^-(\rho^-) e^+ \nu$

For $D^0 \to \pi^- l^+ \nu$, we use the formula (2) with the replacement of V_{cs} by V_{cd}. The obtained branching fractions are presented in Table 6. We find that the branching fractions of Model II in Fig. 9 are about twice those of WSB and Model I. Therefore, in case that we determine the value of $F_1^{D\pi}(0)$ from an experimentally measured branching fraction of $D^0 \to \pi^- l^+ \nu$, the value of $F_1^{D\pi}(0)$ determined with Model II will be about $1/\sqrt{2}$ times its value determined with WSB or Model I. The values of the ratio $\mathcal{B}(D^0 \to \pi^- l^+ \nu)/\mathcal{B}(D^0 \to K^- l^+ \nu)$ we get from the branching fractions in Table 4 and 6 is in agreement with its experimental result $0.101 \pm 0.020 \pm 0.003$ [10].

For $D^0 \to \rho^- l^+ \nu$, we use the formula (8) with the replacement of V_{cs} by V_{cd}. The obtained branching fractions are presented in Table 7. We find in Table 7 that the results of WSB and Model II are the same, and they are much different from the results of Model I. Therefore, like the $D^0 \to K^{*-} l^+ \nu$ case,

Table 8: The obtained branching fractions and their ratio for $\bar{B}^0 \to D^{(*)+} l^- \bar{\nu}$.

	$\mathcal{B}(\bar{B}^0 \to D^{(*)+} e^- \bar{\nu})$	$\mathcal{B}(\bar{B}^0 \to D^{(*)+} \mu^- \bar{\nu})$	$\mathcal{B}(\bar{B}^0 \to D^{(*)+} \tau^- \bar{\nu})$
D^+	$(1.85^{-0.47}_{+0.56}) \times 10^{-2}$	$(1.84^{-0.47}_{+0.56}) \times 10^{-2}$	$(0.52^{-0.07}_{+0.07}) \times 10^{-2}$
Ratio	1	$0.996^{+0.001}_{-0.001}$	$0.278^{+0.049}_{-0.035}$
D^{*+}	$4.76^{-0.47}_{+0.50}$	$4.74^{-0.46}_{+0.50}$	$1.22^{-0.06}_{+0.06}$
Ratio	1	$0.996^{+0.002}_{-0.000}$	$0.256^{+0.014}_{-0.013}$

the property of the form factor $A_1(q^2)$ determines the branching fractions, and the results are not sensitive to the other form factors ($A_0(q^2)$, $A_2(q^2)$ and $V(q^2)$).

5 $\bar{B}^0 \to D^{(*)+} l^- \bar{\nu}$

We use the form factors in (9) with the following $\mathcal{F}_D(y)$ which are experimentally determined [11]:

$$\mathcal{F}(y) = \mathcal{F}(1)[1 - \rho^2(y-1)] \tag{14}$$

$$\text{with} \quad \begin{cases} \rho_D^2 = 0.59 \pm 0.25\,, & |V_{cb}|\mathcal{F}_D(1) \times 10^2 = 3.37 \ \text{ for } \mathcal{F}_D(y) \\ \rho_{D^*}^2 = 0.84 \pm 0.15\,, & |V_{cb}|\mathcal{F}_{D^*}(1) \times 10^2 = 3.51 \text{ for } \mathcal{F}_{D^*}(y)\,. \end{cases}$$

By using the form factors in (9) with $\mathcal{F}_D(y)$ in (14), we obtain from (2) the spectra of $\bar{B}^0 \to D^+ l^- \bar{\nu}$ presented in Figure 3, where we find that the spectrum for muon drops down near $q^2 = 0$. (The spectrum for electron also drops down near the very end of $q^2 = 0$.) We present the obtained branching fractions and their ratio in Table 8, which gives $\mathcal{B}(\bar{B}^0 \to D^+ \tau^- \bar{\nu})/\mathcal{B}(\bar{B}^0 \to D^+ e^- \bar{\nu}) = 0.278^{+0.049}_{-0.035}$ in good agreement with the ALEPH estimation 0.29 ± 0.08 [12]. Since the HQET provides good informations about the heavy to heavy form factors, our results for the exclusive B to D semileptonic decays are very reliable.

By using the form factors in (9) with $\mathcal{F}_{D^*}(y)$ in (14), we obtain from (8) the spectra of $\bar{B}^0 \to D^{*+} l^- \bar{\nu}$ presented in Figure 4. We present the obtained branching fractions and their ratio in Table 8, where we find that $\mathcal{B}(\bar{B}^0 \to D^{*+} \tau^- \bar{\nu})/\mathcal{B}(\bar{B}^0 \to D^{*+} e^- \bar{\nu}) = 0.256^{+0.014}_{-0.013}$ which is significantly smaller than the ALEPH estimation 0.37 ± 0.08 [12].

6 Conclusion

We studied the heavy meson exclusive semileptonic decays with the lepton mass effects into consideration. In D meson decays we studied the sensitivity to form factors models. The results show that the decays to pseudoscalar meson are sensitive to the property of the form factor $F_1(q^2)$, and the decays to vector meson are sensitive to the form factor $A_1(q^2)$ and not on the other ones ($A_0(q^2)$, $A_2(q^2)$ and $V(q^2)$). In B meson decays we studied $\bar{B}^0 \to D^{(*)+}l^-\bar{\nu}$ using experimentally measured form factors in connection with the HQET, and found the $d\Gamma/dq^2$ spectra and branching fractions for $l = e$, μ and τ. We think that the improvements in understanding the form factors are important not only for the extractions of the CKM matrix elements, but also for the investigation of the internal structures of hadrons, and phenomenological analyses such as the one in this work are useful for those improvements.

Acknowledgments

This work was supported by Non-Directed-Research-Fund, Korea Research Foundation 1997, by the Basic Science Research Institute Program, Ministry of Education, Project No. BSRI-97-2414, by the Korea Science and Engineering Foundation, Grant 985-0200-002-2, and by the Research Fund of Kangnung National University 1997.

References

1. N. Cabibbo, Phys. Rev. Lett. **10**, 531 (1963); M. Kobayashi and T. Maskawa, Prog. Theor. Phys. **49**, 652 (1973).
2. J.G. Körner and G.A. Schuler, Phys. Lett. B **231**, 306 (1989); Z. Phys. C **46**, 93 (1990).
3. A. Khodjamirian, R. Rückl and C.W. Winhart, Phys. Rev. D **58**, 054013 (1998).
4. N. Isgur and M.B. Wise, Phys. Lett. B **232**, 113 (1989); **237**, 527 (1990).
5. M. Wirbel, B. Stech and M. Bauer, Z. Phys. C **29**, 637 (1985); **34**, 103 (1987); M. Bauer and M. Wirbel, Z. Phys. C **42**, 671 (1989).
6. UKQCD Collaboration, D.R. Burford et al., Nucl. Phys. B **447**, 425 (1995); L.D. Debbio, Phys. Lett. B **416**, 392 (1998).
7. D.S. Hwang and B.-H. Lee, hep-ph/9801421 (1998), to appear in Eur. Phys. J. C.
8. E687 Collaboration, P.L. Frabetti et al., Phys. Lett. B **364**, 127 (1995).
9. A. Ryd, Review of Form Factor Measurements, in *Seventh International Symposium on heavy Flavor Physics*, Santa Barbara, July, 1997.
10. E687 Collaboration, P.L. Frabetti et al., Phys. Lett. B **382**, 312 (1996).
11. CLEO Collaboration, B. Barish et al., Phys. Rev. D **51**, 1014 (1995); M. Athanas et al., Phys. Rev. Lett. **79**, 2208 (1997).
12. ALEPH Collaboration, D. Buskulic et al., Phys. Lett. B **343**, 444 (1995); **395**, 373 (1997).

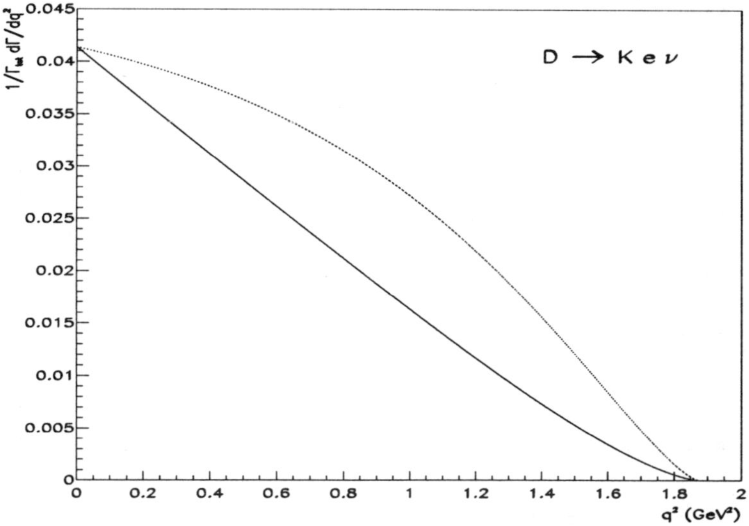

Figure 1: $(1/\Gamma_{tot})(d\Gamma/dq^2)$ of $D^0 \to K^- e^+ \nu$ for three models: solid line for WSB, dashed for Model I (solid and dashed lines overlap in this figure), and dotted for Model II.

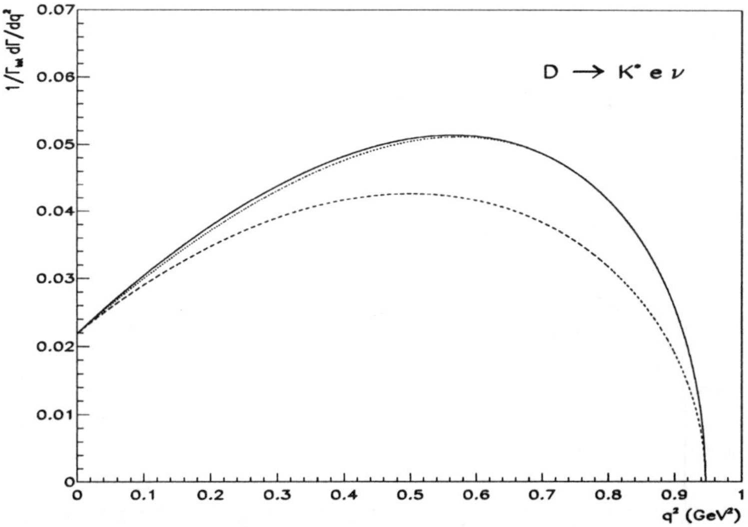

Figure 2: $(1/\Gamma_{tot})(d\Gamma/dq^2)$ of $D^0 \to K^{*-} l^+ \nu$ for three models: solid line for WSB, dashed for Model I, and dotted for Model II (the solid and dotted lines almost overlap in this figure).

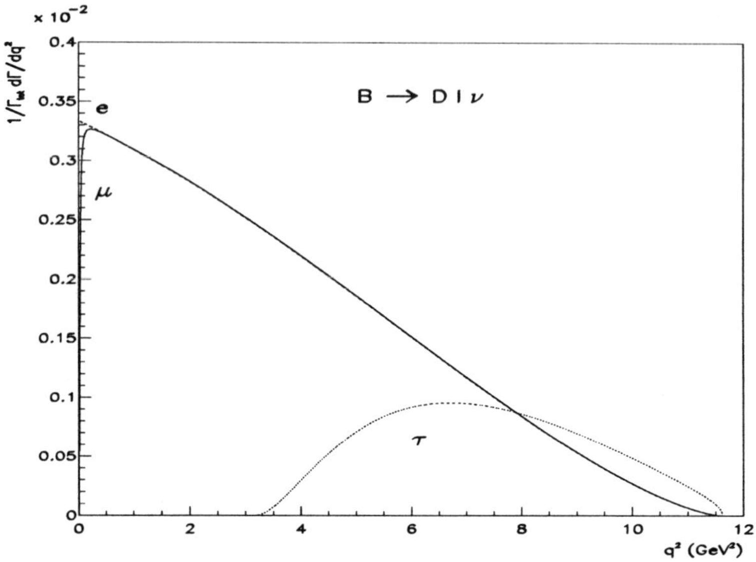

Figure 3: $(1/\Gamma_{tot})(d\Gamma/dq^2)$ of $\bar{B}^0 \to D^+ l^- \bar{\nu}$ for $l = e$, μ and τ.

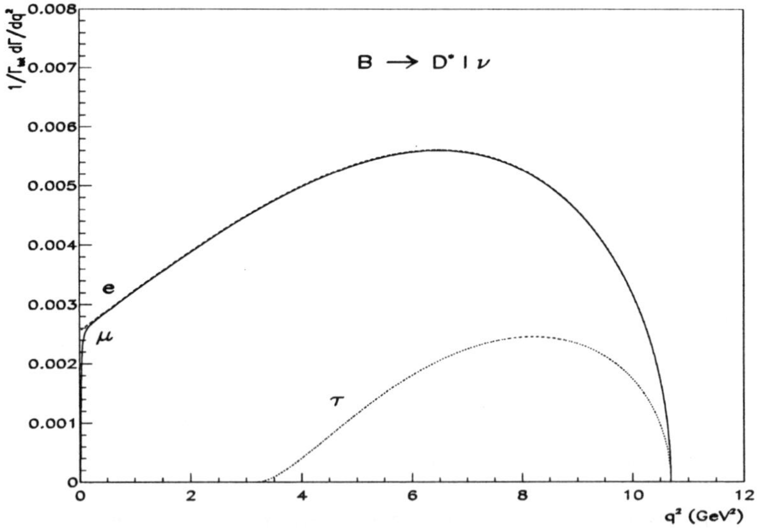

Figure 4: $(1/\Gamma_{tot})(d\Gamma/dq^2)$ of $\bar{B}^0 \to D^{*+} l^- \bar{\nu}$ for $l = e$, μ and τ.

PQCD ANALYSIS OF INCLUSIVE HEAVY HADRON DECAYS

Hoi-Lai Yu
Institute of Physics, Academia Sinica, Taipei, Taiwan 115, Republic of China

We develop the perturbative QCD formalism for inclusive heavy hadron decays. We obtain simultaneously a low semileptonic branching ratio $BR(B \to Xl\bar{\nu})$ without increasing the average charm yield $\langle n_c \rangle$ per B decay, a small lifetime ratio $\tau(\Lambda_b)/\tau(B_d)$, and the correct absolute decay widths of the B meson and of the Λ_b baryon.

We shall work in the rest frame of the B meson, and choose the following light-cone components for relevant momenta,

$$P_B = (P_B^+, P_B^-, \mathbf{0}_\perp) , \quad p_\ell = (p_\ell^+, 0, \mathbf{0}_\perp) , \quad p_\nu = (p_\nu^+, p_\nu^-, \mathbf{p}_{\nu\perp}) , \tag{1}$$

with $P_B^+ = P_B^- = M_B/\sqrt{2}$ and $p_\nu^2 = 0$. The independent variables are identified as p_ℓ^+, p_ν^- and p_ν^+, and their relations to E_ℓ, q^2 and q_0 are $E_\ell = p_\ell^+/\sqrt{2}$, $q^2 = 2p_\ell^+ p_\nu^-$ and $q_0 = (p_\ell^+ + p_\nu^+ + p_\nu^-)/\sqrt{2}$, respectively. We define $P_b = P_B - p$ as the b quark momentum, which satisfies $P_b^2 \approx M_b^2$, M_b being the b quark mass. p is the kicks from the light components inside the B meson, which has a large plus component p^+ and small transverse components $\mathbf{p}_\perp$. The b quark decays into a u quark with momentum $P_u = P_B - p - q$. We have distinguished the B meson momentum P_B from the b quark momentum P_b here.

It is more convenient to employ the scaling variables

$$x = \frac{2E_\ell}{M_B} , \quad y = \frac{q^2}{M_B^2} , \quad y_0 = \frac{2q_0}{M_B} , \tag{2}$$

instead of the dimensionful ones E_ℓ, q^2 and q_0. Note that the scaling variables are defined in terms of the B meson mass M_B, since we formulate the factorization according to the B meson kinematics. Using the momentum configurations defined in eq. (1), the phase space is

$$0 \le x \le 1, \quad 0 \le y \le x, \quad \frac{y}{x} + x \le y_0 \le y + 1 . \tag{3}$$

In the end point region with $x \to 1$ ($p_\ell^+ \to M_B/\sqrt{2}$) and $y \to 0$ ($p_\nu^- \to 0$), we have $y_0 \to 1$ ($p_\nu^+ \to 0$) and $p \to 0$. The u quark then has a large minus component $P_u^- = (1 - y/x)M_B/\sqrt{2}$ but a very small plus component $P_u^+ = (1 - y_0 - y/x)M_B/\sqrt{2}$, and thus a very small invariant $P_u^2 = M_B^2(1 - y_0 + y)$, which forms an on-shell jet subprocess. The u quark travels a long distance of $\mathcal{O}(1/\Lambda_{\rm QCD})$ before hadronized. Besides, the B meson is dominated by soft dynamics, which is the origin of the soft function. Therefore, the important contributions are factorized into the soft (S), jet (J) and hard (H) subprocesses.

The factorization formula for the inclusive semileptonic decay $B \to X_u \ell \nu$ is written as

$$\frac{1}{\Gamma_\ell^{(0)}} \frac{d^3\Gamma}{dx dy dy_0} = M_B^2 \int_{z_{\rm min}}^{z_{\rm max}} dz \int d^2\mathbf{p}_\perp \times S(z, \mathbf{p}_\perp, \mu) J(z, P_u^-, \mathbf{p}_\perp, \mu) H(z, P_u^-, \mathbf{p}_\perp, \mu) , \quad (4)$$

with the momentum fraction z defined by $z = P_b^+/P_B^+ = 1 - p^+/P_B^+$ and $\Gamma_\ell^{(0)} = \frac{G_F^2}{16\pi^3}|V_{ub}|^2 M_B^5$. μ in eq. (4) is the renormalization and factorization scale. The triple differential decay rate is, of course, μ independent. Note that in the region $y \to x \sim 1$ the outgoing u quark becomes soft and eq. (4) fails. We shall show that contributions from this dangerous region are suppressed by phase space. The upper limit of z takes the value $z_{\rm max} = 1$ in our analysis. If performing the factorization according to the b quark kinematics, one must assume $z_{\rm max} = M_B/M_b$, which is greater than 1, in order to fill up the kinematic window. It has been explained [2] that $z_{\rm max} > 1$ is not allowed in perturbation theory, and is thus of nonperturbative origin. From the kinematic constraints in eq. (3) and the on-shell condition of the u-quark jet, the lower limit of z should be $z_{\rm min} = x$, instead of $z_{\rm min} = 0$.

The tree-level expressions for the convolution factors J and H are given by

$$\begin{aligned} J^{(0)} &= \delta(P_u^2) \\ &= \delta\left(M_B^2\left[1 - y_0 + y - (1-z)(1-\frac{y}{x}) - \frac{2\mathbf{p}_\perp \cdot \mathbf{p}_{\nu\perp}}{M_B^2} - \frac{\mathbf{p}_\perp{}^2}{M_B^2}\right]\right) , \\ H^{(0)} &\propto (P_b \cdot p_\nu)(p_\ell \cdot P_u) = ((P_B - p)\cdot p_\nu)(p_\ell \cdot P_u) \\ &\propto (x-y)\left(y_0 - x - (1-z)\frac{y}{x} + \frac{2\mathbf{p}_\perp \cdot \mathbf{p}_{\nu\perp}}{M_B^2}\right) . \end{aligned} \quad (5)$$

Equation (4) can be regarded as an expression at the intermediate stage in the derivation of conventional factorization theorems. If the $\mathbf{p}_\perp$ dependence in J and H is negligible, the variable $\mathbf{p}_\perp$ in S can be integrated over, and eq. (4) reduces to the conventional factorization formula.

Suppose we consider higher-order corrections to eq. (4) from a gluon crossing the final state cut, and route the loop momentum ℓ through, say, the jet subprocess. Without losing generality, we approximate $J(p^+ + \ell^+, P_u^- + \ell^-, \mathbf{p}_\perp +$

$\ell_\perp) \approx J(p^+, P_u^-, \mathbf{p}_\perp + \ell_\perp)$ according to the kinimatic relations $\ell^+ < p^+$, $\ell^- < P_u^-$ and $\ell_\perp \approx \mathbf{p}_\perp$. Hence, the loop integral cannot be performed unless the dependence of J on transverse momentum is known. This difficulty can be removed by Fourier transform,

$$J(p^+, P_u^-, \mathbf{p}_\perp + \ell_\perp) = \int \frac{d^2\mathbf{b}}{(2\pi)^2} \tilde{J}(p^+, P_u^-, \mathbf{b}) e^{i(\mathbf{p}_\perp + \ell_\perp)\cdot \mathbf{b}} , \tag{6}$$

where the impact parameter b (Fourier conjugate variable of $p_\perp$) measures the transverse distance travelled by the jet. Using eq. (6), the $\ell_\perp$ dependence is decoupled from the jet function, and the factor $e^{i\ell_\perp \cdot \mathbf{b}}$ is absorbed into the loop integral, which can then be performed. Therefore, an extra factor $e^{i\ell_\perp \cdot \mathbf{b}}$ is associated with each gluon crossing the final state cut in our formalism.

To further simplify the factorization formula, we neglect those terms involving $\mathbf{p}_{\nu\perp}$ in J and H. This is a good approximation for $x \to 0$ and $x \to 1$, since for $x \to 0$ contributions from transverse momenta are not important, and for $x \to 1$ the magnitude $p_{\nu\perp} = M_B\sqrt{(y_0 - x - y/x)y/x}$ vanishes. Equation (4) then becomes

$$\begin{aligned} \frac{1}{\Gamma_\ell^{(0)}} \frac{d^3\Gamma}{dx dy dy_0} &= M_B^2 \int_x^1 dz \int \frac{d^2\mathbf{b}}{(2\pi)^2} \\ &\quad \times \tilde{S}(z, \mathbf{b}, \mu) \tilde{J}(z, P_u^-, \mathbf{b}, \mu) H(z, P_u^-, \mu) \, . \end{aligned} \tag{7}$$

Large radiative corrections spoil the perturbation theory and must be organized. One has to resum these large corrections up to next-to-leading logarithms. The first step in resummation is to map out the leading regions of radiative corrections. In the collinear region with the loop momentum ℓ parallel to P_u and in the soft region with $\ell \to 0$ we can eikonalize the heavy b-quark line. With the eikonal approximation, the b-quark propagator is expressed as $1/v \cdot \ell$ to the order $1/M_B$ with $v = (1, 1, \mathbf{0}_\perp)$. Hence, the factorization in eq. (7) is in fact valid to $\mathcal{O}(1/M_B)$. The physics involved in this approximation is that a soft gluon or a gluon moving parallel to P_u can not explore the details of the b quark, and its dynamics can be factorized. This is consistent with the HQEFT, where the b quark is treated as a classical relativistic particle carrying color source. Since the large mass M_B does not appear in the eikonal propragator, the only large scale in J is P_u^-. After eikonalization procedures, one obtain the Sudakav exponent that resumes large double logarithms arise from both collinear and soft regions,

$$S = 2 \int_{1/b}^{m_B} \frac{dp}{p} \int_{1/b}^{p} \frac{d\mu}{\mu} A(\alpha_s(\mu)) \, , \tag{8}$$

$$A = C_F \frac{\alpha_s}{\pi} + \left[\frac{67}{9} - \frac{\pi^2}{3} - \frac{10}{27} n_f \right] \left(\frac{\alpha_s}{\pi} \right)^2 , \tag{9}$$

with $C_F = 4/3$ the color factor.

Having summed up the double logarithms, we concentrate on the single logarithms in $\tilde{S}$, H and the initial condition $\tilde{J}(b, \mu)$. Since both the differential decay rate and the Sudakov exponent $s(P_u^-, b)$ are RG invariant, we have the following RG equations:

$$\begin{aligned} \mathcal{D}\tilde{J}(b,\mu) &= -2\gamma_q \tilde{J}(b,\mu) , \\ \mathcal{D}\tilde{S}(b,\mu) &= -\gamma_S \tilde{S}(b,\mu) , \\ \mathcal{D}H(P_u^-,\mu) &= (2\gamma_q + \gamma_S) H(P_u^-,\mu) , \end{aligned} \tag{10}$$

with

$$\mathcal{D} = \mu \frac{\partial}{\partial \mu} + \beta(g) \frac{\partial}{\partial g} . \tag{11}$$

$\gamma_q = -\alpha_s/\pi$ is the quark anomalous dimension in axial gauge, and $\gamma_S = -(\alpha_s/\pi) C_F$ is the anomalous dimension of $\tilde{S}$. Integrating eq. (10), we obtain the evolution of all the convolution factors,

$$\begin{aligned} \tilde{J}(z, P_u^-, b, \mu) &= \exp\left[-2s(P_u^-, b) - 2\int_{1/b}^{\mu} \frac{d\bar{\mu}}{\bar{\mu}} \gamma_q(\alpha_s(\bar{\mu})) \right] \tilde{J}(z, b, 1/b) , \\ \tilde{S}(z, b, \mu) &= \exp\left[-\int_{1/b}^{\mu} \frac{d\bar{\mu}}{\bar{\mu}} \gamma_S(\alpha_s(\bar{\mu})) \right] f(z, b, 1/b) , \\ H(z, P_u^-, \mu) &= \exp\left[-\int_{\mu}^{P_u^-} \frac{d\bar{\mu}}{\bar{\mu}} [2\gamma_q(\alpha_s(\bar{\mu})) + \gamma_S(\alpha_s(\bar{\mu}))] \right] \\ &\times \; H(z, P_u^-, P_u^-) . \end{aligned} \tag{12}$$

We shall neglect the intrinsic b dependence of the distribution function f below. If Sudakov suppression in the large-b region is strong, we may drop the evolution of f and $\tilde{J}$ in b, which is proportional to $\alpha_s(1/b)$. Hence, we assume $f(z, b, 1/b) = f(z)$, $\tilde{J}(z, b, 1/b) = \tilde{J}^{(0)}(z, b)$, the Fourier transform of the tree-level expression in eq. (5), and $H(z, P_u^-, P_u^-) = H^{(0)}(z, P_u^-)$.

Substituting eq. (12) into (7), we derive the factorization formula for the inclusive semileptonic B meson decay,

$$\begin{aligned} \frac{1}{\Gamma_\ell^{(0)}} \frac{d^3\Gamma}{dx dy dy_0} &= M_B^2 \int_x^1 dz \int_0^\infty \frac{b db}{2\pi} f(z) \tilde{J}^{(0)}(z, b) H^{(0)}(z, P_u^-) \\ &\times \exp[-S(P_u^-, b)] . \end{aligned} \tag{13}$$

The complete Sudakov exponent S is given by

$$S(P_u^-, b) = 2s(P_u^-, b) - \frac{5}{3\beta_1}\ln\frac{\hat{P}_u^-}{\hat{b}} , \tag{14}$$

with $\hat{P}_u^- = \ln(P_u^-/\Lambda)$, which combines all the exponents in eq. (12) and includes both leading and next-to-leading logarithms. It is straightforward to observe from eq. (14) that the Sudakov form factor e^{-S} falls off quickly at large $b \sim 1/\Lambda$, where $\alpha_s(1/b) > 1$ and perturbation theory fails. Hence, the Sudakov form factor guarantees that main contributions to the factorization formula come from the small b, or short-distance, region, and the perturbative treatment is indeed self-consistent.

Since all the double logarithmic corrections have been absorbed into the jet subprocess, the soft function $\tilde{S}$ contains only soft single logarithms. These single logarithms can be summed by solving the RG equation $\mathcal{D}\tilde{S} = -\gamma_S\tilde{S}$ as shown in eq. (10). The solution has been given in eq. (12),

$$\tilde{S}(z, b, \mu) = \exp\left[-\int_{1/b}^{\mu}\frac{d\bar{\mu}}{\bar{\mu}}\gamma_S(\alpha_s(\bar{\mu}))\right] f(z, b) , \tag{15}$$

where the initial condition $f(z)$ for the evolution of $\tilde{S}$ must be determined phenomenologically. From the definition of $\tilde{S}$, it is obvious that f depends only on the properties of the bound state $|B\rangle$, but not on the particular short distance subprocess. Therefore, f is a process-independent universal function describing the distribution of the b quark inside a B meson. Since the intrinsic b dependence (the perturbative b dependence has been collected into the exponent S) is not known yet, we take the ansatz $f(z, b) = f(z)\exp(-\Sigma(b))$, which leads to $\Sigma \to 0$ as $b \to 0$ according to the definition $f(z, b = 0) \equiv f(z)$. It is also natrual to assume $\Sigma > 0$ for all b from the viewpoint that the b quark is bounded inside the B meson. Hence, the intrinsic b depedence provides further suppression. The nonperturbative function $f(z)$, identified as the B-meson distribution function, can be expressed as the matrix element of the b quark fields, whose first three moments are[3,4]

$$\begin{aligned}
&\int_0^1 f(z)dz = 1 , \\
&\int_0^1 f(z)(1-z)dz = \frac{\bar{\Lambda}}{m_B} + O(\Lambda_{\rm QCD}^2/m_B^2) , \\
&\int_0^1 f(z)(1-z)^2 dz = \frac{1}{m_B^2}\left(\bar{\Lambda}^2 - \frac{\lambda_1}{3}\right) + O(\Lambda_{\rm QCD}^3/m_B^3) .
\end{aligned} \tag{16}$$

Though the exponent Σ is unknown, we can, however, extract its leading behavior by means of the IR renormalon analysis. Note that the perturbative Sudakov factor e^{-S} in Eq. (15) becomes unreliable as the transverse distance b approaches $1/\Lambda_{QCD}$. Near this end point, $\alpha_s(1/b)$ diverges, and IR renormalon contributions are significant. We reexpress the RG result of the evolution of the distribution function, which is contained in the second term of S, as

$$W = \exp\left[4\pi C_F \int \frac{d^4 l}{(2\pi)^4} \frac{v_\mu v_\nu}{(v\cdot l)^2} 2\pi\delta(l^2)\alpha_s(l_T^2) e^{i\mathbf{l}_T\cdot\mathbf{b}} N^{\mu\nu}\right] . \quad (17)$$

The tensor $N^{\mu\nu} = g^{\mu\nu} - (n^\mu l^\nu + l^\mu n^\nu)/(n\cdot l) + n^2 l^\mu l^\nu/(n\cdot l)^2$ comes from the gluon propagator in axial gauge $n\cdot A = 0$.

Substituting the identity $\alpha_s(l_T^2) = \pi\int_0^\infty d\sigma \exp[-2\sigma\beta_1 \ln(l_T/\Lambda_{QCD})]$, $\beta_1 = (33-2n_f)/12$, into Eq. (17), and performing the loop integral for $n \propto (-1, 1, \mathbf{0})$, we obtain

$$W = \exp\left[C_F \int_0^\infty d\sigma \left(\frac{b\Lambda_{QCD}}{2}\right)^{2\sigma\beta_1} \frac{\Gamma(-\sigma\beta_1)}{\Gamma(1+\sigma\beta_1)}\right] . \quad (18)$$

It is easy to observe that the pole of $\Gamma(-\sigma\beta_1)$ at $\sigma\to 0$ gives the perturbative anomalous dimension of the distribution function appearing in Eq. (8). The extra poles at $\sigma \to 1/\beta_1, 2/\beta_1,\ldots$, then correspond to the IR renormalons, giving corrections of powers b^2, b^4,..., respectively. These renormalons generate singularities, which must be compensated by the nonperturbative power correctons in order to have a well-defined perturbative expansion. Using the "minimal" ansatz [2], *ie* picking up only the leading (left-most) renormalon contribution, we parametrize the exponent $\Sigma(b)$ by $\Sigma(b) = c' m_B^2 b^2$, corresponding to the fact that the power corrections start at $O(b^2)$. Certainly, other parametrizations consistent with this fact are equally good. We postulate the following two-parameter distribution function for the B meson,

$$f_B(z) = N\frac{z(1-z)^2}{[(z-a)^2+\epsilon z]^2}\theta(1-z) , \quad (19)$$

The charged lepton spectrum for the decay $B \to X_u \ell\nu$ from the naive quark model is obtained by simply choosing $f(z) = \delta(1-z)$ and ignoring the transverse momentum dependence in $J^{(0)}$ and $H^{(0)}$. A simple calculation leads to

$$\frac{1}{\Gamma_\ell^{(0)}}\frac{d\Gamma}{dx} = \frac{x^2}{6}(3-2x) . \quad (20)$$

This spectrum does not fall off at the end point of the spectrum, contradicting the observed behavior of the inclusive semileptonic decays of B mesons. The discrepancy implies that the tree-level analysis is not appropriate, especially in the end-point region where PQCD corrections are important.

The spectrum from the parton model without Sudakov suppression is obtained by adopting $H^{(0)} = (x-y)[y_0 - x - (1-z)y/x]$ and $P_u^2 = M_B^2[1 - y_0 + y - (1-z)(1-y/x)]$. With integration over y_0, we derive

$$\frac{1}{\Gamma_\ell^{(0)}}\frac{d\Gamma}{dx} = \int_0^x dy \int_x^1 dz f(z)(x-y)(y+z-x)\,. \tag{21}$$

At last, including Sudakov suppression and all other single logarithms, we arrive at the charged lepton spectrum of the $B \to X_u \ell\nu$ decay that takes into account both large perturbative and nonperturbative corrections,

$$\begin{aligned}\frac{1}{\Gamma_\ell^{(0)}}\frac{d\Gamma}{dx} &= M_B \int_0^x dy \int_0^{1/\Lambda} db \int_x^1 dz f(z)(x-y)\xi\Big[(z+y-x)J_1(\xi M_B b) \\ &\quad - \frac{2}{M_B b}\xi J_2(\xi M_B b) + \xi^2 J_3(\xi M_B b)\Big] e^{-S(P_u^-, b)}\,,\end{aligned} \tag{22}$$

with $\xi = \sqrt{(x-y)(z/x-1)}$ and J_1, J_2, J_3 are the Bessel functions of order 1,2 and 3 respectively.

We stress that our results do not violate the conclusion from HQEFT and also respect global quark-hadron duality, if they were interpreted properly. To confirm this, we identify $P_b = (M_b^2/2P_B^-, P_B^-, \mathbf{0})$ as the momentum carried by a free b quark in factorization theorems, where the minus component P_b^- has been set to P_B^-. That is, the free b quark is not at rest inside the B meson. Reexpress eq. (21) into a form similar to that in [3]:

$$\frac{1}{\Gamma_\ell^{(0)}}\frac{d\Gamma}{dx} = F(x)\theta\left(\frac{M_b^2}{M_B^2} - x\right) + F\left(\frac{M_b^2}{M_B^2}\right) M(x)\,, \tag{23}$$

with $F(x) = x^2(3-2x)/6$ being the quark-model prediction derived from the conventional approaches and

$$F\left(\frac{M_b^2}{M_B^2}\right) M(x) = \int_0^x dy \int_x^1 dz f(z)(x-y)(y+z-x) - F(x)\theta\left(\frac{M_b^2}{M_B^2} - x\right)\,. \tag{24}$$

The step function in eq. (23) specifies the maximal E_ℓ in the decay of a free b quark with the above momentum P_b. The function $M(x)$, representing nonperturbative corrections to the b quark decay, coincides with the shape function $S(x)$ defined in [3].

We shall show that the contribution from $M(x)$ to the total decay rate is indeed of $\mathcal{O}(1/M_B^2)$. Integrating eq. (24) over x, we obtain

$$F\left(\frac{M_b^2}{M_B^2}\right)\int_0^1 M(x)dx = \frac{1}{12}\int_0^1 dz z^4 f(z) - \int_0^{M_b^2/M_B^2} F(x)dx \; . \tag{25}$$

An arbitrary structure function f, which possesses the same moment as in eq. (16), can be expanded in terms of δ-functions:

$$f(z) = \delta(1-z) - \frac{\bar{\Lambda}}{M_B}\delta'(1-z) + \mathcal{O}(\bar{\Lambda}^2/M_B^2) \; . \tag{26}$$

Inserting eq. (26) into (25), we justify straightforwardly that the nonperturbative correction

$$\begin{aligned} F\left(\frac{M_b}{M_B}\right)\int_0^1 M(x)dx &= \int_{M_b^2/M_B^2}^1 F(x)dx - \frac{1}{12}\frac{\bar{\Lambda}}{M_B}\int_0^1 dz z^4\delta'(1-z) \\ + \;\; \mathcal{O}(\bar{\Lambda}^2/M_B^2) &= \frac{1}{3}\frac{\bar{\Lambda}}{M_B} - \frac{1}{3}\frac{\bar{\Lambda}}{M_B} + \mathcal{O}(\bar{\Lambda}^2/M_B^2) \end{aligned} \tag{27}$$

vanishes at $\mathcal{O}(1/M_B)$ as concluded in [3]. In summary, the quark-model contribution from the window between $x = M_b^2/M_B^2$ and $x = 1$, with a width of $\mathcal{O}(1/M_B)$, cancels the $\mathcal{O}(1/M_B)$ correction from the structure function, such that the nonperturbative correction is of $\mathcal{O}(1/M_B^2)$. Our result provides an explicit dynamical demonstration the validity of global quark-hadron duality.

For the nonleptonic modes $b \to c\bar{u}d$ and $b \to c\bar{c}s$, which involve more complicated strong interactions, additional subprocesses need to be included except for those in the semileptonic cases. These modes involve three scales: the W boson mass M_W, at which the matching condition of the effective Hamiltonian for nonleptonic decays to the original Hamiltonian is defined, the characteristic scale t of the hard subamplitude, which reflects the specific dynamics of heavy hadron decays, and the b quark transverse momentum $p_\perp$, or equivalently $1/b$, which serves as the factorization scale mentioned above. To analyze the nonleptonic decays, we have proposed a three-scale factorization formalism [5] that embodies both effective field theory and the standard PQCD factorization theorems. The hard gluon exchanges among quarks, which generate logarithms of M_W, are factorized into a *harder* function [5]. Collinear divergences and soft divergences exist in radiative corrections simultaneously, when the light final-state quarks, such as $\bar{u}$, d, and s, become energetic. The resultant double logarithms $\ln^2(\bar{p}b)$, $\bar{p}$ being the jet momentum defined later, are absorbed into the corresponding jet functions.

The various logarithmic corrections are summed to all orders using the resummation technique and renormalization-group (RG) equations, whose results are the evolutions among the scales M_W, t, and $1/b$. The summation of the logarithms $\ln(M_W/t)$ is identified as the Wilson coefficient of the effective Hamiltonian, which describes the evolution from the characteristic scale M_W of the harder function to the characteristic scale t of the hard subamplitude. The summation of the logarithms $\ln(tb)$ leads to the evolution from t to $1/b$. The double logarithms $\ln^2(\bar{p}b)$ in the jet functions are resummed to give a Sudakov factor [6], which suppresses the long-distance contributions from the large b region, and thus improves the applicability of PQCD.

With the above ingredients, we have calculated the branching ratios of the modes $b \to cl\bar{\nu}$, $b \to c\bar{u}d$ and $b \to c\bar{c}s$ involved in the inclusive B meson decays [4]. It was observed that the decay rate of the single-charm mode $b \to c\bar{u}d$ is enhanced by the QCD evolution effects more than the double-charm mode $b \to c\bar{c}s$ is. In this way the semileptonic branching ratio $B_{\rm SL} \equiv BR(B \to Xl\bar{\nu})$ is suppressed without increasing the average charm yield $\langle n_c \rangle$ per B decay. The controversy of the small $B_{\rm SL}$ in inclusive B meson decays, which cannot be explained by the quark model, is then resolved. This solution differs from the usual attempts in the literature, in which the $b \to c\bar{c}s$ mode is increased more.

We shall further show that our formalism provides a possible explanation of the low lifetime ratio $\tau(\Lambda_b)/\tau(B_d)$. In the current experimental status, we treat the moment of the Λ_b baryon disribution function as a free parameter. Different dynamics between the Λ_b baryon and B meson decays are then reflected by the moments of their corresponding distribution functions. It will be found that with reasonable choices of the parameter, we obtain a larger (more than 10%) difference in the Λ_b baryon and B meson lifetimes. Our predictions for the absolute decay widths of the B meson and of the Λ_b baryon are also consistent with the data.The factorization formulas for the nonleptonic decay rates are quoted as [4]

$$\begin{aligned}
\frac{\Gamma_{\rm NL}}{\Gamma_0} &= \frac{M_B^2}{2}\int dxdydy_0 \int_{z_{\rm min}}^{1} dz \int_0^\infty bdb \left[\frac{N_c+1}{2}c_+^2(t) + \frac{N_c-1}{2}c_-^2(t)\right] \\
&\quad \times H(x,y,y_0,z)\prod_i \tilde{J}_i(x,y,y_0,z,b)\prod_j \tilde{J}_j(x,y,y_0,z,b) f_B(z,b) \\
&\quad \times \exp\left[-\sum_j s(\bar{p}_j,b) - s_{JS}(t,b)\right] . \qquad (28)
\end{aligned}$$

The functions $\tilde{J}_i$, $i = c$ and $\bar{c}$ for the $b \to c\bar{c}s$ mode and $i = c$ for the $b \to c\bar{u}d$

mode, and the light-quark jet functions $\tilde{J}_j$, $j = s$ for the $b \to c\bar{c}s$ mode and $j = \bar{u}$, d for the $b \to c\bar{u}d$ mode, represent the final-state cuts . The B meson distribution function f_B, being universal, is the same as that employed for the semileptonic decays.

Several reasonable choices of $\lambda_1(\Lambda_b)$ are made, and results for $M_c = 1.6$ GeV, $M_D = 1.869$ GeV and $M_\tau = 1.7771$ GeV are listed in the table below, where the hard parameter κ is varied to reflect the theoretical uncertainty. Results of $\tau(B_d)$, $\tau(\Lambda_b)$, $\tau(\Lambda_b)/\tau(B_d)$, $B_{\rm SL}$, and $\langle n_c\rangle$ for various $\lambda_1(\Lambda_b)$.

(a) $\lambda_1(\Lambda_b) = -0.40$ GeV^2 ($\bar{\Lambda}(\Lambda_b) = 1.05$ GeV)

κ	$\tau(B_d) \times 10^{12}$ s	$\tau(\Lambda_b) \times 10^{12}$ s	$\tau(\Lambda_b)/\tau(B_d)$	$B_{\rm SL}$	$\langle n_c\rangle$
1.0	$1.73^{+0.37}_{-0.28}$	$1.64^{+0.35}_{-0.26}$	0.95	12.52	1.195
1.5	$1.56^{+0.33}_{-0.25}$	$1.49^{+0.32}_{-0.24}$	0.96	11.26	1.196
2.0	$1.43^{+0.30}_{-0.23}$	$1.36^{+0.29}_{-0.22}$	0.96	10.30	1.194

(b) $\lambda_1(\Lambda_b) = -0.71$ GeV^2 ($\bar{\Lambda}(\Lambda_b) = 1.01$ GeV)

κ	$\tau(B_d) \times 10^{12}$ s	$\tau(\Lambda_b) \times 10^{12}$ s	$\tau(\Lambda_b)/\tau(B_d)$	$B_{\rm SL}$	$\langle n_c\rangle$
1.0	$1.73^{+0.37}_{-0.28}$	$1.56^{+0.33}_{-0.25}$	0.90	12.52	1.195
1.5	$1.56^{+0.33}_{-0.25}$	$1.41^{+0.30}_{-0.23}$	0.91	11.26	1.196
2.0	$1.43^{+0.30}_{-0.23}$	$1.29^{+0.27}_{-0.21}$	0.91	10.30	1.194

(c) $\lambda_1(\Lambda_b) = -1.10$ GeV^2 ($\bar{\Lambda}(\Lambda_b) = 0.97$ GeV)

κ	$\tau(B_d) \times 10^{12}$ s	$\tau(\Lambda_b) \times 10^{12}$ s	$\tau(\Lambda_b)/\tau(B_d)$	$B_{\rm SL}$	$\langle n_c\rangle$
1.0	$1.73^{+0.37}_{-0.28}$	$1.52^{+0.32}_{-0.24}$	0.88	12.52	1.195
1.5	$1.56^{+0.33}_{-0.25}$	$1.38^{+0.29}_{-0.22}$	0.88	11.26	1.196
2.0	$1.43^{+0.30}_{-0.23}$	$1.26^{+0.27}_{-0.20}$	0.89	10.30	1.194

1. H-n. Li and H.L. Yu, Phys. Rev. Lett. **74**, 4388 (1995);
   H-n. Li and H.L. Yu, Phys. Lett. **B353**, 301 (1995).
2. G.P. Korchemsky and G. Sterman, Phys. Lett. **B340**, 96 (1994).
3. M. Neubert, Phys. Rev. D49, 4623 (1994).
4. H-n. Li and H.L. Yu, Phys. Rev. D53, 4970 (1996).
5. C. H. V. Chang, H-n. Li, Phys. Rev. D **55**, 5577 (1997).
6. C. H. V. Chang, D. Chang, W. F. Chang, H-n. Li and H. L. Yu, hep-ph/**9803360**, to appear in Phys. Rev. D.

Rare B and K decays: Model independent test and new physics

T. Morozumi
Dept. of Physics, Hiroshima University, 739-8526 Higashi-Hiroshima, Japan.

Rare K and B decays are useful for testing FCNC of the SM and new physics search. With high statics experiments of semi-leptonic B and K rare decays, the detailed study on the various distributions may be possible. In the first part of my talk, we introduce a model independent study on rare B decay $B \to X_s \ell^+ \ell^-$. In the second part of my talk, rare K decay $K_L \to \pi_0 \nu \bar{\nu}$ is studied in a left-right symmetric model with right-handed neutrinos.

1 Contents

- Model independent study on $B \to X_s \ell^+ \ell^-$. [1]
- $K_L \to \pi_0 \nu \bar{\nu}$ and CP property. The prediction in a left-right model. [2 3]

2 Model independent study on $B \to X_s \ell^+ \ell^-$

Presently, the experimental upper bound on $B \to X_s \ell^+ \ell^-$ and the prediction of the SM are:

$$\begin{aligned} Br.(B \to X_s \ell^+ \ell^-)_{EXP} &< 5.7 \times 10^{-5}, \\ Br.(B \to X_s \ell^+ \ell^-)_{SM.} &= 8.4 \pm 2.3 \times 10^{-6}. \end{aligned} \tag{1}$$

In the amplitude of $B \to X_s \ell^+ \ell^-$, the effects of new physics are generally parametrized as:

$$\begin{aligned} \mathcal{M} &= \mathcal{M}_{\rm SM} + \mathcal{M}_{\rm NEW}, \\ \mathcal{M}_{\rm NEW} &\simeq \sum_{i=1}^{10} C_i O_i, \end{aligned} \tag{2}$$

where O_is are local four fermi operators (See below) and C_is are their coefficients. We study two measurements: the branching ratio and the forward and backward asymmetry, $B(C_i)$ and $A(C_i)$. By studying how the branching ratio and the forward and backward asymmetry depend on each coefficient C_i, we can plot the correlation (B, A). It is useful for discriminating various interactions. The most general four fermi interactions are written as,

$$\mathcal{M}_{\rm NEW} = \frac{G_F \alpha}{\sqrt{2}\pi} V_{ts}^* V_{tb} \; [\; C_{LL}\; \bar{s}_L \gamma_\mu b_L \; \bar{l}_L \gamma^\mu l_L + C_{LR}\; \bar{s}_L \gamma_\mu b_L \; \bar{l}_R \gamma^\mu l_R$$

$$+ C_{RL}\ \bar{s}_R\gamma_\mu b_R\ \bar{l}_L\gamma^\mu l_L + C_{RR}\ \bar{s}_R\gamma_\mu b_R\ \bar{l}_R\gamma^\mu l_R$$
$$+ C_{LRLR}\ \bar{s}_L b_R\ \bar{l}_L l_R + C_{RLLR}\ \bar{s}_R b_L\ \bar{l}_L l_R$$
$$+ C_{LRRL}\ \bar{s}_L b_R\ \bar{l}_R l_L + C_{RLRL}\ \bar{s}_R b_L\ \bar{l}_R l_L$$
$$+ C_T\ \bar{s}\sigma_{\mu\nu} b\ \bar{l}\sigma^{\mu\nu} l + iC_{TE}\ \bar{s}\sigma_{\mu\nu} b\ \bar{l}\sigma_{\alpha\beta} l\ \epsilon^{\mu\nu\alpha\beta}], \quad (3)$$

Using the four fermi interaction, we can compute the differential branching ratio and the forward and backward asymmetry. By integrating the double differential rate over the dilepton $(mass)^2$ from $1GeV^2$ to $8GeV^2$ (below the J/ψ resonance), we plot the correlations (B, A) in Fig.17 and 18 of Ref.(1). Our observation is that there is strong sensitivity on C_{LL} which comes from the interference between C_{LL} and the corresponding standard model amplitude $C_9 - C_{10}$. The tensor and the scalar interaction increase the branching ratio but the asymmetry is not changed.

2.1 $K_L \to \pi_0\nu\bar{\nu}$and CP property

$K_L \to \pi_0\nu\bar{\nu}$ is CP violating process in the SM. This is because in the effective hamiltonian of the SM, the decay occurs through the four fermi interaction $\bar{s}_L\gamma_\mu d_L\bar{\nu}_L\gamma^\mu\nu_L$. In this case, the CP of the final state is even. Thus, the decay mode is CP violating process. However, in the models with right-handed neutrino, it is possible to have scalar type spinor bilinear for neutrino and anti-neutrino. In the case, the final state is CP odd. Therefore the decay is CP conserving. Here we give a concrete example such that the CP conserving contribution due to the scalar interaction contribute to the process in the context of a left-right model. In the model, there is CP conserving amplitude because of ν_R (RH neutrino). There is an enhancement factor in the corresponding hadronic matrix element. Futhermore, pion energy spectrum in the low energy region is distorted from the SM one by the presence of the scalar type interaction. To illustrate this point, we start with the following effective lagrangian:

$$\mathcal{L}_{eff} = -\frac{\kappa G_F}{\sqrt{2}}\{C_{SM}\bar{s}\gamma_\mu(1-\gamma_5)d\bar{\nu}\gamma^\mu(1-\gamma_5)\nu + C_S\bar{s}d\bar{\nu}\nu\} + h.c.. \quad (4)$$

The amplitude for $K_L \to \pi^0\nu\bar{\nu}$ is:

$$Amp. = -\frac{\kappa G_F}{\sqrt{2}}\{(pC_{SM} - qC^*_{SM}) < \pi^0|\bar{s}\gamma_\mu d|K^0 > \bar{\nu}\gamma^\mu(1-\gamma_5)\nu$$
$$+ (pC_S + qC^*_S) < \pi^0|\bar{s}d|K^0 > \bar{\nu}\nu\} \quad (5)$$

In the CP conserving limit for the $K^0\bar{K}^0$ mixing, i.e., $p = q$, the SM contribution is proportional to ImC_{SM} while the contribution of the scalar interaction is proportional to ReC_S, which signals the SM contribution is CP violating and the scalar type interaction is CP conserving. We can relate vector form factor and scalar form factor,

$$
\begin{aligned}
<\pi^0|\bar{s}d|K^0> &= p_-{}^{\mu} <\pi^0|\bar{s}\gamma_\mu d|K^0> \\
&= \frac{f_+ M_K^2}{m_d - m_s}(1 - \delta_\pi^2 + \xi\hat{t}), \\
<\pi^0|\bar{s}\gamma^\mu d|K^0> &= f_+ p_+^\mu + f_- p_-^\mu, \qquad \xi = \frac{f_-}{f_+},
\end{aligned}
\tag{6}
$$

where $\delta_\pi = \frac{M_\pi}{M_K}$, $p_- = p_K - p_\pi$ and $\hat{t} = p_-^2/{M_k}^2$. The matrix element of the scalar interaction is enhanced by a factor of $\frac{M_K}{m_s}$ compared to that of the vector form factor. Now we turn to the estimation of C_S in a left-right model.

2.2 A brief introduction to the left-right model

The gauge group of the model is $SU(2)_L \times SU(2)_R \times U(1)$ and the ordinary quarks, isosinglet quarks and the relevant Higgs scalars are assigned to the following gauge group representations:[4].

$$
\begin{aligned}
\psi_L^i &= \begin{pmatrix} u^i \\ d^i \end{pmatrix}_L : (2,1,1/3), \\
\psi_R^i &= \begin{pmatrix} u^i \\ d^i \end{pmatrix}_R : (1,2,1/3), \\
& U_{L,R}^i : (1,1,4/3), \quad D_{L,R}^i : (1,1,-2/3), \\
& \phi_L : (2,1,1), \quad \phi_R : (1,2,1),
\end{aligned}
\tag{7}
$$

where $i = 1,2,3$. In the present model, quark and charged lepton mass is generated through the seesaw type mass matrix (See for details Ref.(4).) and for the mixing angle between iso-singlet and doublet quarks satisfies the following equation.

$$
c_L c_R m + s_L s_R M = 0, \tag{8}
$$

where s_L (s_R) are the sin of the mixing angles for left(right)-handed fermions. We denote $c_L(c_R)$ for the cos of the angle. $m(M)$ is doublet(singlet) fermion mass. Calculating the box diagram in which charged $SU(2)_L$ and $SU(2)_R$ gauge bosons; i.e., W_L and W_R are exchanged, the corresponding amplitude C_S is given by:

$$
C_S = 2V_{cs}^* V_{cd}\frac{1}{M_{WR}^2}\{m_c c_L^c c_R^c m_\tau c_L^\tau c_R^\tau d(x_c, y_\tau) + M_c s_L^c s_R^c m_\tau c_L^\tau c_R^\tau d(X_c, y_\tau)
$$

$$+ m_c c_L^c c_R^c M_\tau s_L^\tau s_R^\tau d(x_c, Y_\tau) + M_c s_L^c s_R^c M_\tau s_L^\tau s_R^\tau d(X_c, Y_\tau)$$
$$= 2\lambda \frac{m_c m_\tau}{M_{WR}^2} \{d(x_c, y_\tau) - d(X_c, y_\tau) - d(x_c, Y_\tau) + d(X_c, Y_\tau)\}, \quad (9)$$

where d is a function of internal fermions. Here only the contribution of the charm flavors and tau flavors is important. $x_c(X_c)$ are the $(mass)^2$ of charm (isosinglet charm) quarks normalized by M_{WL}^2. $y_\tau(Y_\tau)$ are the $(mass)^2$ of tau (isosinglet tau) leptons normalized by M_{WL}^2. On the otherhand, the SM contribution is dominated by top quark loop and is given by:

$$C_{SM} = V_{ts}^* V_{td} D_{IL}(x_t), \quad (10)$$

where D is a Inami-Lim function. The typical values for ImC_{SM} and ReC_{LR} are given as:

$$ImC_{SM} = 5.2 \times 10^{-3} \lambda\eta, \quad ReC_S = 7.5 \times 10^{-3} \lambda\beta, \quad \beta = \frac{M_{WL}^2}{M_{WR}^2}, \quad (11)$$

where λ is Cabbibo angle and 0.22 and η is CP violating parameter of Kobayashi-Maskawa in Wolfenstein parametrization. The pion energy spectrum is given by:

$$\frac{dB}{dx_\pi} \simeq \sqrt{x_\pi^2 - 4\delta^2}\{(x_\pi^2 - 4\delta^2)|pC_{SM} - qC_{SM}|^2 \times N_f$$
$$+ 3\hat{t}(\frac{M_K}{m_d - m_s})^2 (1 - \delta^2 + \xi\hat{t})^2 |(pC_S + qC_S^*)|^2\}, \quad (12)$$

where $N_f = 3$, $x_\pi = \frac{2E_\pi}{M_K}$, and $\delta = \frac{M_\pi}{M_K}$. On the standard model part, the contribution from three flavors of neutrinos is summed up while on the left-right model part, only tau neutrino contribution is important. The ratio of the CP conserving left-right contribution versus CP violation SM like contribution is given by,

$$\frac{LR}{SM} = O(\frac{M_K \beta}{m_s \eta})^2. \quad (13)$$

For example, if $\eta = 0.25$, the branching ratio and the pion energy spectrum of $K_L \to \pi_0 \nu \bar{\nu}$ may be sensitive to $500(GeV) < M_{WR} < 1(TeV)$. In Fig.12 and Fig.13 in Ref.(2).

3 Summary

We summarize our results here.

- A model independent analysis on $B \to X_s l^+ l^-$ is presented and the effects of ten local four fermi interactions on the branching ratio and the forward and backward asymmetry are studied

- The correlation plotts (B, A) are shown and they are useful to discriminate various interaction.

- In the future, we may pin down new physics once the deviation from the SM is found.

- CP conserving amplitude through scalar type interaction $\bar{s}d\bar{\nu}\nu$ may contribute to the process, $K_L \to \pi_0 \nu \bar{\nu}$

- The matrix element of the scalar interaction is enhanced by a factor of $\frac{M_K}{m_s}$ and because of the CP conserving amplitude. Also pion energy spectrum is enhanced at low energy region. In a left-right symmetric model, the coefficient of the scalar interaction is computed. It is shown that pion energy spectrum is sensitive to $M_{WR} < 1(TeV)$. If $M_{WR} = 500GeV$ and $\eta = 0.25$, 30% increase of the branching ratio is expected.

Acknowledgments

I would like to thank H. Y. Cheng, W. How, G. Lin, H. Yu and the organizers of the meeting. I also would like to thank S. Fukae, C.S.Kim, Y. Kiyo, M. Tanimoto, P. Parada, M. N. Rebelo and T. Yoshikawa on discussions and collaborations. The works are supported by Grant-in-Aid for Scientific Research on Priority Areas (Physics of CP violation) and by Joint international Scientific Research (08044089).

References

1. S. Fukae, C.S. Kim, T. Morozumi, and T. Yosikawa hep-ph/9807254.
2. Y. Kiyo, T. Morozumi, M. Tanimoto, P. Parada, and M.N. Rebelo ,hep-ph/9809333.
3. Y. Kiyo, T. Morozumi, and M. Tanimoto, hep-ph/9805307, in the proceedings of the international workshop on Fermion Masses and CP violation March5-6, 1998 (Hiroshima) 254.
4. T. Morozumi, T. Satoh, M. N. Rebelo and M. Tanimoto, *Phys. Lett.* B **410**, 233 (1997).

RADIATIVE LEPTONIC $B_{u,c}$ DECAYS AND LEPTON POLARIZATION ASYMMETRIES

C. Q. GENG, C. C. LIH
Department of Physics, National Tsing Hua University Hsinchu, Taiwan, ROC

W. M. ZHANG
Department of Physics, National Cheng Kung University Tainan, Taiwan, ROC

We study the radiative leptonic decays of $B^+_{u,c} \to l^+\nu_l\gamma$ within the light front framework. We show that the decay rates of $B^+_u \to l\nu_l\gamma$ ($l = e, \mu$) are larger than that of the corresponding purely leptonic modes. We give a general analysis on the lepton polarization asymmetries in the decays. Especially, we discuss the transverse component of the τ lepton asymmetries in the tau modes due to CP violation in theories beyond the standard model to search T violating effect.

1 Introduction

It is known that the purely leptonic B_q decays of $B_q \to l\nu_l$ ($q = u, c$) could be used to determine the weak mixing element of $|V_{qb}|$ in the Cabibbo-Kobayashi-Maskawa (CKM) matrix[1] as well as the value of the B_q meson decay constant f_{B_q}.[2] The decay rates of these purely leptonic modes are given by

$$\Gamma(B_q \to l\bar{\nu}_l) = \frac{G_F^2}{8\pi}|V_{qb}|^2 f_{B_q}^2 \left(\frac{m_l^2}{M_{B_q}^2}\right) M_{B_q}^3 \left(1 - \frac{m_l^2}{M_{B_q}^2}\right)^2 . \quad (1)$$

However, the rates for $B_q \to e\bar{\nu}_e$ and $\mu\bar{\nu}_\mu$ in Eq. (1) are helicity suppressed with the suppression factors of $m_l^2/M_{B_q}^2$ for $l = e$ and μ, respectively, and explicitly, one has that

$$\begin{aligned} Br(B_u^+ \to e^+\nu_e, \mu^+\nu_\mu, \tau^+\nu_\tau) &\simeq 5.1 \times 10^{-12}\,,\ 2.3 \times 10^{-7}\,,\ 5.1 \times 10^{-5}\,, \\ Br(B_c^+ \to e^+\nu_e, \mu^+\nu_\mu, \tau^+\nu_\tau) &\simeq 1.4 \times 10^{-9}\,,\ 6.2 \times 10^{-5}\,,\ 1.5 \times 10^{-2}\,, \end{aligned} \quad (2)$$

by taking $|V_{u(c)b}| = 3 \times 10^{-3}$ (0.04), $f_{B_{u(c)}} = 180$ (360) MeV, $\tau_{B^+_{u(c)}} \simeq 1.62$ $(0.52) \times 10^{-12}\,s$, and $M_{B_{u(c)}} = 5.2$ (6.3) GeV.[3] Clearly, it is difficult to measure these decays, especially for the light charged lepton modes. Although there are no suppressions for the τ channels, it is hard to observe the decays experimentally because of the low efficiency.

Recently, there has been a considerable amount of theoretical attention[4–12] to the radiative leptonic B decays. These decays receive two types of contributions: internal-bremsstrahlung (IB) and structure-dependent (SD) [13]. The

IB contributions are still helicity suppressed [4], while the SD ones contain the electro-magnetic coupling constant α but they are free of the helicity suppression. Therefore, the radiative decay rates of $B_q^+ \to l^+\nu_l\gamma$ ($l = e, \mu$) could have an enhancement with respect to the purely leptonic modes of $B_q^+ \to l^+\nu_l$ due to the SD contributions.

With the possible large branching ratios in the radiative leptonic decays, the $B_u^+ \equiv B^+$ modes could be measured in the two B factories at KEK and SLAC, while that for the B_c^+ could be done at LHC where about 2×10^8 B_c mesons are estimated to be produced. [14]

In this talk, we first present the branching ratios for the decays of $B_q \to l\nu_l\gamma$. We will use the light front formulation [15,16] to evaluate the hadronic matrix elements. We then discuss the lepton polarization asymmetries. We shall concentrate on the transverse parts (P_T) of the asymmetries in the tau decay modes of $B^+ \to \tau^+\nu_\tau\gamma$ and $B_c^+ \to \tau^+\nu_\tau\gamma$, which are related to the T-odd triple correlation $\vec{s}_\tau \cdot (\vec{p}_\tau \times \vec{p}_\gamma)$ with $\vec{s}_\tau$ being the tau lepton spin vector and $\vec{p}_i$ ($i = \tau$ and γ) the momenta of the τ and photon in the rest frame of the B-meson. As in the case of the radiative $K_{\mu 2}^+$ decay [17], the CKM phase does not affect the lepton transverse polarization and therefore, a non-zero value of P_T could be a signature of physics beyond the standard model. To illustrate our results, we will discuss the polarizations in some typical non-standard CP violation theories, such as the left-right symmetric and three-Higgs doublet models.

2 Branching ratios of $B_q^+ \to l^+\nu_l\gamma$

The decay amplitudes for radiative leptonic B decays of $B_q^+ \to l^+\nu_l\gamma$ can be written into two parts

$$M(B_q^+ \to l^+\nu_l\gamma) = M_{IB} + M_{SD}\,, \tag{3}$$

where M_{IB} and M_{SD} represent the IB and SD parts, and are given by

$$\begin{aligned} M_{IB} &= ie\frac{G_F}{\sqrt{2}} V_{qb} f_B m_l \epsilon_\mu^* B^\mu\,, \\ M_{SD} &= -i\frac{G_F}{\sqrt{2}} V_{qb} \epsilon_\alpha L_\beta H^{\alpha\beta}\,, \end{aligned} \tag{4}$$

respectively, with

$$\begin{aligned} B^\mu &= \bar{u}(p_\nu)(1+\gamma_5)\Big(\frac{p^\mu}{p\cdot q} - \frac{2p_l^\mu + \not{q}\gamma^\mu}{2p_l\cdot q}\Big)v(p_l, s_l), \\ L_\beta &= \bar{u}(p_\nu)\gamma_\beta(1-\gamma_5)v(p_l, s_l), \end{aligned}$$

$$H^{\alpha\beta} = e\frac{F_A}{M_B}(-g^{\alpha\beta}p\cdot q + p^\alpha q^\beta) + ie\frac{F_V}{M_B}\epsilon^{\alpha\beta\mu\nu}q_\mu p_\nu, \tag{5}$$

where ϵ_μ is the photon polarization vector, p, p_l, p_ν, and q are the four momenta of B_q^+, l^+, ν and γ, respectively, s_l is the polarization vector of the l^+, f_B is the B-mason decay constant, and $F_{A(V)}$ is the form factor of the vector (axial-vector) current. The factors f_B and $F_{V,A}$ are defined as follows:

$$\begin{aligned} < 0|\bar{q}\gamma^\mu\gamma_5 b|B(p)> &= -if_B p^\mu\,, \\ < \gamma(q)|\bar{q}\gamma^\mu\gamma_5 b|B(p)> &= e\frac{F_A}{M_B}\left[\epsilon^{*\mu}(p\cdot q) - (\epsilon^*\cdot p)q^\mu\right]\,, \\ < \gamma(q)|\bar{q}\gamma^\mu b|B(p)> &= ie\frac{F_V}{M_B}\epsilon^{\mu\alpha\beta\gamma}\epsilon^*_\alpha p_\beta q_\gamma\,. \end{aligned} \tag{6}$$

Numerically, $f_{B_u(B_c)} = 0.18\ (0.36)\ GeV$, and F_A and F_V can be found in the light front quark model at the time-like momentum transfers in which the physical accessible kinematic region is $0 \le p^2 \le p^2_{\max}$ with one loop level [9,12].

For the IB amplitude, the charged B meson emits leptons via the axial-vector current, and the photon is radiated from the external charged particles. We note that the amplitude, M_{IB}, is proportional to the ratio m_ℓ/M_B and therefore it is helicity suppressed for the light charged lepton modes. The SD amplitude is governed by the vector and axial vector form factors, in which the photon is emitted from intermediate states. Gauge invariance leaves only two form factors, $F_{V,A}$, undetermined in the SD part.

For $B_q \to \gamma$ transitions, it is known that as the recoil momentum increases, we have to start considering relativistic effects seriously. In particular, at the maximum recoil point, there is no reason to expect that the non-relativistic quark model is still applicable. A consistent treatment of the relativistic effect of the quark motion and spin in a bound state is a main issue of the relativistic quark model. The light front quark model [18,19] is the widely accepted relativistic quark model in which a consistent and relativistic treatment of quark spins and the center-of-mass motion can be carried out. The results of F_A and F_V in the entire range of momentum transfer p^2 are shown in Figures 1 and 2 for $B_u \to \gamma$ and $B_c \to \gamma$, respectively.

In the standard model, using the values shown in the figures, we get the integrated branching ratios as [9,12]

$$\begin{aligned} Br(B_u^+ \to e^+\nu_e\gamma, \mu^+\nu_\mu\gamma, \tau^+\nu_\tau\gamma) &\simeq 1.3\times10^{-6}\,,\ 1.3\times10^{-6}\,,\ 1.8\times10^{-6}\,, \\ Br(B_c^+ \to e^+\nu_e\gamma, \mu^+\nu_\mu\gamma, \tau^+\nu_\tau\gamma) &\simeq 2.0\times10^{-5}\,,\ 2.0\times10^{-5}\,,\ 8.3\times10^{-5}\,. \end{aligned} \tag{7}$$

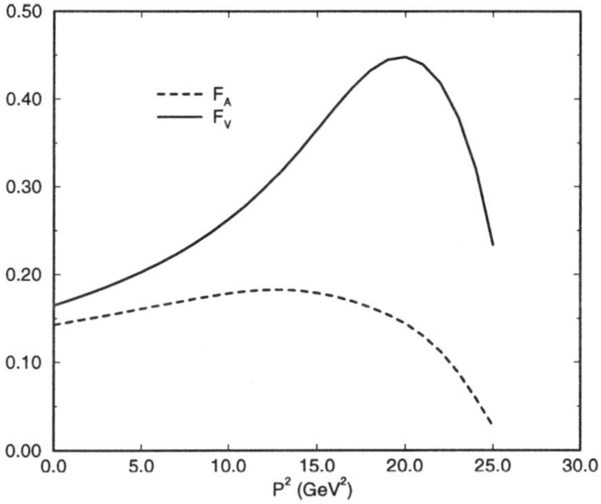

Figure 1: The values of the form factors F_V (solid curve) and F_A (dashed curve) as functions of the momentum transfer p^2 for $B_u \to \gamma$.

From Eqs. (2) and (7), we find that

$$R^l_{B_u} = \frac{\Gamma(B_u^+ \to l^+\nu_l\gamma)}{\Gamma(B_u^+ \to l^+\nu_l)} \simeq 2.5 \times 10^5\,,\ 5.6\,,$$
$$R^l_{B_c} = \frac{\Gamma(B_c^+ \to l^+\nu_l\gamma)}{\Gamma(B_c^+ \to l^+\nu_l)} \simeq 1.4 \times 10^4\,,\ 0.3\,, \tag{8}$$

for $l = e$ and μ, respectively. We note that the current experimental limits are 2.0×10^{-4} for $B_u^+ \to e^+\nu_e\gamma$ and 5.2×10^{-5} for $B_u^+ \to \mu^+\nu_\mu\gamma$.

3 Lepton polarization asymmetries in $B_q^+ \to l^+\nu_l\gamma$

For a general investigation of lepton polarization asymmetries in $B_q^+ \to l^+\nu_l\gamma$ including some new physics, we first carry out the most general four-fermion interactions given by

$$\mathcal{L} = -\frac{G_F}{\sqrt{2}} V_{qb}\bar{q}\gamma^\alpha(1-\gamma_5)b\bar{\nu}\gamma_\alpha(1-\gamma_5)l + G_s\bar{q}b\bar{\nu}(1+\gamma_5)l + G_P\bar{q}\gamma_5 b\bar{\nu}(1+\gamma_5)l$$
$$+G_V\bar{q}\gamma^\alpha b\bar{\nu}\gamma_\alpha(1-\gamma_5)l + G_A\bar{q}\gamma^\alpha\gamma_5 b\bar{\nu}\gamma_\alpha(1-\gamma_5)l + H.c.\,, \tag{9}$$

where G_F is the Fermi constant, V_{qb} is the CKM mixing element, $q = u$ and c, corresponding to $B_u^+ \equiv B^+$ and B_c^+ mesons, $l = e$, μ, and τ, and G_S, G_P,

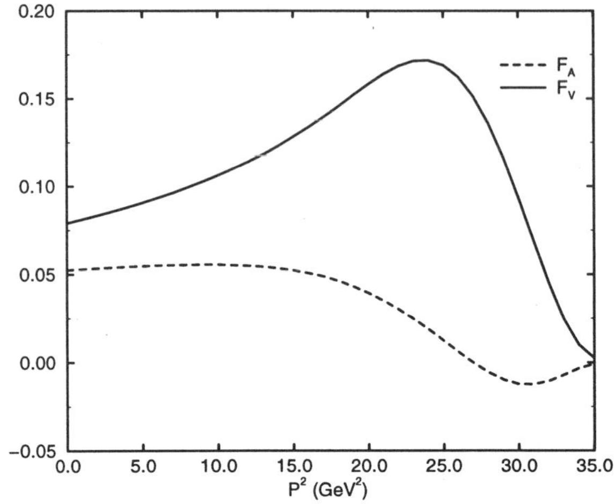

Figure 2: Same as Figure 1 but for $B_u \to \gamma$.

G_V, and G_A denote form factors for the scalar, pseudoscalar, vector, and axial vector interactions, arising from new physics, respectively. In the standard model, G_i $(i = S, P, V, A)$ vanish and only the first term in Eq. (9) is present.

We now examine the probability of the process $B_q^+ \to l^+\nu_l\gamma$ as a function of the four momenta of the particles and the polarization vector s_μ of the lepton. We write the components of s_μ in term of $\vec{\eta}$, the unit vector along the lepton spin in its the rest frame, given by

$$s_0 = \frac{\vec{p_l}\cdot\vec{\eta}}{m_l}\,, \vec{s} = \vec{\eta} + \frac{s_0}{E_l + m_l}\vec{p_l}\,. \tag{10}$$

In the B_q^+ rest frame, the partial decay rate is found to be

$$d\Gamma = \frac{1}{(2\pi)^3}\frac{1}{8M_B} \mid M \mid^2 dE_\gamma dE_l\,, \tag{11}$$

and

$$|M|^2 = A_0(x,y) + (A_L\vec{e}_L + A_N\vec{e}_N + A_T\vec{e}_T)\cdot\vec{\eta}\,, \tag{12}$$

where $\vec{e}_i$ $(i = L, N, T)$ are the unit vectors along the longitudinal, normal and transverse components of the tau polarization, defined by

$$\vec{e}_L = \frac{\vec{p_l}}{|\vec{p_l}|},$$

$$\vec{e}_N = \frac{\vec{p_l} \times (\vec{q} \times \vec{p_l})}{|\vec{p_l} \times (\vec{q} \times \vec{p_l})|},$$
$$\vec{e}_T = \frac{\vec{q} \times \vec{p_l}}{|\vec{q} \times \vec{p_l}|}, \tag{13}$$

respectively. The quantities of A_0, A_L, A_N, A_T can be calculated directly and are given by

$$\begin{aligned} A_0(x,y) = & \frac{1}{2}e^2 G_F^2 V_{qb}^2 (1-\lambda) \left\{ \frac{4m_l^2 |f_B|^2}{\lambda x^2} \left[x^2 + 2(1-r_l)\left(1 - x - \frac{r_l}{\lambda}\right)\right] \right. \\ & + M_B^4 x^2 \left[|F_V + F_A|^2 \frac{\lambda^2}{1-\lambda}\left(1 - x - \frac{r_l}{\lambda}\right) + |F_V - F_A|^2 (y - \lambda)\right] \\ & - 4 M_B m_l^2 \left[Re[f_B (F_V + F_A)^*] \left(1 - x - \frac{r_l}{\lambda}\right) \right. \\ & \left.\left. - \ Re[f_B(F_V - F_A)^*] \frac{1 - y + \lambda}{\lambda}\right]\right\}, \end{aligned} \tag{14}$$

$$\begin{aligned} A_L(x,y) = & -e^2 G_F^2 V_{qb}^2 \frac{(1-\lambda)}{2\lambda\sqrt{y^2 - 4r_l}} \left\{ \frac{4m_l^2 |f_B|^2}{\lambda x^2} \left[x(\lambda y - 2r_l)(x + y - 2\lambda) \right.\right. \\ & \left. -(y^2 - 4r_l)(\lambda x + 2r_l - 2\lambda)\right] - M_B^4 \lambda x^2 \left[|F_V + F_A|^2 \frac{\lambda}{1-\lambda} \right. \\ & \left. \left(\lambda y - 2r_l\right)(1 - x - \frac{r_l}{\lambda}) + |F_V - F_A|^2 \left(y^2 - \lambda y - 2r_l\right)\right] \\ & - 4 M_B m_l^2 \left[Re\{f_B (F_V + F_A)^*\} \lambda \left(1 - x - \frac{r_l}{\lambda}\right)(2 - 2x - y) \right. \\ & \left.\left. + Re\{f_B (F_V - F_A)^*\} \left((1-y)(y-\lambda) + 2r_l - \lambda\right)\right]\right\}, \end{aligned} \tag{15}$$

$$\begin{aligned} A_N(x,y) = & \ e^2 G_F^2 V_{qb}^2 \frac{(1-\lambda)\sqrt{\lambda y - \lambda^2 - r_l}}{M_B \lambda \sqrt{y^2 - 4r_l}} \left\{ \frac{4m_l^3 |f_B|^2}{\lambda x}(x + y - 2\lambda) \right. \\ & - M_B^4 m_l \lambda x^2 \left[|F_V + F_A|^2 \frac{\lambda}{1-\lambda}\left(1 - x - \frac{r_l}{\lambda}\right) + |F_V - F_A|^2 \right] \\ & - 2 M_B^3 m_l \left[Re\{f_B (F_V + F_A)^*\} \left(\frac{(r_l - \lambda)(1 - x - r_l)}{1 - \lambda} \right.\right. \\ & \left.\left.\left. + \lambda x (1 - x)\right) - Re\{f_B (F_V - F_A)^*\}(y - 2r_l)\right]\right\}, \end{aligned} \tag{16}$$

$$A_T(x,y) = -2e^2 G_F^2 V_{qb}^2 M_B^2 m_l \frac{1-\lambda}{\lambda} \sqrt{\lambda y - \lambda^2 - r_l}\, \{Im[f_B(F_V + F_A)^*]$$

$$\frac{\lambda}{1-\lambda}\left(1-x-\frac{r_l}{\lambda}\right)+Im[f_B(F_V-F_A)^*]\Big\}\,, \tag{17}$$

where $\lambda = (x+y-1-r_l)/x$, $r_l = m_l^2/M_B^2$, and $x = 2p\cdot q/p^2 = 2E_\gamma/M_B$ and $y = 2p\cdot p_l/p^2 = 2E_l/M_B$ are normalized energies of the photon and l, respectively. If we define the longitudinal, normal and transverse lepton polarization asymmetries by

$$P_i(x,y) = \frac{d\Gamma(\vec{e}_i) - d\Gamma(-\vec{e}_i)}{d\Gamma(\vec{e}_i) + d\Gamma(-\vec{e}_i)}\,,\ (i = L, N, T)\,, \tag{18}$$

we find that

$$P_i(x,y) = \frac{A_i(x,y)}{A_0(x,y)}\,,\ (i = L, N, T)\,. \tag{19}$$

From Eqs. (12), (13) and (18), it is easily seen that the asymmetries of P_L and P_N are even quantities under time-reversal transformation, while P_T is an odd one. We note that both P_N and P_T depend on the lepton mass as seen from Eqs. (17) and (19).

In the standard model, we find that

$$\begin{aligned} &P_N(B^+_{u,c} \to l^+\nu_l\gamma) \simeq 0\,, \\ &P_L(B^+_{u,c} \to l^+\nu_l\gamma) \simeq 100\% \end{aligned} \tag{20}$$

for the $l = e$ and μ in the allowed phase space. For the tau modes, we obtain that

$$P_L(B^+_{u,c} \to \tau^+\nu_\tau\gamma)|_{Average} \simeq 22\%\,,\ 43\%\,. \tag{21}$$

Since f_B and $F_{V,A}$ are all real in the standard model, the transverse lepton polarizations in the decays of $B_q^+ \to l^+\nu_l\gamma$ are zero.

We now study transverse lepton polarization asymmetries in non-standard models where some of G_i in Eq. (9) exist and we shall focus on the tau modes since the transverse polarizations are very small for the light lepton modes of $B_q^+ \to l^+\nu_l\gamma$ $(l = e,\mu)$ due to $m_l \ll M_{B_q}$. It is clear that, in order to have a non-zero T violating P_T for τ, the coupling constants $G_{(P,A,V)}$ have to be present and furthermore at least one of them is complex to provide the CP violating phase. However, it is interesting to note that there is no contribution to the transverse τ polarization if the interaction beyond the standard model has only left-handed vector current, because there is no relative phase between the amplitudes of M_{IB} and M_{SD}.

With new physics, the factors f_B and $F_{A,V}$ in Eq. (5) need to be replaced as follows[17]:

$$\begin{aligned} f_B &\to f_B(1+\Delta_P+\Delta_A)\,, \\ F_A &\to F_A(1+\Delta_A)\,, \\ F_V &\to F_V(1-\Delta_V)\,, \end{aligned} \tag{22}$$

with

$$\Delta_{(P,A,V)} = \frac{\sqrt{2}}{G_F V_{qb}} \left(\frac{G_P M_B^2}{(m_b+m_u)m_\tau}, G_A, G_V \right) . \tag{23}$$

In Eq. (22), the factors of $\Delta_{(P,A,V)}$ could be complex numbers due to new physical phases arising from G_P, G_V, G_A, respectively. We rewrite $P_T(x,y)$ in Eq. (18) as

$$P_T(x,y) = P_T^V(x,y) + P_T^A(x,y) \tag{24}$$

with

$$\begin{aligned} P_T^V(x,y) &= \sigma_V(x,y)[Im(\Delta_A+\Delta_V)]\,, \\ P_T^A(x,y) &= [\sigma_V(x,y)-\sigma_A(x,y)]Im(\Delta_P)\,, \end{aligned} \tag{25}$$

where

$$\begin{aligned} \sigma_V(x,y) &= 2e^2 G_F^2 V_{qb}^2 M_B^2 m_\tau f_B F_V \frac{\sqrt{\lambda y - \lambda^2 - r_\tau}}{\rho_0(x,y)} \\ &\quad \times \left(\frac{-1+\lambda}{\lambda} - \left(1 - x - \frac{r_\tau}{\lambda}\right)\right), \\ \sigma_A(x,y) &= 2e^2 G_F^2 V_{qb}^2 M_B^2 m_\tau f_B F_A \frac{\sqrt{\lambda y - \lambda^2 - r_\tau}}{\rho_0(x,y)} \\ &\quad \times \left(\frac{-1+\lambda}{\lambda} + \left(1 - x - \frac{r_\tau}{\lambda}\right)\right). \end{aligned} \tag{26}$$

One can show from Eq. (26) that the values of $\sigma_V(x,y)$ and $\sigma_V(x,y)-\sigma_A(x,y)$ in $B_{u,c}^+ \to \tau^+\nu_\tau\gamma$ are between 0.1 and 0.8, respectively.

To illustrate the estimations of P_T, we consider the left-right symmetric and three-Higgs-doublet models and study new CP-violation phases outside the CKM mechanism.

• *Left-right symmetric models*

In models with $SU(2)_L \times SU(2)_R \times U(1)_{B-L}$ gauge symmetries, the quark level four-fermion interaction is given by

$$\mathcal{L}_{RL} = -2\sqrt{2}G_F(\frac{g_R}{g_L})K_{qb}^{R^*}\,\xi\bar{b}\gamma_\mu P_R q\bar{\nu}\gamma^\mu P_L\mu\,, \tag{27}$$

which contributes to $B_q^+ \to \tau^+\nu_\tau\gamma$, where $g_{L,R}$ are coupling constants for $SU(2)_{L,R}$, ξ is the left-right mixing angle, and $K_{qb}^{R^*}$ is the right-handed CKM (RCKM) matrix. In general, the gauge coupling constants g_L and g_R are unequal to each other and they are free parameters. Compare $\mathcal{L}$ in Eq. (2) with $\mathcal{L}_{RL}$ in Eq. (27), we find that

$$G_A = G_V = -\frac{G_F}{\sqrt{2}}(\frac{g_R}{g_L})K_{qb}^{R^*}\,\xi\,. \tag{28}$$

Because we have the relations between $\Delta_{A,V}$ and $G_{A,V}$ in Eq. (23), we get

$$\Delta_A = \Delta_V = -\frac{K_{qb}^{R^*}\,\xi}{V_{qb}}(\frac{g_R}{g_L})\,. \tag{29}$$

From the definition of the τ transverse polarization, we obtain

$$P_T = 2\sigma_V(\frac{g_R\xi}{g_L})\ Im(K_{qb}^{R^*}). \tag{30}$$

In a class of the specific models [20], it is found that

$$\xi\frac{g_R}{g_L} < 3.3 \times 10^{-2} \tag{31}$$

for $M_R > 549\ GeV$. With the value in Eq. (31) and the assumption of $2\sigma_V Im(K_{qb}^{R^*}) \sim 1$, we get that

$$P_T(B_q^+ \to \tau^+\nu_\tau\gamma) < 3.3\%\,, \tag{32}$$

for both $q = u$ and c modes.

• *Three-Higgs-Doublet Models*

In a multi-Higgs doublet model (MHDM), the spontaneous CP violation (SCPV) arises from complex vaccum expectation values. In the Weinberg three-Higgs-doublet model (THMD) [21], CP phase occurs in the charged-Higgs-boson mixing and it contributes to FCNC at tree level. We shall concentrate on this phase and assume that the CKM matrix is real. With fermion mass eigenstates, the Yukawa interaction of physical charged scalars is given by

$$\mathcal{L} = (2\sqrt{2}G_F)^{1/2} \sum_{i=1}^{2} (\alpha_i \bar{U}_L K M_D D_R + \beta_i \bar{U}_R M_U K D_L + \gamma_i \bar{\nu}_L M_E E_R) H_i^+ + h.c. , \tag{33}$$

where M_D, M_U, and M_E are the diagonal mass matries of down, up-type quarks and charged leptons, respectively, K is the mixing matrix for quark sector, H_i^+ $(i = 1, 2)$ denote the two charged physical scalars and α_i, β_i, γ_i are complex coupling constants. Comparing Eq. (33) with Eq. (9), we find that the pseudoscalar coupling constant G_P has the following form

$$G_P = \sqrt{2} G_F V_{qb} m_\tau \sum_{i=1}^{2} \frac{\gamma_i}{m_i^2} (m_q \beta_i^* - m_b \alpha_i^*) . \tag{34}$$

From Eq. (23), one has that

$$Im\Delta_P = \frac{M_{B_q}^2}{m_b + m_q} \sum_{i=1}^{2} Im \frac{\gamma_i}{m_i^2} (m_q \beta_i^* - m_b \alpha_i^*) . \tag{35}$$

Assuming that $400\ GeV < m_{H_1} \equiv m_H \ll m_{H_2}$, the experimental limit on the inclusive process $B \to X \tau \nu_\tau$ decay gives the strongest bound [22,23]

$$\frac{|Im \alpha_1^* \gamma_1|}{m_H^2} < 0.23\ (GeV)^{-2} , \tag{36}$$

which leads to

$$|Im\Delta_P|_{u,c} < 5.8 ,\ 6.7 \tag{37}$$

where we have ignored the terms relating to m_q in Eq. (35). From Eq. (25), with $|\sigma_V - \sigma_A| \sim 0.1$ and the limit in Eq. (37), the transverse τ polarization asymmetries of $B_{u,c}^+ \to \tau^+ \nu_\tau \gamma$ are found to be

$$P_T(B_{u,c}^+ \to \tau^+ \nu_\tau \gamma) < 0.58 ,\ 0.67 . \tag{38}$$

4 Conclusions

We have studied the radiative decays of $B_{u,c}^+ \to l \nu_l \gamma$ within the light front framework. We have shown that the contributions from the diagrams with photon emission from the external quarks for $B^+_{u,c} \to l\nu_l\gamma$ with $l = e$ and μ are no longer helicity suppressed. We have found that, in the standard model, the branching ratios of $B^+_{u(c)} \to e^+\nu_e\gamma$, $B^+_{u(c)} \to \mu^+\nu_\mu\gamma$ and $B^+_{u(c)} \to \tau^+\nu_\tau\gamma$ are 1.3×10^{-6} (2.0×10^{-5}), 1.3×10^{-6} (2.0×10^{-5}), and 1.8×10^{-6} (8.3×10^{-5}), respectively. We have also examined the lepton polarization asymmetries. We have found that the average longitudinal lepton polarizations of $B^+_{u,c} \to l\nu_l\gamma$ with $l = e$ and μ are close to 1, while that of $B^+_{u,c} \to \tau^+\nu_\tau\gamma$ are 0.22 and 0.43, respectively. For the T-odd transverse lepton polarization asymmetries in the decays, we have demonstrated that they are zero in the standard model but can be large in the non-standard CP violation theories for the tau modes. Explicitly, we have found that $P_T(B^+_{u(c)} \to \tau^+\nu_\tau\gamma)$ can be as large as 3.3×10^{-2} (3.3×10^{-2}) and 0.58 (0.67) in the left-right symmetric and three-Higgs doublet models. Experimentally, to observe the above tau transverse polarization asymmetries in the non-standard models at the $n\sigma$ level, one needs at least 5.0×10^8 (1.1×10^7) and 1.7×10^6 (2.7×10^4) $B^+_{u(c)}$ decays, which could be accessible in the B factories and LHC, respectively. Clearly, a measurement of such effect is a clean signature of CP violation beyond the standard model.

Acknowledgments

This work was supported in part by the National Science Council of Republic of China under contract number NSC-87-2112-M-007-040.

References

1. N. Cabibbo, *Phys. Rev. Lett.* **10** (1963) 531; M. Kobayashi and T. Maskawa, *Prog. Theor. Phys.* **49** (1973) 652.
2. For review see A.J. Buras, M.K. Harlander, "Heavy flavors", Eds. A.J. Buras, M. Lindner, (World Scientific, Singapore) p. 58; A. Ali, *Nucl. Phys.* B, Proc. Suppl. **39BC**, 408 (1995); S. Playfer and S. Stone, *Int. J. Mod. Phys.* **A10**, 4107 (1995).
3. Particle Data Group, *Phys. Rev.* **D54**, 1 (1996).
4. G. Burdman, T. Goldman and D. Wyler, *Phys. Rev.* **D51**, 111 (1995).
5. G. Eilam, I. Halperin and R. Mendel, *Phys. Lett.* **B361**, 137 (1995).
6. D. Atwood, G. Eilam and A. Soni Mod. Phys. Lett. **A11**, 1061 (1996).
7. P. Colangelo, F. De Fazio, G. Nardulli, *Phys. Lett.* **B372**, 331 (1996); **B386**, 328 (1996).
8. T. Aliev, A. Özpineci and M. Savci, *Phys. Lett.* **B393**, 143 (1997); *Phys. Rev.* **D55**, 7059 (1997).

9. C. Q. Geng, C. C. Lih, and Wei-Min Zhang, *Phys. Rev.* **D57**, 5697 (1998).
10. C.-H. Chang, J.-P. Cheng C.-D. Lü, **hep-ph**/9712325.
11. T.M. Aliev, M. Savcı, **hep-ph**/9804407.
12. C. Q. Geng, C. C. Lih, and Wei-Min Zhang, in preparation.
13. T. Goldmann and W. J. Wilson, *Phys. Rev.* **D15**, 709 (1977), and references therein.
14. Chao–Hsi Chang, Yu–Qi Chen, *Phys. Rev.* **D48**, 4086 (1993); K. Cheung, *Phys. Rev. Lett.* **71**, 3413 (1993).
15. Wei-Min Zhang, *Chin. J. Phys.* **31**, 717 (1994).
16. Chi-Yee Cheung, Wei-Min Zhang and Guey-Lin Lin, *Phys. Rev.* **D52**, 2915 (1995).
17. C. H. Chen, C. Q. Geng, and C. C. Lih, *Phys. Rev.* **D56**, 6856 (1997).
18. M. Terent'ev, *Sov. J. Phys.* **24**, 106 (1976); V. Berestetsky and M. Terent'ev, *ibid* **24**, 547 (1976); **25**, 347 (1977); P. Chung, F. Coester and W. Polyzou, *Phys. Lett.* **B205**, 545 (1988).
19. W. Jaus, *Phys. Rev.* **D41**, 3394 (1990); **44**, 2851 (1991); P. J. O'Donnell and Q. P. Xu, *Phys. Lett.* **B325**, 219; **336**, 113 (1994).
20. G. Barenboim, J. Bernabéu, J. Prades, and M. Raidal, *Phys. Rev.* **D55**, 4213 (1997).
21. S. Weinberg, *Phys. Rev. Lett.* **37**, 657 (1976).
22. M.Kobayashi, T. T. Lin and Y. Okada, *Prog. Thror. Phys.* **B396**, 81 (1996).
23. Y. Grossman, *Nucl. Phys.* **B426**, 355 (1994).

CP VIOLATION IN B DECAYS AND STRATEGIES FOR EXTRACTING CKM PHASES

ROBERT FLEISCHER
Theory Division, CERN
CH-1211 Geneva 23, Switzerland
E-mail: Robert.Fleischer@cern.ch

A brief review of CP violation in the B-meson system and of strategies to determine the angles α, β and γ of the unitarity triangle of the CKM matrix is given. Both general aspects and some recent developments are discussed, including a critical look at the "benchmark" modes, CP violation in penguin decays, B_s decays in the light of the width difference $\Delta\Gamma_s$, charged B decays, and strategies to probe the CKM angle γ with $B \to \pi K$ modes.

1 Setting the Scene

At present, CP violation is one of the least experimentally explored phenomena of the Standard Model, and is very promising in the search for indications of "new physics" at future experiments. In order to accomplish this task, it is crucial to have CP-violating processes available that can be analysed in a reliable way within the framework of the Standard Model, where CP violation is closely related to the Cabibbo–Kobayashi–Maskawa matrix (CKM matrix),[1] connecting the electroweak eigenstates of the d, s and b quarks with their mass eigenstates. As far as CP violation is concerned, the central feature is that – in addition to three generalized Cabibbo-type angles – also a *complex phase* is needed in the three-generation case to parametrize the CKM matrix. This complex phase is the origin of CP violation within the Standard Model.

A closer look shows that CP-violating observables are proportional to the following combination of CKM matrix elements:[2]

$$J_{\rm CP} = \pm\,{\rm Im}\left(V_{i\alpha}V_{j\beta}V_{i\beta}^{*}V_{j\alpha}^{*}\right) \quad (i \neq j,\, \alpha \neq \beta)\,, \tag{1}$$

representing a measure of the "strength" of CP violation in the Standard Model. Since $J_{\rm CP} = \mathcal{O}(10^{-5})$, CP violation is a small effect. In many scenarios of new physics,[3] several additional complex couplings are present, leading to new sources of CP violation.

Concerning phenomenological applications, the parametrization

$$\hat{V}_{\rm CKM} = \begin{pmatrix} 1-\frac{1}{2}\lambda^2 & \lambda & A\lambda^3 R_b\, e^{-i\gamma} \\ -\lambda & 1-\frac{1}{2}\lambda^2 & A\lambda^2 \\ A\lambda^3 R_t\, e^{-i\beta} & -A\lambda^2 & 1 \end{pmatrix} + \mathcal{O}(\lambda^4) \tag{2}$$

with $\lambda = 0.22$, $A \equiv |V_{cb}|/\lambda^2 = 0.81 \pm 0.06$, $R_b \equiv |V_{ub}/(\lambda V_{cb})| = \sqrt{\rho^2 + \eta^2} = 0.36 \pm 0.08$, and $R_t \equiv |V_{td}/(\lambda V_{cb})| = \sqrt{(1-\rho)^2 + \eta^2} = \mathcal{O}(1)$ turns out to be very useful. It is a modification of the Wolfenstein parametrization,[4] exhibiting not only the hierarchy of the CKM elements, but also the dependence on the angles $\beta = \beta(\rho, \eta)$ and $\gamma = \gamma(\rho, \eta)$ of the usual "non-squashed" unitarity triangle of the CKM matrix.[5]

Although the discovery of CP violation[6] goes back to 1964, so far this phenomenon has been observed only within the neutral K-meson system, where it is described by two complex quantities, called ε and ε', which are defined by the following ratios of decay amplitudes:

$$\frac{A(K_{\rm L} \to \pi^+\pi^-)}{A(K_{\rm S} \to \pi^+\pi^-)} = \varepsilon + \varepsilon', \quad \frac{A(K_{\rm L} \to \pi^0\pi^0)}{A(K_{\rm S} \to \pi^0\pi^0)} = \varepsilon - 2\,\varepsilon'. \tag{3}$$

While $\varepsilon = (2.280 \pm 0.013) \times e^{i\frac{\pi}{4}} \times 10^{-3}$ parametrizes "indirect" CP violation, originating from the fact that the mass eigenstates of the neutral kaon system are not CP eigenstates, the quantity $\text{Re}(\varepsilon'/\varepsilon)$ measures "direct" CP violation in $K \to \pi\pi$ transitions. The CP-violating observable ε plays an important role to constrain the unitarity triangle[7] and implies – using reasonable assumptions about certain hadronic parameters – in particular a positive value of the Wolfenstein parameter η. Despite enormous experimental efforts, the question of whether $\text{Re}(\varepsilon'/\varepsilon) \neq 0$ could not yet be answered. However, in the near future, this issue should be clarified by improved measurements at CERN and Fermilab, as well as by the KLOE experiment at DAΦNE. Unfortunately, the calculations of $\text{Re}(\varepsilon'/\varepsilon)$ are very involved and suffer at present from large hadronic uncertainties.[7] Consequently, this observable will not allow a powerful test of the CP-violating sector of the Standard Model, unless the hadronic matrix elements of the relevant operators can be brought under better control. Probably the major goal of a possible future observation of $\text{Re}(\varepsilon'/\varepsilon) \neq 0$ would hence be the unambiguous exclusion of "superweak" models of CP violation.[8]

In order to test the Standard Model description of CP violation, the rare decays $K_{\rm L} \to \pi^0\nu\overline{\nu}$ and $K^+ \to \pi^+\nu\overline{\nu}$ are more promising and may allow a determination of $\sin(2\beta)$ with respectable accuracy.[9] Yet the kaon system by itself cannot provide the whole picture of CP violation. Consequently, it is essential to study CP violation outside this system. In this respect, the B-meson system appears to be most promising, which is also reflected by the tremendous experimental efforts at the future B-factories. There are of course also other interesting systems to explore CP violation and to search for signals of new physics, for instance the D-meson system, where sizeable mixing or CP-violating effects would signal new physics because of the tiny Standard Model "background". In the following, we shall focus on B decays.

2 The Central Target: CP Violation in the B System

With respect to testing the Standard Model description of CP violation, the major role is played by non-leptonic B decays, which can be divided into three decay classes: decays receiving both tree and penguin contributions, pure tree decays, and pure penguin decays. There are two types of penguin topologies: gluonic (QCD) and electroweak (EW) penguins related to strong and electroweak interactions, respectively. Because of the large top-quark mass, also the latter operators play an important role in several processes.[10]

To analyse non-leptonic B decays theoretically, one uses low-energy effective Hamiltonians, which are calculated by making use of the operator product expansion, yielding transition matrix elements of the following structure:

$$\langle f|\mathcal{H}_{\text{eff}}|i\rangle \propto \sum_k C_k(\mu)\langle f|Q_k(\mu)|i\rangle\,. \tag{4}$$

The operator product expansion allows us to separate the short-distance contributions to this transition amplitude from the long-distance ones, which are described by perturbative Wilson coefficient functions $C_k(\mu)$ and non-perturbative hadronic matrix elements $\langle f|Q_k(\mu)|i\rangle$, respectively. As usual, μ denotes an appropriate renormalization scale.

In the case of $|\Delta B| = 1$, $\Delta C = \Delta U = 0$ transitions, which will be of particular interest for the following discussion, we have

$$\mathcal{H}_{\text{eff}} = \mathcal{H}_{\text{eff}}(\Delta B = -1) + \mathcal{H}_{\text{eff}}(\Delta B = -1)^\dagger, \tag{5}$$

where

$$\mathcal{H}_{\text{eff}}(\Delta B = -1) = \frac{G_{\text{F}}}{\sqrt{2}}\left[\sum_{j=u,c} V_{jq}^* V_{jb}\left\{\sum_{k=1}^{2} Q_k^{jq}\, C_k(\mu) + \sum_{k=3}^{10} Q_k^{q}\, C_k(\mu)\right\}\right]. \tag{6}$$

Here $\mu = \mathcal{O}(m_b)$, Q_k^{jq} are four-quark operators, the label $q \in \{d, s\}$ corresponds to $b \to d$ and $b \to s$ transitions, and k distinguishes between current–current ($k \in \{1,2\}$), QCD ($k \in \{3,\ldots,6\}$) and EW ($k \in \{7,\ldots,10\}$) penguin operators. The evaluation of such low-energy effective Hamiltonians has been reviewed in Ref. 11, where the four-quark operators are given explicitly and numerical values for their Wilson coefficient functions can be found.

2.1 CP Asymmetries in Decays of Neutral B Mesons

A particularly simple and interesting situation arises, if we restrict ourselves to decays of neutral B_q mesons ($q \in \{d, s\}$) into CP self-conjugate final states

$|f\rangle$, satisfying the relation $(\mathcal{CP})|f\rangle = \pm|f\rangle$. In this case, the corresponding time-dependent CP asymmetry can be expressed as

$$a_{\rm CP}(t) \equiv \frac{\Gamma(B_q^0(t) \to f) - \Gamma(\overline{B_q^0}(t) \to f)}{\Gamma(B_q^0(t) \to f) + \Gamma(\overline{B_q^0}(t) \to f)} =$$
$$\mathcal{A}_{\rm CP}^{\rm dir}(B_q \to f)\cos(\Delta M_q t) + \mathcal{A}_{\rm CP}^{\rm mix-ind}(B_q \to f)\sin(\Delta M_q t)\,, \tag{7}$$

where the direct CP-violating contributions have been separated from the mixing-induced CP-violating contributions, which are characterized by

$$\mathcal{A}_{\rm CP}^{\rm dir}(B_q \to f) \equiv \frac{1-|\xi_f^{(q)}|^2}{1+|\xi_f^{(q)}|^2} \quad \text{and} \quad \mathcal{A}_{\rm CP}^{\rm mix-ind}(B_q \to f) \equiv \frac{2\,{\rm Im}\,\xi_f^{(q)}}{1+|\xi_f^{(q)}|^2}\,, \tag{8}$$

respectively. Here direct CP violation refers to CP-violating effects arising directly in the corresponding decay amplitudes, whereas mixing-induced CP violation is due to interference between B_q^0–$\overline{B_q^0}$ mixing and decay processes. Note that Eq. (7) has to be modified in the B_s case for $t \gtrsim 1/|\Delta\Gamma_s|$ because of the expected sizeable width difference $\Delta\Gamma_s$.[12] In general, the observable[10]

$$\xi_f^{(q)} = \mp\, e^{-i\phi_{\rm M}^{(q)}} \frac{\sum\limits_{j=u,c} V_{jr}^* V_{jb} \langle f|\mathcal{Q}^{jr}|B_q^0\rangle}{\sum\limits_{j=u,c} V_{jr} V_{jb}^* \langle f|\mathcal{Q}^{jr}|\overline{B_q^0}\rangle}\,, \tag{9}$$

where $\mathcal{Q}^{jr} \equiv \sum\limits_{k=1}^{2} Q_k^{jr} C_k(\mu) + \sum\limits_{k=3}^{10} Q_k^{jr} C_k(\mu)$, and where

$$\phi_{\rm M}^{(q)} = \begin{cases} 2\beta \text{ for } q = d \\ \;0 \;\text{ for } q = s \end{cases} \tag{10}$$

denotes the weak B_q^0–$\overline{B_q^0}$ mixing phase, suffers from large uncertainties, which are introduced by the hadronic matrix elements in Eq. (9). There is, however, a very important special case. If the decay $B_q \to f$ is dominated by a single CKM amplitude, these matrix elements cancel, and $\xi_f^{(q)}$ takes the simple form

$$\xi_f^{(q)} = \mp \exp\left[-i\left(\phi_{\rm M}^{(q)} - \phi_{\rm D}^{(f)}\right)\right], \tag{11}$$

where $\phi_{\rm D}^{(f)}$ is a weak decay phase, which is given as follows ($r \in \{d,s\}$):

$$\phi_{\rm D}^{(f)} = \begin{cases} -2\gamma \text{ for dominant } \bar{b} \to \bar{u}\,u\,\bar{r} \text{ CKM amplitudes in } B_q \to f \\ \;\;0 \;\;\text{ for dominant } \bar{b} \to \bar{c}\,c\,\bar{r} \text{ CKM amplitudes in } B_q \to f. \end{cases} \tag{12}$$

Probably the most important applications of this well-known formalism are the decays $B_d \to J/\psi\, K_S$ and $B_d \to \pi^+\pi^-$. If one goes through the relevant Feynman diagrams contributing to the former channel (for a detailed discussion, see Ref. 10), one finds that it is dominated by the $\bar{b} \to \bar{c}\, c\, \bar{s}$ CKM amplitude. Consequently, the decay phase vanishes, and we have

$$\mathcal{A}_{\rm CP}^{\rm mix-ind}(B_d \to J/\psi\, K_S) = +\sin[-(2\beta - 0)]\,. \tag{13}$$

Since Eq. (11) applies with excellent accuracy to the decay $B_d \to J/\psi\, K_S$ – the point is that penguins enter essentially with the same weak phase as the leading tree contribution – it is usually referred to as the "gold-plated" mode to determine the CKM angle β.[13] First attempts to measure $\sin(2\beta)$ through the CP-violating asymmetry (13) have recently been performed by the OPAL and CDF collaborations. Their results are as follows:[14]

$$\sin(2\beta) = \begin{cases} 3.2^{+1.8}_{-2.0} \pm 0.5 & \text{OPAL Collaboration} \\ 1.8 \pm 1.1 \pm 0.3 & \text{CDF Collaboration;} \end{cases} \tag{14}$$

they favour the Standard Model expectation of a *positive* value of this quantity. The presently allowed range arising from the usual fits of the unitarity triangle is given by $0.36 \lesssim \sin(2\beta) \lesssim 0.80$.[15] In the B-factory era, an experimental uncertainty of $\Delta \sin(2\beta)|_{\rm exp} = 0.08$ seems to be achievable.

In the case of the decay $B_d \to \pi^+\pi^-$, mixing-induced CP violation would measure $-\sin(2\alpha)$ through

$$\mathcal{A}_{\rm CP}^{\rm mix-ind}(B_d \to \pi^+\pi^-) = -\sin[-(2\beta + 2\gamma)] = -\sin(2\alpha)\,, \tag{15}$$

if there were no penguin contributions present. However, we have to deal with such topologies, leading to hadronic corrections to Eq. (15) that were analysed by many authors.[16-18] Last year, the CLEO collaboration reported the observation of several exclusive B-meson decays into two light pseudoscalar mesons.[19] However, $B \to \pi\pi$ modes have not yet been seen and the upper limits for the corresponding branching ratios are not "favourable". The recent CLEO results indicate moreover that we have in fact to worry about the penguin corrections to Eq. (15).

There are various methods on the market to control the penguin uncertainties in a quantitative way. Unfortunately, these strategies are usually rather challenging in practice. The best known approach was proposed by Gronau and London.[20] It makes use of the $SU(2)$ isospin relations

$$\sqrt{2}\, A(B^+ \to \pi^+\pi^0) = A(B_d^0 \to \pi^+\pi^-) + \sqrt{2}\, A(B_d^0 \to \pi^0\pi^0) \tag{16}$$

$$\sqrt{2}\, A(B^- \to \pi^-\pi^0) = A(\overline{B_d^0} \to \pi^+\pi^-) + \sqrt{2}\, A(\overline{B_d^0} \to \pi^0\pi^0)\,, \tag{17}$$

which can be represented in the complex plane as two triangles. The sides of these triangles are determined through the corresponding branching ratios, while their relative orientation can be fixed by measuring the CP-violating observable $\mathcal{A}_{\rm CP}^{\rm mix-ind}(B_d \to \pi^+\pi^-)$. Following these lines, it is in principle possible to extract a value of α taking into account the QCD penguin contributions. It should be noted that EW penguins cannot be controlled using this isospin strategy. Their effect is, however, expected to be rather small in this case, leading to $|\Delta\alpha_{\rm EW}/\sin\alpha| \lesssim 4°$.[21] An attempt to analyse other isospin-breaking effects, which may affect the isospin relations (16) and (17) through $\pi^0 - \eta, \eta'$ mixing, was recently performed in Ref. 22. Unfortunately, the Gronau–London approach suffers from an experimental problem, since the measurement of $B_d \to \pi^0\pi^0$ decays is very difficult. Theoretical estimates based on the "factorization" hypothesis (for a critical look at this concept, see Ref. 23) give branching ratios at the 10^{-6} level.[24] However, upper bounds on the combined branching ratio $\mathrm{BR}(B_d \to \pi^0\pi^0)$, i.e. averaged over the decay and its charge conjugate, may already lead to interesting upper bounds on the QCD penguin uncertainty affecting the determination of α.[18,25]

An alternative to the Gronau–London strategy to extract the CKM angle α is provided by $B \to \rho\pi$ modes.[26] Here the isospin triangle relations (16) and (17) are replaced by pentagonal relations, and the corresponding approach is rather complicated. The $SU(3)$ flavour symmetry offers also ways to determine α. For example, it is possible to extract this angle with the help of a triangle construction by measuring in addition to the $B_d \to \pi^+\pi^-$ observables those of $B_d \to K^0\overline{K^0}$ decays.[27] The latter would also be interesting to obtain insights into certain final-state interaction processes.[28] A measurement of both the direct and the mixing-induced CP asymmetries in $B_d \to \pi^+\pi^-$, together with the $B^+ \to \pi^+K^0$ branching ratio, would provide another step towards the control of the QCD penguin uncertainties.[29] Several strategies to constrain and determine α along these lines were recently proposed in Ref. 18. As sketched above, a solid extraction of this CKM angle is unfortunately quite difficult and could be out of reach for the first generation of B-factory experiments.

A decay appearing frequently in the literature as a tool to determine the CKM angle γ is $B_s \to \rho^0 K_{\rm S}$. In this case, however, penguins are expected to lead to serious problems – even more serious than in $B_d \to \pi^+\pi^-$ – so that this mode appears to be the "wrong" way to extract γ.[10] Moreover, the rapid B_s^0–$\overline{B_s^0}$ oscillations, as well as the small expected branching ratio at the 10^{-7} level, make experimental studies of $B_s \to \rho^0 K_{\rm S}$ very difficult. It should be kept in mind, however, that this channel may be in better shape to probe γ, if the concept of "colour suppression" should not work in this case. A recent model calculation within a perturbative framework can be found in Ref. 30.

2.2 CP Violation in Penguin Modes as a Probe of New Physics

In order to test the Standard Model description of CP violation, penguin-induced modes play an important role. Because of the loop suppression of these "rare" processes, it is possible – and indeed it is the case in several specific model calculations – that new-physics contributions to these decays are of similar magnitude as those of the Standard Model.[3]

An important example is the $b \to s$ penguin mode $B_d \to \phi K_S$. The corresponding branching ratio is expected to be of $\mathcal{O}(10^{-5})$ and may be large enough to investigate this channel at the future B-factories.[24] In contrast to the $b \to d$ penguin case,[31] the corresponding decay amplitude does not contain a sizeable CP-violating weak phase within the Standard Model. Consequently, direct CP violation in $B_d \to \phi K_S$ is tiny, and mixing-induced CP violation measures simply the weak B_d^0–$\overline{B_d^0}$ mixing phase, implying the relation

$$\mathcal{A}_{\rm CP}^{\rm mix-ind}(B_d \to J/\psi\, K_S) = \mathcal{A}_{\rm CP}^{\rm mix-ind}(B_d \to \phi\, K_S) = -\sin(2\beta)\,, \qquad (18)$$

which represents an interesting probe of new-physics contributions to $b \to s$ decay processes. The theoretical accuracy of this relation is limited by certain neglected terms that are CKM-suppressed by $\mathcal{O}(\lambda^2 R_b)$, and may lead to direct CP violation in $B_d \to \phi K_S$. Simple model calculations performed at the perturbative quark level indicate asymmetries of at most $\mathcal{O}(1\%)$.[10,17] However, the impact of long-distance effects is hard to quantify (for a recent attempt, see Ref. 32). The importance of $B_d \to \phi K_S$ and similar modes, such as $B_d \to \eta' K_S$, to search for new-physics effects in $b \to s$ flavour-changing neutral current processes was emphasized by several authors.[10,17,33]

Studies of CP-violating effects in inclusive $B \to X_s \gamma$ decays, which can be analysed reliably in QCD by means of the operator product expansion,[34] also play an important role in the search for new physics. Within the Standard Model, direct CP violation in $B \to X_s \gamma$ is very small, i.e. below the 1% level, whereas it may well be as large as 50% in new-physics scenarios with enhanced chromomagnetic dipole operators.[35]

2.3 The B_s System in the Light of $\Delta\Gamma_s$

In the B_s system, very rapid B_s^0–$\overline{B_s^0}$ oscillations are expected, requiring an excellent vertex resolution system. Studies of CP violation in B_s decays are therefore regarded as being very difficult. An alternative route to investigate CP-violating effects may be provided by the width difference $\Delta\Gamma_s$.[36] Because of this width difference, already *untagged* data samples of B_s decays may exhibit CP-violating effects.[12] Several "untagged strategies" to extract the CKM angle γ were proposed, using for example angular distributions in $B_s \to K^{*+}K^{*-}$, $K^{*0}\overline{K^{*0}}$ or $B_s \to D^*\phi$, $D_s^{*\pm}K^{*\mp}$ decays.[37-39]

The B_s system provides interesting probes also for physics beyond the Standard Model. Important examples are the decays $B_s \to D_s^+ D_s^-$ and $B_s \to J/\psi\,\phi$. The latter is the counterpart of the "gold-plated" mode $B_d \to J/\psi\, K_S$ to measure β and is very promising for experiments performed at future hadron machines. These transitions are dominated by a single CKM amplitude and allow – in principle even from their untagged data samples[37] – the extraction of a CP-violating weak phase $\phi_{\rm CKM} \equiv 2\lambda^2\eta$, which is expected to be of $\mathcal{O}(0.03)$ within the Standard Model. Consequently, an extracted value of $\phi_{\rm CKM}$ that is much larger than this Standard Model expectation would signal new-physics contributions to B_s^0–$\overline{B_s^0}$ mixing.[3] Suggestions for efficient determinations of the observables of the $B_s \to D_s^{*+}D_s^{*-}$ and $B_s \to J/\psi\,\phi$ angular distributions, as well as of $\Delta\Gamma_s$ and of the B_s mass difference ΔM_s, were given in Ref. 40.

A time-dependent study of $B_s \to J/\psi\,\phi$ decays is also interesting to resolve a discrete ambiguity in the determination of the CKM angle β.[41] Such ambiguities are a typical feature of the strategies to extract CKM phases sketched above; they could complicate the search for new physics considerably. Several strategies were recently proposed to deal with these problems.[42]

2.4 CP Asymmetries in Decays of Charged B Mesons

Since mixing effects are not present in the charged B-meson system, the measurement of a non-vanishing CP asymmetry in a charged B decay would give us unambiguous evidence for direct CP violation, thereby ruling out "superweak" models.[8] Such CP asymmetries arise from interference between decay amplitudes with both different CP-violating weak and CP-conserving strong phases. Whereas the weak phases are related to the CKM matrix, the strong phases are induced by strong final-state interaction effects and introduce in general severe theoretical uncertainties into the calculation. An interesting situation arises, however, in $B^\pm \to \rho^\pm\rho^0(\omega) \to \rho^\pm\pi^+\pi^-$ decays, where $\rho^0(\omega)$ denotes the $\rho^0-\omega$ interference region.[43] In this case, experimental data on $e^+e^- \to \pi^+\pi^-$ processes can be used to constrain the hadronic uncertainties affecting the corresponding CP asymmetry, which is related to $\sin\alpha$ and may well be as large as $\mathcal{O}(20\%)$ at the ω invariant mass. Direct CP violation in three-body decays such as $B^\pm \to K^\pm\pi^+\pi^-$, which involve various intermediate resonances, was recently considered in Ref. 44. Here the Dalitz plot distributions may provide information on the CKM angle γ (see also Ref. 45).

In order to extract angles of the unitarity triangle from decays of charged B mesons, amplitude relations – either exact or approximate ones based on flavour symmetries – play an important role. A review of these methods can be found in Ref. 10. The "prototype" is the approach to determine γ with the help of triangle relations between the $B^{\pm} \to DK^{\pm}$ decay amplitudes proposed by Gronau and Wyler.[46] Unfortunately, the corresponding triangles are expected to be very "squashed". Moreover, one has to deal with additional experimental problems,[47] so that this approach is very difficult from a practical point of view. More refined variants and different ways to combine the information provided by $B \to DK$ decays to probe the CKM angle γ were proposed by several authors.[47,48] The CLEO collaboration has recently observed the colour-allowed decay $B^- \to D^0 K^-$ and its charge conjugate, which is the first detection of a B decay originating from $b \to c\,\bar{u}\,s$ quark subprocesses.[49]

2.5 Strategies to Probe the CKM Angle γ with $B \to \pi K$ Modes

The decays $B^+ \to \pi^+ K^0$, $B_d^0 \to \pi^- K^+$ and their charge conjugates, which were observed by the CLEO collaboration last year,[19] play an important role to probe the CKM angle γ at future B-factories.[21,50] So far, only results for the combined branching ratios $\mathrm{BR}(B^{\pm} \to \pi^{\pm} K)$ and $\mathrm{BR}(B_d \to \pi^{\mp} K^{\pm})$ have been published at the 10^{-5} level with large experimental uncertainties. In order to obtain information on γ, the ratio

$$R \equiv \frac{\mathrm{BR}(B_d \to \pi^{\mp} K^{\pm})}{\mathrm{BR}(B^{\pm} \to \pi^{\pm} K)} \tag{19}$$

of the combined $B \to \pi K$ branching ratios, and the "pseudo-asymmetry"

$$A_0 \equiv \frac{\mathrm{BR}(B_d^0 \to \pi^- K^+) - \mathrm{BR}(\overline{B_d^0} \to \pi^+ K^-)}{\mathrm{BR}(B^+ \to \pi^+ K^0) + \mathrm{BR}(B^- \to \pi^- \overline{K^0})} \tag{20}$$

play a key role. Making use of the $SU(2)$ isospin symmetry of strong interactions, the following amplitude relations can be derived (for a detailed discussion, see Ref. 28):

$$A(B^+ \to \pi^+ K^0) \equiv P\,, \quad A(B_d^0 \to \pi^- K^+) = -\,[P + T + P_{\rm ew}]\,, \tag{21}$$

where the "penguin" amplitude P is *defined* by the $B^+ \to \pi^+ K^0$ decay amplitude, $P_{\rm ew} \equiv -|P_{\rm ew}|e^{i\delta_{\rm ew}}$ is essentially due to electroweak penguins, and $T \equiv |T|e^{i\delta_T}e^{i\gamma}$ is usually referred to as a "tree" amplitude. However, because of a subtlety in the implementation of the isospin symmetry, the amplitude T does not only receive contributions from colour-allowed tree-diagram-like topologies, but also from penguin and annihilation topologies.[28,51] The general expressions for the amplitudes P, T and $P_{\rm ew}$ are given in Ref. 51. Let us just note here that we have

$$P \propto \left[1 + \rho\, e^{i\theta} e^{i\gamma}\right], \tag{22}$$

where $\rho\, e^{i\theta}$ is a measure of the strength of certain rescattering effects (θ is a CP-conserving strong phase). The quantities $r \equiv |T|/\sqrt{\langle |P|^2\rangle}$ and $\epsilon \equiv |P_{\rm ew}|/\sqrt{\langle |P|^2\rangle}$ with $\langle |P|^2\rangle \equiv (|P|^2 + |\overline{P}|^2)/2$, as well as the strong phase differences $\delta \equiv \delta_T - \delta_{tc}$ and $\Delta \equiv \delta_{\rm ew} - \delta_{tc}$, where δ_{tc} measures the strong phase of the difference of the penguin topologies with internal top and charm quarks, turn out to be very useful to parametrize the observables R and A_0.[51]

If both R and A_0 are measured, contours in the $\gamma - r$ plane can be fixed,[51] corresponding to a mathematical implementation of a simple triangle construction.[21] In order to determine the CKM angle γ, the quantity r, i.e. the magnitude of the "tree" amplitude T, has to be fixed. At this step, a certain model dependence enters. Using arguments based on "factorization", one comes to the conclusion that a future theoretical uncertainty of r as small as $\mathcal{O}(10\%)$ may be achievable.[50,52] However, since the properly defined amplitude T does not only receive contributions from colour-allowed "tree" topologies, but also from penguin and annihilation processes,[28,51] it may be shifted sizeably from its "factorized" value so that $\Delta r = \mathcal{O}(10\%)$ may be too optimistic.

Interestingly, it is possible to derive bounds on γ that do *not* depend on r at all[53] (for a discussion of $B_s \to K\overline{K}$ decays, see Ref. 39). If we eliminate the strong phase δ in the ratio R of combined $B \to \pi K$ branching ratios with the help of the pseudo-asymmetry A_0 and treat r as a "free" variable, while keeping ρ, θ and ϵ, Δ fixed, we find that R takes the following minimal value:[51]

$$R_{\rm min} = \kappa \sin^2\gamma + \frac{1}{\kappa}\left(\frac{A_0}{2\sin\gamma}\right)^2. \tag{23}$$

In this expression, rescattering and EW penguin effects are described by

$$\kappa = \frac{1}{w^2}\left[1 + 2\,(\epsilon\, w)\cos\Delta + (\epsilon\, w)^2\right] \quad \text{with} \quad w = \sqrt{1 + 2\,\rho\cos\theta\cos\gamma + \rho^2}. \tag{24}$$

An allowed range for γ is related to $R_{\rm min}$, since values of γ implying $R_{\rm exp} < R_{\rm min}$, where $R_{\rm exp}$ denotes the experimentally determined value of R, are excluded. In the "original" bounds on γ derived in Ref. 53, no information provided by A_0 has been used, i.e. both r and δ were kept as "free" variables, and the special case $\rho = \epsilon = 0$, i.e. $\kappa = 1$, has been assumed, implying $\sin^2\gamma < R_{\rm exp}$. A particularly interesting situation arises, if R is measured to be smaller than 1. In this case, the information on γ provided by the $B \to \pi K$ decays is complementary to the presently allowed range of $41^\circ \lesssim \gamma \lesssim 134^\circ$, which arises from the usual fits of the unitarity triangle,[15] since a certain interval around 90° can be excluded.[53] A measurement of $A_0 \neq 0$ allows us to exclude in addition a range around 0° and 180°.[51] Unfortunately, the present data do not yet provide an answer to the question of whether $R < 1$. The results reported by the CLEO collaboration last year give $R = 0.65 \pm 0.40$,[19] whereas a very recent, updated analysis yields $R = 1.0 \pm 0.4$.[54]

The theoretical accuracy of these bounds is limited both by rescattering processes of the kind $B^+ \to \{\pi^0 K^+, \pi^0 K^{*+}, \ldots\} \to \pi^+ K^0$,[55-57] and by EW penguin effects.[50,56] An important implication of the rescattering effects may be a sizeable CP asymmetry, A_+, in the decay $B^+ \to \pi^+ K^0$. The rescattering effects can be controlled in the bounds on γ through experimental data. A first step towards this goal is provided by the CP asymmetry A_+. In order to go beyond it, $B^\pm \to K^\pm K$ decays – the $SU(3)$ counterparts of $B^\pm \to \pi^\pm K$ – play a key role,[51,58] allowing us to include the rescattering processes in the contours in the $\gamma - r$ plane and the associated constraints on γ completely (for alternative strategies, see Refs. 28 and 57). Since the "short-distance" expectation for the combined branching ratio $\mathrm{BR}(B^\pm \to K^\pm K)$ is $\mathcal{O}(10^{-6})$,[24] experimental studies of $B^\pm \to K^\pm K$ appear to be difficult. These modes have not yet been observed, and only upper limits for $\mathrm{BR}(B^\pm \to K^\pm K)$ are available.[19,54] However, rescattering effects may enhance this quantity significantly, and could thereby make $B^\pm \to K^\pm K$ measurable at future B-factories.[51,58] Another important indicator of large FSI effects is provided by $B_d \to K^+ K^-$ decays,[59] for which stronger experimental bounds already exist.[19,54]

In the case of the decays $B^+ \to \pi^+ K^0$ and $B_d^0 \to \pi^- K^+$, EW penguins contribute only through "colour-suppressed" topologies; estimates based on the "factorization" hypothesis typically give values for ϵ at the 1% level.[24] These crude estimates may, however, underestimate the role of the EW penguins.[50,56] An improved description is possible by using the general expressions for the EW penguin operators and performing appropriate Fierz transformations; on the way to controlling the corresponding uncertainties in the bounds on γ with the help of experimental data, the decay $B^+ \to \pi^+ \pi^0$ provides an important first step.[51]

3 Conclusions and Outlook

In conclusion, we have seen that certain non-leptonic B decays offer a fertile ground to test the Standard Model description of CP violation. Detailed experimental studies of these modes at B-factories are just ahead of us and may bring unexpected results, which could guide us to the physics lying beyond the Standard Model. The coming years will certainly be very exciting!

Acknowledgements

I would like to thank the organizers, in particular Professors Hai-Yang Cheng, Guey-Lin Lin and Hoi-Lai Yu, for inviting me to this stimulating workshop, for their kind hospitality, and for covering my local expenses. The travel support from the *Deutsche Forschungsgemeinschaft* is also gratefully acknowledged.

References

1. N. Cabibbo, *Phys. Rev. Lett.* **10**, 531 (1963); M. Kobayashi and T. Maskawa, *Progr. Theor. Phys.* **49**, 652 (1973).
2. C. Jarlskog, *Phys. Rev. Lett.* **55**, 1039 (1985); *Z. Phys.* **C29**, 491 (1985).
3. For reviews, see Y. Grossman, Y. Nir and R. Rattazzi, in *Heavy Flavours II*, eds. A.J. Buras and M. Lindner (World Scientific, Singapore, 1998) [hep-ph/9701231]; M. Gronau and D. London, *Phys. Rev.* **D55**, 2845 (1997); Y. Nir and H.R. Quinn, *Annu. Rev. Nucl. Part. Sci.* **42**, 211 (1992); R. Fleischer, CERN-TH/97-241 (1997) [hep-ph/9709291], to appear in the proceedings of the 7th International Symposium on Heavy Flavor Physics, Santa Barbara, California, 7-11 July 1997; L. Wolfenstein, *Phys. Rev.* **D57**, 6857 (1998).
4. L. Wolfenstein, *Phys. Rev. Lett.* **51**, 1945 (1983).
5. L.L. Chau and W.-Y. Keung, *Phys. Rev. Lett.* **53**, 1802 (1984); C. Jarlskog and R. Stora, *Phys. Lett.* **B208**, 268 (1988).
6. J.H. Christenson, J.W. Cronin, V.L. Fitch and R. Turlay, *Phys. Rev. Lett.* **13**, 138 (1964).
7. For a review, see A.J. Buras and R. Fleischer, in *Heavy Flavours II*, eds. A.J. Buras and M. Lindner (World Scientific, Singapore, 1998) [hep-ph/9704376].
8. L. Wolfenstein, *Phys. Rev. Lett.* **13**, 562 (1964).
9. G. Buchalla and A.J. Buras, *Phys. Rev.* **D54**, 6782 (1996); *Phys. Lett.* **B333**, 221 (1994); A.J. Buras, *Phys. Lett.* **B333**, 476 (1994).
10. For a review, see R. Fleischer, *Int. J. Mod. Phys.* **A12**, 2459 (1997).
11. G. Buchalla, A.J. Buras and M.E. Lautenbacher, *Rev. Mod. Phys.* **68**, 1125 (1996).
12. I. Dunietz, *Phys. Rev.* **D52**, 3048 (1995).
13. A.B. Carter and A.I. Sanda, *Phys. Rev. Lett.* **45**, 952 (1980); *Phys. Rev.* **D23**, 1567 (1981); I.I. Bigi and A.I. Sanda, *Nucl. Phys.* **B193**, 85 (1981).
14. OPAL Collaboration (K. Ackerstaff *et al.*), CERN-EP/98-001 (1998) [hep-ex/9801022]; CDF Collaboration (F. Abe *et al.*), FERMILAB-PUB-

98/189-E [hep-ex/9806025].
15. A.J. Buras, TUM-HEP-299/97 (1997) [hep-ph/9711217], to appear in the proceedings of the 7th International Symposium on Heavy Flavor Physics, Santa Barbara, California, 7-11 July 1997.
16. See, for instance, M. Gronau, *Phys. Lett.* **B300**, 163 (1993); J.P. Silva and L. Wolfenstein, *Phys. Rev.* **D49**, R1151 (1994); R. Aleksan *et al.*, *Phys. Lett.* **B356**, 95 (1995); F. DeJongh and P. Sphicas, *Phys. Rev.* **D53**, 4930 (1996); M. Ciuchini *et al.*, *Nucl. Phys.* **B501**, 271 (1997); P.S. Marrocchesi and N. Paver, *Int. J. Mod. Phys.* **A13**, 251 (1998).
17. A. Ali, G. Kramer and C.-D. Lü, DESY 98-056 (1998) [hep-ph/9805403].
18. J. Charles, LPTHE-Orsay 98-35 (1998) [hep-ph/9806468].
19. CLEO Collaboration (R. Godang *et al.*), *Phys. Rev. Lett.* **80**, 3456 (1998).
20. M. Gronau and D. London, *Phys. Rev. Lett.* **65**, 3381 (1990).
21. R. Fleischer, *Phys. Lett.* **B365**, 399 (1996).
22. S. Gardner, UK/TP 98-02 (1998) [hep-ph/9806423].
23. A. Buras and L. Silvestrini, TUM-HEP-315-98 (1998) [hep-ph/9806278].
24. For a recent study, see A. Ali, G. Kramer and C.-D. Lü, DESY 98-041 (1998) [hep-ph/9804363].
25. Y. Grossman and H.R. Quinn, *Phys. Rev.* **D58**, 017504 (1998).
26. H. Lipkin, Y. Nir, H. Quinn and A. Snyder, *Phys. Rev.* **D44**, 1454 (1991); A. Snyder and H. Quinn, *Phys. Rev.* **D48**, 2139 (1993).
27. A.J. Buras and R. Fleischer, *Phys. Lett.* **B360**, 138 (1995).
28. A.J. Buras, R. Fleischer and T. Mannel, CERN-TH/97-307 (1997) [hep-ph/9711262], to appear in *Nucl. Phys.* **B**.
29. R. Fleischer and T. Mannel, *Phys. Lett.* **B397**, 269 (1997).
30. B.F.L. Ward, UTHEP-97-1001 (1997) [hep-ph/9806310].
31. R. Fleischer, *Phys. Lett.* **B341**, 205 (1994); A.J. Buras and R. Fleischer, *Phys. Lett.* **B341**, 379 (1995).
32. Y. Grossman, G. Isidori and M.P. Worah, *Phys. Rev.* **D58**, 057504 (1998).
33. Y. Grossman and M.P. Worah, *Phys. Lett.* **B395**, 241 (1997); M. Ciuchini *et al.*, *Phys. Rev. Lett.* **79**, 978 (1997); D. London and A. Soni, *Phys. Lett.* **B407**, 61 (1997); G. Barenboim, J. Bernabeu and M. Raidal, *Phys. Rev. Lett.* **80**, 4625 (1998); D. Guetta, WIS-98/9 (1998) [hep-ph/9805274].
34. For a recent analysis, see A.L. Kagan and M. Neubert, CERN-TH/98-99 (1998) [hep-ph/9805303].
35. A.L. Kagan and M. Neubert, CERN-TH/98-1 (1998) [hep-ph/9803368].
36. For a recent calculation of $\Delta\Gamma_s$, see M. Beneke *et al.*, CERN-TH/98-261

(1998) [hep-ph/9808385].
37. R. Fleischer and I. Dunietz, *Phys. Rev.* **D55**, 259 (1997).
38. R. Fleischer and I. Dunietz, *Phys. Lett.* **B387**, 361 (1996).
39. R. Fleischer, CERN-TH/97-281 (1997) [hep-ph/9710331], to appear in *Phys. Rev.* **D**.
40. A.S. Dighe, I. Dunietz and R. Fleischer, CERN-TH/98-85 (1998) [hep-ph/9804253], to appear in *Eur. Phys. J.* **C**.
41. A.S. Dighe, I. Dunietz and R. Fleischer, *Phys. Lett.* **B433**, 147 (1998).
42. Y. Grossman, Y. Nir and M.P. Worah, *Phys. Lett.* **B407**, 307 (1997); Y. Grossman and H.R. Quinn, *Phys. Rev.* **D56**, 7259 (1997); Y. Grossman, B. Kayser, and Y. Nir, *Phys. Lett.* **B415**, 90 (1997); J. Charles *et al.*, *Phys. Lett.* **B425**, 375 (1998).
43. R. Enomoto and M. Tanabashi, *Phys. Lett.* **B386**, 413 (1996); S. Gardner, H.B. O'Connell and A.W. Thomas, *Phys. Rev. Lett.* **80**, 1834 (1998).
44. R. Enomoto, Y. Okada and Y. Shimizu, *Phys. Lett.* **B433**, 109 (1998).
45. I. Bediaga *et al.*, [hep-ph/9804222].
46. M. Gronau and D. Wyler, *Phys. Lett.* **B265**, 172 (1991); see also I. Dunietz, *Phys. Lett.* **B270**, 75 (1991).
47. D. Atwood, I. Dunietz and A. Soni, *Phys. Rev. Lett.* **78**, 3257 (1997).
48. M. Gronau, *Phys. Rev.* **D58**, 037301 (1998); Z.-Z. Xing, LMU-98-06 (1998) [hep-ph/9804434]; M. Gronau and J.L. Rosner, FERMILAB-PUB-98-227-T (1998) [hep-ph/9807447]; J.-H. Jang and P. Ko, KAIST-14/98 (1998) [hep-ph/9807496].
49. CLEO Collaboration (M. Athanas *et al.*), *Phys. Rev. Lett.* **80** (1998) 5493.
50. M. Gronau and J.L. Rosner, *Phys. Rev.* **D57**, 6843 (1998).
51. R. Fleischer, CERN-TH/98-60 (1998) [hep-ph/9802433], to appear in *Eur. Phys. J.* **C**.
52. F. Würthwein and P. Gaidarev, CALT-68-2153 (1997) [hep-ph/9712531].
53. R. Fleischer and T. Mannel, *Phys. Rev.* **D57**, 2752 (1998).
54. J. Alexander, plenary talk given at the 29th International Conference on High Energy Physics – ICHEP 98, Vancouver, Canada, 23-29 July 1998.
55. L. Wolfenstein, *Phys. Rev.* **D52**, 537 (1995); J.-M. Gérard and J. Weyers, UCL-IPT-97-18 (1997) [hep-ph/9711469]; D. Atwood and A. Soni, *Phys. Rev.* **D58**, 036005 (1998).
56. M. Neubert, *Phys. Lett.* **B424**, 152 (1998).
57. A.F. Falk *et al.*, *Phys. Rev.* **D57**, 4290 (1998).
58. R. Fleischer, CERN-TH/98-128 (1998) [hep-ph/9804319], to appear in *Phys. Lett.* **B**.
59. M. Gronau and J.L. Rosner, EFI-98-23 (1998) [hep-ph/9806348].

Onwards from Fast η' in B Decays: Large CP Violation from New Physics?

George W.S. Hou [a]
Brookhaven National Lab, Physics Department, Bldg. 510A
Upton, NY 11973, USA
E-mail: wshou@quark.phy.bnl.gov

Suppressed in SM, the best place for direct CP violation search is in the already observed charmless $b \to s$ modes. In SM with FSI phases, however, a_{CP} in $K\pi$ modes could be as large as 20–30% but have sign correlations between $K^-\pi^+/K^-\pi^0$ and $\bar{K}^0\pi^-$. New Physics effects such as enhanced color dipole bsg coupling could give $a_{CP} \sim$ 40–60% in $K\pi$ and ϕK modes that are distinguishable from FSI.

1 Motivation

1997 can be called the Year of the Strong Penguin: a handful of two-body charmless B decays were observed [1] by CLEO for the first time, giving firm indication that strong penguins are dominant. Something completely unexpected also emerged in η' modes: Not only exclusive modes are very sizable, semi-inclusive $B \to \eta' + X_s$ with fast η' was found [2] to be close to 10^{-3}.

We concern ourselves here with direct CP violation in these modes. The a_{CP} reach is only $\sim$ 100% at present. But in 2–5 years this could go down to 30% to 10%. The modes that are already observed so far would certainly be the most sensitive probes. What physics do they and can they probe?

What has been observed so far are charmless $b \to s$ decays. The ordering $K\pi > \pi\pi$ clearly indicates that strong penguin > tree. We recall that in SM, a_{CP} for pure penguin $b \to s$ transitions are suppressed by the factor

$$\mathrm{Im}\,(V_{us}V_{ub}^*)/|V_{cs}V_{cb}^*| \simeq \eta\lambda^2 < 1.7\%, \tag{1}$$

so $a_{CP} > 10\%$ in pure penguin modes would imply New Physics! We therefore have a discovery window in the next few years for beyond SM (BSM) effects. The question then is: What BSM? *Is large a_{CP} possible in $b \to s$ modes?* Rather than trying to be exhaustive, we wish to demonstrate that CP asymmetries in $b \to s$ transitions *can indeed be large* with simple extensions of SM, and sometimes even within SM.

New Physics will be illustrated with *large color dipole bsg* coupling

$$-\frac{G_F}{\sqrt{2}}\frac{g}{16\pi^2}V_{tb}V_{ts}^*\, c_8\, \bar{s}\sigma_{\mu\nu}G^{\mu\nu}m_b\,(1+\gamma_5)b. \tag{2}$$

[a] Permanent address: Dept. of Physics, National Taiwan Univ., Taipei, Taiwan, R.O.C.

In SM one finds $c_8^{\rm SM}(m_b) \simeq -0.3$ which leads to $b \to sg$ (with g on-shell) $\sim 0.2\%$, a small rate that is usually neglected. But because $b \to sg$ just does not give tangible signatures, our experimental knowledge of the strength of c_8 is actually rather poor. In fact, the long-standing "deficit" in B decay semileptonic branching ratio ($\mathcal{B}_{s.l.}$) and charm counting (n_C) point towards the possibility of sizable $b \to sg$ in Nature.[3,4] If $b \to sg \approx 10\%$, which implies $c_8 \sim 2$, $\mathcal{B}_{s.l.}$ and n_C can each be lowered by that amount and the problems would go away. A recent CLEO bound[5] gives $b \to sg < 7\%$ at 90% C.L., which comes indirectly from the study of $B \to D\bar{D}K + X$ decay. But even if one takes this seriously there is still much room, and $b \to sg$ at 1–5% would be very hard to rule out. What we stress here is: if c_8 is large in Nature, it must come from New Physics and carry naturally a KM-independent CP phase.

The idea of a large bsg color dipole coupling and its associated new CP phase has been applied to $B \to \eta' + X_s$ decay. We have advocated that the $g^*g\eta'$ anomaly coupling mechanism[6,7] is needed to account for the energetic η' (or equivalently, the recoil m_{X_s}) spectrum. Then, with new CP phase σ in $c_8 \cong 2e^{i\sigma}$ interfering with absorptive parts from usual c_{3-6} penguin coefficients, a_{CP} in inclusive m_{X_s} spectrum could be at 10% level.[7]

We explore here the general impact of a large color dipole coupling on CP asymmetries in charmless $b \to s$ decays. If $b \to sg$ rate is really of order 1–10% in Nature, even if this rate itself is hard to measure, other charmless $b \to s$ decays must be affected through interference effects.

2 Model of Unconstrained CP Phase

To have large $b \to sg$ and evade $b \to s\gamma$ constraint at the same time, one needs additional source for radiating gluons but not photons. Gluinos ($\tilde{g}$) fit the bill nicely. In SUSY one usually sets soft squark mass terms to be "universal" to suppress FCNC and to reduce the number of parameters. But it has been shown[4,8] that nonuniversal soft $m_{\tilde{d}_j}$ masses could give large c_8 without violating the $b \to s\gamma$ constraint. In previous studies, however, the possibility of new CP phases were not considered.

As an existence proof, let us consider a minimal model of $\tilde{s} - \tilde{b}$ mixing, the simplest would be LL mixing[9] which mimics SM couplings, but one could also have RR or LR mixing models.[10] The phase of d_i quarks are fixed by gauge interaction, and there is just one mixing angle θ and one phase ϕ. Since this mixing involves only the second and third generations, one evades low energy bounds that involve first generation quarks, such as neutron edm, the K system, and even B_d-$\bar{B}_d$ mixing. This is a natural model that is tailor made for generating large effects in $b \to s$ transitions (as well as B_s mixing)!

3 Direct CP Violation in Inclusive $b \to s\bar{q}q$ Decay

The theory of inclusive decays are cleaner since one can use the quark/parton language. The absorptive parts arrising from short distance perturbative rescattering [11] can be used and one is insensitive to long distance phases. However, experimental detection poses a challenge, unless partial reconstruction techniques can be made to work.

Since penguins dominate charmless $b \to s$ decays, one is interested in CP violation in pure penguin processes such as $b \to s\bar{d}d$ and $s\bar{s}s$. But since these rates and asymmetries occur at $\mathcal{O}(\alpha_S^2)$, care [12] has to be taken in treating CP violation in the $b \to s\bar{u}u$ mode, which has the distinction of receiving also the tree contribution. Although the tree amplitude alone does not lead to CP violation, while tree–penguin interference occurs only at $\mathcal{O}(\alpha_S)$, to be consistent with treating pure penguin CP asymmetries, one needs to take into account the absorptive part carried by the gluon propagator (bubble graph) associated with the penguin. This $\mathcal{O}(\alpha_S^2)$ tree–penguin interference term is needed to maintain CPT and unitarity in rates and hence a_{CP}.

The above discussion has been stated in terms of "full" theory (exact loop calculation) to lowest relevant order in α_S. Since QCD corrections are important and relatively well developed by now, we adopt an operator language in computing inclusive rates. We start from the effective Hamiltonian

$$H_{\rm eff} = \frac{4G_F}{\sqrt{2}}\left[V_{ub}V_{us}^*(c_1O_1 + c_2O_2) - V_{ib}V_{is}^*\, c_j^iO_j\right], \tag{3}$$

with i summed over u, c, t and j over 3 to 8. The operators are defined as

$$\begin{aligned}
&O_1 = \bar{u}_\alpha\gamma_\mu Lb_\beta\, \bar{s}_\beta\gamma^\mu Lu_\alpha, \quad O_2 = \bar{u}\gamma_\mu Lb\, \bar{s}\gamma^\mu Lu,\\
&O_{3,5} = \bar{s}\gamma_\mu Lb\, \bar{q}\gamma^\mu L(R)q, \quad O_{4,6} = \bar{s}_\alpha\gamma_\mu Lb_\beta\, \bar{q}_\beta\gamma^\mu L(R)q_\alpha,\\
&\tilde{O}_8 = \frac{\alpha_s}{4\pi}\, \bar{s}i\sigma_{\mu\nu}T^a\frac{m_bq^\nu}{q^2}Rb\, \bar{q}\gamma^\mu T^a q,
\end{aligned} \tag{4}$$

where $\tilde{O}_8$ arises from the dimension 5 color dipole O_8 operator of Eq. (2), and $q = p_b - p_s$. We have neglected electroweak penguins for simplicity. The Wilson coefficients c_j^i are evaluated to NLO order in regularization scheme independent way, for $m_t = 174$ GeV, $\alpha_s(m_Z^2) = 0.118$ and $\mu = m_b = 5$ GeV. Numerically, [13] $c_{1,2} = -0.313,\ 1.150$, $c_{3,4,5,6}^t = 0.017,\ -0.037,\ 0.010,\ -0.045$, and $c_8^{\rm SM} = c_8^t - c_8^c = -0.299$. We note that $c_{1,2}$ are resummations of series starting at $\mathcal{O}(\alpha_S^0)$, while c_{3-6} start at $\mathcal{O}(\alpha_S^1)$ which is reflected in their relative smallness. However, one power of α_S is factored out by convention in defining O_8, hence c_8 starts at $\mathcal{O}(\alpha_S^0)$ and its size is comparable to c_1 within SM. One has to keep track of the *relevant leading order* in α_S when comparing with "full theory" approach discussed earlier.

To get absorptive parts, we add $c_{4,6}^{u,c}(q^2) = -Nc_{3,5}^{u,c}(q^2) = -P^{u,c}(q^2)$ for u, c quarks in the loop, where

$$P^{u,c}(q^2) = \frac{\alpha_s}{8\pi} c_2 \left(\frac{10}{9} + G(m_{u,c}^2, q^2) \right),$$

and

$$G(m^2, q^2) = 4 \int x(1-x) dx \ln \frac{m^2 - x(1-x)q^2}{\mu^2}.$$

To respect CPT/unitarity at $\mathcal{O}(\alpha_S^2)$, for c_{3-8}^t at $\mu^2 = q^2 < m_b^2$, we substitute

$$\mathrm{Im}\, c_8 = \frac{\alpha_s}{4\pi} c_8 \sum_{u,d,s,c} \mathrm{Im}\, G(m_i^2, q^2),$$

$$\mathrm{Im}\, c_{4,6}^t = -N \mathrm{Im}\, c_{3,5}^t = \frac{\alpha_s}{8\pi} \left[c_3^t \mathrm{Im}\, G(m_s^2, q^2) + (c_4^t + c_6^t) \sum_{u,d,s,c} \mathrm{Im}\, G(m_i^2, q^2) \right],$$

when interfering with the tree amplitude. We note that the use of operator language can be misleading at this stage since the absorptive parts are not resummed while the Wilson coefficients are. One could easily lose track of α_S counting that is needed for maintaining CPT/unitarity if one thinks too naively in effective theory language.

Having made all these precautions, we can square amplitudes in a straightforward manner to obtain rates and arrive at the asymmetries. Since at lower order one has $b \to sg$ decay, the $|c_8|^2$ term has a $\log q^2$ pole behavior. We regulate it by simply cutting it off at 1 GeV. Fig. 1(a) gives the rates for $b \to sd\bar{d}$ (solid) and $\bar{b} \to \bar{s}d\bar{d}$ (dashed) vs. y= q^2/m_b^2. The SM result does not show a prominent low q^2 tail since $b \to sg$ is small, and the asymmetry comes mostly from below $c\bar{c}$ threshold. For larger q^2 the a_{CP} is GIM suppressed.[12] The SM asymmetry is indeed tiny. For new physics enhanced $c_8 = 2e^{i\sigma}$, we consider the cases for $\sigma = \pi/4$, $\pi/2$ and $3\pi/4$. Besides a very prominent low q^2 tail since $b \to sg$ is now $\sim 10\%$, the salient feature is the rather large rate asymmetries above $c\bar{c}$ threshold. The reason is because the new physics σ phase now evades the SM constraint of Eq. (1), and the c_8 amplitude interferes with standard c_{3-6} penguins which carry the absorptive parts due to (perturbative) $c\bar{c}$ rescattering, but the $u\bar{u}$ rescattering absorptive part is suppressed by V_{ub}. Note that for $\sigma = \pi/4$, $3\pi/4$ one has constructive, destructive interference, respectively. For the latter case, the overall rate is close to SM but the asymmetries are much larger, reaching 30% for large q^2!

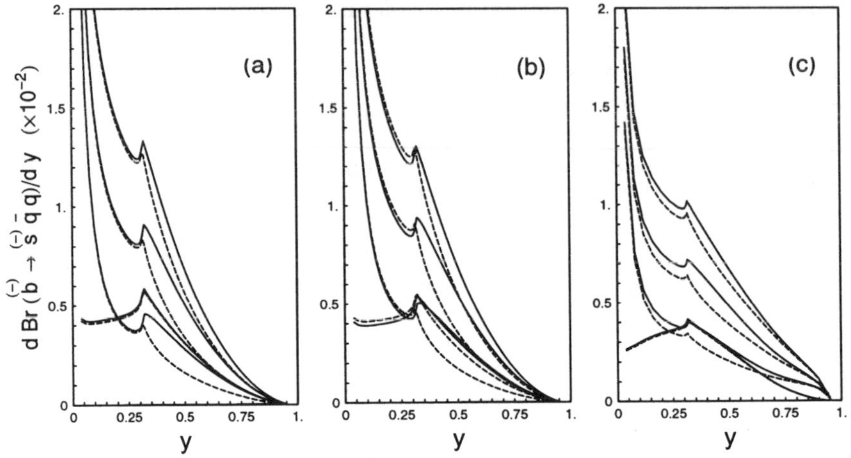

Figure 1: Inclusive rate vs. y= q^2/m_b^2 for (a) $b \to s\bar{d}d$, (b) $s\bar{u}u$ and (c) $s\bar{s}s$ decays (solid) and $\bar{b}$ decays (dashed). Curves with prominent small y tail are for $c_8 = 2e^{i\sigma}$ with $\sigma = \pi/4$ (top), $\pi/2$ (middle), $3\pi/4$ (bottom), while the other is SM result.

For $b \to s\bar{u}u$ the tree diagram also contributes, and one has to include the absorptive part in gluon propagator as discussed earlier. Because of this, the rate asymmetries in SM occur both below and above $c\bar{c}$ threshold, as can be seen in Fig. 1(b). Each are larger than the $b \to s\bar{d}d$ case but are of opposite sign hence they tend to cancel each other. [12] If c_8 is enhanced, however, the dominant mechanism is again interference between c_8 and the usual penguins, hence the results are similar to the $b \to s\bar{d}d$ case. For $b \to s\bar{s}s$ mode, one has to take into account identical particle effects. As seen in Fig. 1(c), this leads to the peculiar shapes at large and small q^2, and the asymmetry is now smeared over all q^2. But otherwise it is similar to the $b \to s\bar{d}d$ case.

The integrated inclusive results are summarized in Table 1.

Table 1: Inclusive Br (in 10^{-3})/a_{CP} (in %) for SM and for $c_8 = 2e^{i\sigma}$.

	SM	$\sigma =$ 0	$i\pi/4$	$i\pi/2$	$i3\pi/4$	$i\pi$
$b \to sdd$	2.6/0.8	8.5/0.4	7.6/3.4	5.2/6.5	2.9/8.1	1.9/0.5
$b \to s\bar{u}u$	2.4/1.4	8.1/-0.2	7.5/2.6	5.5/5.6	3.2/8.1	2.0/3.5
$b \to s\bar{s}s$	2.0/0.9	6.9/0.4	6.2/3.2	4.4/6.0	2.6/7.1	1.8/0.4

4 Direct CP Violation in Exclusive Charmless Hadronic Modes

The exclusive modes are more accessible experimentally, as evidenced by the handful of observed modes. Unfortunately, the theory is not clean. One has to evaluate all possible hadronic matrix elements of operators in Eq. (3). Faced with CLEO data, it has become popular[14] to use $N_{\rm eff}$ rather than the value of 3 as dictated by QCD. Although it is a measure of deviation from factorization, it becomes in reality a new process dependent fit parameter. One is still subject to usual approximations and inprecise knowledge of form factors and the q^2 value to take. CP asymmetries are especially sensitive to the latter. At the rate level, the $K\pi$ modes are approximately manageable, but the $\eta' K$ and ωK modes seem high while the ϕK mode seems low. Thus, even introducing $N_{\rm eff}$ as a new parameter, there are problems everywhere already in rate. A new development[15] in 1998 is that the $B^- \to K^-\pi^0$ mode has been observed, while the $B^- \to \bar{K}^0\pi^-$ rate came down considerably. One has now three measured $K\pi$ modes and their rates are all around 1.4×10^{-5}.

We shall not discuss the η' modes here since it must have a large contribution from anomaly mechanism and is rather difficult to treat. But we do wish to explore whether an enhanced c_8 could improve agreement with experiment. Before we do so, however, we point out that the $K\pi$ modes offer a rather interesting subtlety: they in general have two isospin components and exhibit larger a_{CP} within SM, and they are very sensitive to final state interaction (FSI) phases. As shown in Ref. [12] but now put in terms of the angle γ, in the absence of FSI phases one finds for $B^- \to K^-\pi^0$ mode

$$a_{uu} \propto \frac{\#_1 \sin\gamma}{|\#_2 - \#_3 \cos\gamma|^2}, \tag{5}$$

where $\#_1$ comes from interference, while $\#_2$ and $\#_3$ come from penguin and tree $b \to s\bar{u}u$ amplitudes, respectively. $\#_3$ and the dispersive part of $\#_2$ have the same sign. At the time of Ref. [12], $\cos\gamma < 0$ was favored, while $\sin\gamma$ was smaller than today, hence a_{uu} was not very large. The present[16] preferred value is $\gamma \sim 64^\circ$, however, and one now has destructive rather than constructive interference. Hence, one not only gains from $\sin\gamma \sim 0.9$ in the numerator, there is also extra enhancement from the denomenator of Eq. (5), and a_{uu} as large as 10% is possible. Furthermore, since one can write $(\bar{s}u)(\bar{u}b) = [(\bar{s}u)(\bar{u}b) + (\bar{s}d)(\bar{d}b)]/2 + [(\bar{s}u)(\bar{u}b) - (\bar{s}d)(\bar{d}b)]/2$, there is in general two isospin components from the tree level O_1 and O_2 operators. These two isospin amplitudes may develope soft FSI phases that are different from each other. If such is the case, it could overrun the perturbatively generated (hence α_S suppressed) absorptive phases, and much larger CP asymmetries can be achieved. While this is good news for CP violation search in general, it is bad news for search of new physics. Can one distiniguish between new physics effects and SM with large FSI phases? The answer is yes, if one compares several modes.

Let us give a little more detail for sake of illustration. We separate the $\bar{B}^0 \to K^-\pi^+$ amplitue into two isospin components, $A = A_{1/2} + A_{3/2}$. Since color allowed amplitudes dominate, $N_{\rm eff} \simeq N = 3$. Defining $v_i = V_{is}^* V_{ib}$ and assuming factorization, we find,

$$A_{1/2} = i\frac{G_F}{\sqrt{2}} f_K F_0 \left(m_B^2 - m_\pi^2\right) \left\{ v_u \left[\frac{2}{3}\left(\frac{c_1}{N} + c_2\right) - \frac{r}{3}\left(c_1 + \frac{c_2}{N}\right)\right] \right.$$
$$\left. -v_j \left[\frac{c_3^j}{N} + c_4^j + \frac{2m_K^2}{(m_b - m_u)(m_s + m_u)}\left(\frac{c_5^j}{N} + c_6^j\right)\right] - v_t \frac{\alpha_s}{4\pi}\frac{m_b^2}{q^2} c_8 \tilde{S}_{\pi K} \right\}, \quad (6)$$

where $F_0 = F_0^{B\pi}(m_K^2)$ is a BSW form factor, $\tilde{S}_{\pi K} \sim -0.76$ is a complicated form factor normalized to F_0, and r is some ratio of $B \to K$ and $B \to \pi$ form factors and a measure of SU(3) breaking. For $A_{3/2}$ one sets c_{3-8}^j to zero and substitutes $2/3$, $-r/3 \longrightarrow 1/3$, $r/3$. The $K^-\pi^0$ mode is analogous, with modifications in $A_{3/2}$ and an overall factor of $1/\sqrt{2}$. The penguins contribute only to $A_{1/2}$, hence the naively pure penguin $B^- \to \bar{K}^0\pi^-$ amplitude has just Eq. (6) with $c_{1,2}$ set to zero. Note that the $c_{5,6}$ effects are sensitive to current quark masses because of effective density-density interaction. The absorptive parts for c_{3-8}^j are evaluated at $q^2 \approx m_b^2/2$ which favors large a_{CP}, but q^2 could be [12] as low as $m_b^2/4$. We plot in Fig. 2(a) and (b) the branching ratio (Br) and a_{CP} vs. angle γ. For $K^-\pi^{+,0}$, a_{CP} peaks at the sizable $\sim 10\%$ just at the currently favored [16] value of $\gamma \simeq 64°$. But for $\bar{K}^0\pi^-$, $a_{CP} \sim \eta\lambda^2$ is very small. We have used $m_s(\mu = m_b) \simeq 120$ MeV since it enhances the rates. With $m_s(\mu = 1\ \text{GeV}) \simeq 200$ MeV, the rates would be a factor of 2 smaller. We find $K^-\pi^+$, $\bar{K}^0\pi^-$, $K^-\pi^0 \sim 1.4,\ 1.6,\ 0.7 \times 10^{-5}$, respectively.

To illustrate the effect of FSI, we now write $A = A_{1/2} + A_{3/2}e^{i\delta}$ and plot in Fig. 2(c) and (d) the Br and a_{CP} vs. δ for $\gamma = 64°$. The rate is not very sensitive to δ which reflects penguin dominance, but a_{CP} can now reach 20%, even 30% for $K^-\pi^0$. We stress that the naively pure penguin $\bar{K}^0\pi^-$ mode is in fact also quite susceptible to FSI phases as it is the isospin partner of $K^-\pi^0$, which definitely receives tree contributions. The $B^- \to \bar{K}^0\pi^-$ mode can receive tree contributions through FSI rescattering. Comparing Fig. 2(b) and (d), a_{CP} in this mode can be *much larger* than the naive factorization result. However, the a_{CP} for $\bar{K}^0\pi^-$ and $K^-\pi^+$ are out of phase, hence, comparing the two cases can give information on the FSI phase δ.

For physics beyond SM such as $c_8 = 2e^{i\sigma}$, there are too many parameters and one needs a strategy. We set $N = 3$ and try to fit observed Br's with the phase σ, then find the preferred a_{CP}. Since the c_8 term now dominates, one

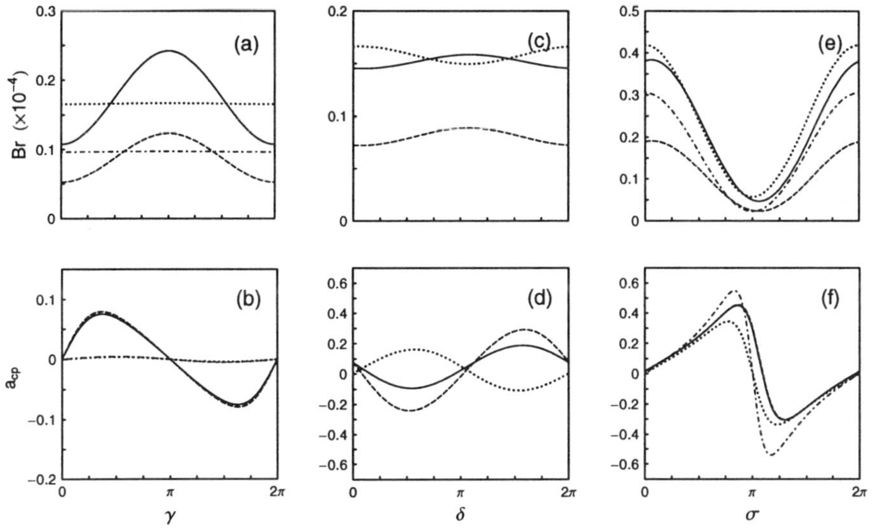

Figure 2: Br and a_{CP} vs. (a), (b) SM unitarity angle γ, (c), (d) FSI phase δ for $\gamma = 64°$, and (e), (f) new physics phase σ for $\gamma = 64°$ and $\delta = 0$. Solid, dotted, dashed and dotdashed lines are for $K^-\pi^+$, $\bar{K}^0\pi^-$, $K^-\pi^0$ and ϕK respectively.

is less sensitive to the FSI phase δ. In fitting Br's, we find that *destructive interference is necessary* which can be understood from the inclusive results of Fig. 1. This means that *large a_{CP}s are preferred!* We plot in Fig. 2(e) and (f) the Br and a_{CP} vs. the new physics phase σ, for $\gamma = 64°$ and $\delta = 0$. The $K^-\pi^+$ and $\bar{K}^0\pi^-$ modes are very close in rate for $\sigma \sim 45° - 180°$, but the $K^-\pi^0$ mode remains a factor of 2 smaller. However, the a_{CP} can now reach 50% for $K^-\pi^+/\bar{K}^0\pi^-$ and 40% for $K^-\pi^0$! These are *truly large asymmetries* and would be easily observed, perhaps even before B Factories turn on (i.e. with CLEO II.V data!). They are in strong contrast to the SM case with FSI phase δ, Fig. 2(d), and can be distinguished.

Genuine pure penguin processes arising from $b \to s\bar{s}s$ give cleaner probes of new physics CP violation effects since they are insensitive to FSI phase. The amplitude for $B^- \to \phi K^-$ decay is

$$A(B \to \phi K) \simeq -i\frac{G_F}{\sqrt{2}} f_\phi m_\phi 2p_B \cdot \varepsilon_\phi F_1(m_\phi^2) \left\{ v_j \left(c_3^j + \frac{c_4^j}{N} + c_5^j \right)\bigg|_{q_2^2} + v_j \left(\frac{c_3^j}{N} + c_4^j + \frac{c_6^j}{N} \right)\bigg|_{q_1^2} + v_t \frac{\alpha_s}{4\pi} \frac{m_b^2}{q_1^2} c_8 \tilde{S}_{\phi K} \right\}. \quad (7)$$

The relevant q^2 is determined by kinematics: $q_1^2 = m_b^2/2$ as before, but for amplitudes without Fierzing $q_2^2 = m_\phi^2$. We have dropped color octet contributions and have checked that they are indeed small. Since the amplitude is pure penguin, c_8 should have no absorptive part. As shown in Fig. 2(a) and (b), the SM rate of $\sim 1 \times 10^{-5}$ is above the CLEO bound of 0.5×10^{-5} while a_{CP} is uninterestingly small. If we allow for new physics enhanced $c_8 = 2e^{i\sigma}$, one again needs destructive interference to match observed rate. The results are plotted in Fig. 2(e) and (f) vs. σ. The rate is lower than the $\bar{K}^0\pi^-/K^-\pi^+$ modes because it is not sensitive to $1/m_s$ and we have used a low m_s value to boost up $B \to K\pi$. The a_{CP} could now reach almost 60%, thanks to the destructive interference preferred by fitting the CLEO limit on rate. We note that the SM asymmetry for $B \to \phi K$ should be of order 1%.

One can now construct an attractive picture. We have noted that recent studies cannot explain the low $B \to \phi K$ upper limit within SM. If c_8 is enhanced by new physics and interferes destructively with SM, $B \to \phi K$ can be brought down to below 5×10^{-6}. The experimentally observed $\bar{K}^0\pi^- \simeq K^-\pi^+$ follows from c_8 dominance, and their rate $\sim 1.4 \times 10^{-5}$ which is 2–3 times larger than the ϕK mode suggests a low m_s value and slight tunings of BSW form factors. Around $\sigma \sim 145°$, the rates are largely accounted for, but a_{CP} for ϕK, $K^-\pi^+/K^-\pi^0$ and $\bar{K}^0\pi^-$ could be *enhanced to the dramatic values* of 55%, 45% and 35%, respectively, and *all of the same sign*. This is certainly distinct from the sign correlations of SM with FSI.

On the down side, within the scenario of strong penguin dominance, which includes the case of enhanced c_8, the $B^- \to K^-\pi^0$ rate is always about a factor of two smaller than the $K^-\pi^+$ mode, and we are unable to accommodate recent CLEO findings.[15,17] We are also barely able to accommodate $B \to \omega K$. Within SM one needs $1/N_{\rm eff} \sim 1$ to be able to account for the large $B \to \omega K \simeq 1.5 \times 10^{-5}$ value, while for $1/N_{\rm eff} \simeq 0$ one can account for only half. Adding new physics induced $c_8 = 2e^{i\sigma}$ effect, we are able to account for Br for both large and small $N_{\rm eff}$, but not for $N = 3$. However, a_{CP} is never more than a few % and hence not very interesting. Since the ωK mode has a single isospin amplitude, it is insensitive to FSI rescattering phases.

5 Discussion and Conclusion

We have emphasized that the $b \to s$ modes that are already observed are the best places for CP search. For discussion on the $B \to K^*\gamma$ and $b \to s\gamma$ modes, the first ever observed penguins, please see the talk of Chua.[10,18] An enhanced c_8 could also lead to difference in $\sin 2\beta$ value measured in $B \to J/\psi K^{*0}$ vs. $J/\psi K_S$ decays. Please see the talk of He for further details.[9,19]

We have used the viable notion of enhanced c_8 with New Physics CP phase for sake of illustration. We recall that B physics has had its share of surprises, such as the long b lifetime, large B_d mixing, and the recent case of prominent $B \to \eta' K$, ωK and inclusive fast $\eta' + X_s$ modes. We therefore must be on our guard that CP violation could be surprisingly large in a variety of B decays. Charmless $b \to s$ transitions offer a unique early window, and *large CP asymmetries may just pop up everywhere* as B Factories turn on!

Seek, and ye shall find! (Faith [b] needed.)

Acknowledgments

I thank Chun-Khiang Chua, Xiao-Gang He, Ben Tseng and Kwei-Chou Yang for collaborative work, and R.O.C. National Science Council for support.

References

1. See e.g. K. Lingel, T. Skwarnicki, and J.G. Smith, hep-ex/9804015.
2. T.E. Browder et al. (CLEO Collab.), Phys. Rev. Lett. **81**, 1786 (1998).
3. B.G. Grządkowski and W.S. Hou, Phys. Lett. B**272**, 383 (1991).
4. A.L. Kagan, Phys. Rev. D**51**, 6196 (1995).
5. T. Coan et al. (CLEO Collab.), Phys. Rev. Lett. **80**, 1150 (1998).
6. D. Atwood and A. Soni, Phys. Lett. B**405**, 150 (1997).
7. W.S. Hou and B. Tseng, Phys. Rev. Lett. **80**, 434 (1998).
8. M. Ciuchini, E. Gabrielli and G. Giudice, Phys. Lett. **B 388** (1996) 353.
9. X.G. He and W.S. Hou, hep-ph/9804293.
10. C.K. Chua, X.G. He and W.S. Hou, hep-ph/9808431.
11. M. Bander, D. Silverman and A. Soni, Phys. Rev. Lett. **43**, 242 (1979).
12. J.-M. Gérard and W.S. Hou, Phys. Rev. Lett. **62**, 855 (1989); Phys. Rev. D**43**, 2909 (1991).
13. N.G. Deshpande and X.G. He, Phys. Lett. B**336**, 471 (1994).
14. See e.g. H.Y. Cheng, hep-ph/9807323; and references therein.
15. M. Artuso et al. (CLEO Collab.), CLEO CONF 98-20.
16. F. Parodi, P. Roudeau and A. Stocchi, hep-ph/9802289.
17. If we ignore $B \to \phi K$ and the notion of enhanced c_8, the near equality of the three observed $K\pi$ modes can be accommodated within SM by inclusion of electroweak penguin effects and tuning γ. See N.G. Deshpande et al., hep-ph/9809329.
18. C.K. Chua, this proceedings.
19. X.G. He, this proceedings.

[b] Large CP violation shall be discovered by those who have faith as such.

Physics with $B \to \eta'$ and $B \to J/\psi$

Xiao-Gang He
Department of Physics, National Taiwan University, Taipei, 10764, Taiwan

In this talk I discuss two topics related to B decays. I will first discuss the Standard Model prediction for $B \to \eta' X_s$, and then discuss how to use experimental information from $B \to J/\psi K^*$ to test CP violation beyond the SM.

I will report some of my recent work on $B \to \eta' X_s$ and $B \to J/\psi K^*$ with my collaborators [1,2,3] in this talk. The processes $B \to \eta' X_s$ and $B \to J/\psi K^*$ have been analyzed in detail by the CLEO collaboration [4,5]. The unexpected large branching ratio of [4] $(6.2 \pm 1.6 \pm 1.3) \times 10^{-4}$ for $B \to \eta' X_s$ has led some to claim that new physics is needed to explain the observation by CLEO [6,7,8]. I will show that in fact the Standard Model prediction is consistent with experimental data when related processes are treated properly [1]. The measurement of the full angular analysis of [5] $B \to J/\psi K^*$ from CLEO has also led to some interesting physics. I will show that information obtained from the angular distribution for this decay provide good test for CP violation beyond the SM [2,3].

1 $B \to \eta' X_s$ in the Standard Model

The recent observation of $B \to \eta' X_s$ [4] and $B \to \eta' K$ [9] decays with high momentum η' has stimulated many theoretical activities [6,7,8,10,11,12,13,14]. One of the mechanisms proposed to account for this decay is $b \to sg^* \to sg\eta'$ [6,10] where the η' meson is produced via the anomalous $\eta' - g - g$ coupling. According to a previous analysis [6], this mechanism within the Standard Model(SM) can only account for 1/3 of the measured branching ratio [4]. There are also other calculations of $B \to \eta' X_s$ based on four-quark operators of the effective weak-Hamiltonian [7,11]. These contributions to the branching ratio, typically 10^{-4}, are also too small to account for $B \to \eta' X_s$, although the four-quark-operator contribution is capable of explaining the branching ratio for the exclusive $B \to \eta' K$ decays [12,13]. These results have inspired proposals for an enhanced $b \to sg$ and other mechanisms arising from physics beyond the SM [6,7,8]. It will be shown here that the SM is in fact consistent with experimental data from CLEO [1].

The quark level effective Hamiltonian for the $B \to \eta' X_s$ decay is given by [15]:

$$
\begin{aligned}
H_{eff}(\Delta B = 1) &= \frac{G_F}{\sqrt{2}}[\sum_{f=u,c} V_{fb}V_{fs}^*(C_1(\mu)O_1^f(\mu) + C_2(\mu)O_2^f(\mu)) \\
&- V_{ts}^*V_{tb}\sum_{i=3}^{6}(C_i^{eff}(\mu)O_i(\mu) + C_8^{eff}(\mu)O_8(\mu))], \qquad (1)
\end{aligned}
$$

The operators are defined in Ref.[15]. C_i are the Wilson Coefficients (WC). The values for the WC's given in Ref.[16,17] will be used in this discussion. The superscript ^{eff} indicate that the matrix element corrections are also included.

Let me first discuss the four-quark operator contributions to $B \to \eta' X_s$. The four-quark operators can induce three types of processes represented by 1) $< \eta'|\bar{q}\Gamma_1 b|B >< X_s|\bar{s}\Gamma_1' q|0 >$, 2) $< \eta'|\bar{q}\Gamma_2 q|0 >< X_s|\bar{s}\Gamma b|B >$, and 3) $< \eta' X_s|\bar{s}\Gamma_3 q|0 >< 0|\bar{q}\Gamma_3'|B >$. Here $\Gamma_i^{(')}$ denotes appropriate gamma matrices. The contribution from 1) gives a "three-body" type of decay, $B \to \eta' s\bar{q}$. The contribution from 2) gives a "two-body" type of decay $b \to s\eta'$. And the contribution from 3) is the annihilation type which is relatively suppressed and will be neglected. Several decay constants and form factors needed in the calculations are listed below:

$$
\begin{aligned}
&< 0|\bar{u}\gamma_\mu\gamma_5 u|\eta' >=< 0|\bar{d}\gamma_\mu\gamma_5 d|\eta' >= i f_{\eta'}^u p_\mu^{\eta'} \\
&< 0|\bar{s}\gamma_\mu\gamma_5 s|\eta' >= i f_{\eta'}^s p_\mu^{\eta'}, \quad < 0|\bar{s}\gamma_5 s|\eta' >= i(f_{\eta'}^u - f_{\eta'}^s)\frac{m_{\eta'}^2}{2m_s}, \\
&f_{\eta'}^u = \frac{1}{\sqrt{3}}(f_1\cos\theta_1 + \frac{1}{\sqrt{2}}f_8\sin\theta_8), \quad f_{\eta'}^s = \frac{1}{\sqrt{3}}(f_1\cos\theta_1 - \sqrt{2}f_8\sin\theta_8), \\
&< \eta'|\bar{q}\gamma_\mu b|B >= F_1^{Bq}(p_\mu^B + p_\mu^{\eta'}) + (F_0^{Bq} - F_1^{Bq})\frac{mB^2 - m_{\eta'}^2}{q^2}q_\mu, \\
&F_{1,0}^{Bq} = \frac{1}{\sqrt{3}}(\frac{1}{\sqrt{2}}\sin\theta F_{1,0}^{B\eta_8} + \cos\theta F_{1,0}^{B\eta_1}). \qquad (2)
\end{aligned}
$$

For the $\eta' - \eta$ mixing associated with decay constants above, I have used the two-angle parameterization. The numerical values of various parameters are obtained from Ref. [18] with $f_1 = 157$ MeV, $f_8 = 168$ MeV, and the mixing angles $\theta_1 = -9.1^0$, $\theta_8 = -22.1^0$. For the mixing angle associated with form factors, I used the one-angle parameterization with[18] $\theta = -15.4^o$, since these form factors were calculated in that formulation [11,7]. In the latter discussion of $b \to sg\eta'$, I shall use the same parameterization in order to compare our results with those of earlier works [10,6]. For form factors, I assume that $F^{B\eta_1} = F^{B\eta_8} = F^{B\pi}$ with dipole and monopole q^2 dependence for F_1 and F_0, respectively. I used the running mass $m_s \approx 120$ MeV at $\mu = 2.5$ GeV and $F^{B\pi} = 0.33$ following Ref.[13].

Using $V_{ts} = 0.038$, $\gamma = 64^0$ and $\mu = 5$ GeV, it is found that the branching ratios in the signal region $p_{\eta'} > 2.2$ GeV ($m_X < 2.35$ GeV) are given by

$$B(b \to \eta' s) = 0.9 \times 10^{-4}, \;\; B(B \to \eta' s\bar{q}) = 0.1 \times 10^{-4}. \tag{3}$$

The branching ratio can reach 2×10^{-4} if all parameters take values in favor of $B \to \eta' X_s$. Clearly the mechanism by four-quark operator is not sufficient to explain the observed $B \to \eta' X_s$ branching ratio.

I now turn to $b \to \eta' sg$ through the QCD anomaly. To see how the effective Hamiltonian in Eq. (1) can be applied to calculate this process, let me rearrange the effective Hamiltonian such that

$$\begin{aligned}\sum_{i=3}^{6} C_i O_i &= (C_3^{eff} + \frac{C_4^{eff}}{N_c})O_3 + (C_5^{eff} + \frac{C_6^{eff}}{N_c})O_5 \\ &-2(C_4^{eff} - C_6^{eff})O_A + 2(C_4^{eff} + C_6^{eff})O_V + C_8^{eff} O_8,\end{aligned} \tag{4}$$

where

$$O_A = \bar{s}\gamma_\mu(1-\gamma_5)T^a b \sum_q \bar{q}\gamma^\mu\gamma_5 T^a q, \;\; O_V = \bar{s}\gamma_\mu(1-\gamma_5)T^a b \sum_q \bar{q}\gamma^\mu T^a q. \tag{5}$$

Since the light-quark bilinear in O_V carries the quantum number of a gluon, one expects [10] O_V give contribution to the $b \to sg^*$ form factors. In fact, by applying the QCD equation of motion : $D_\nu G_a^{\mu\nu} = g_s \sum \bar{q}\gamma^\mu T^a q$, one obtains, $O_V = (1/g_s)\bar{s}\gamma_\mu(1-\gamma_5)T^a b D_\nu G_a^{\mu\nu}$. The effective $b \to sg^*$ vertex can be written as

$$\Gamma_\mu^{bsg} = -\frac{G_F}{\sqrt{2}} V_{ts}^* V_{tb} \frac{g_s}{4\pi^2}(\Delta F_1 \bar{s}(q^2\gamma_\mu - \not{q}\, q_\mu)LT^a b - iF_2 m_b \bar{s}\sigma_{\mu\nu}q^\nu RT^a b). \tag{6}$$

The form factors ΔF_1 and F_2 are defined according to the convention in Ref. [6]. One obtains

$$\Delta F_1 = \frac{4\pi}{\alpha_s}(C_4^{eff}(\mu) + C_6^{eff}(\mu)), \;\; F_2 = -2C_8^{eff}(\mu) \tag{7}$$

Note that the relative sign of ΔF_1 and F_2 obtained agree with those in Refs. [6,7] and [19] which results in a destructive interference.

At the hadronic level the anomalous $\eta' - g - g$ coupling is given by: $a_g(\mu)\cos\theta\epsilon_{\mu\nu\alpha\beta}q^\alpha k^\beta$ with $a_g(\mu) = \sqrt{N_F}\alpha_s(\mu)/\pi f_{\eta'}$, q and k the momenta of the two gluons. Using the WC's at $\mu = 5$ GeV, the branching ratio is found to be $B(b \to sg\eta') = 5.6 \times 10^{-4}$ with a cut on $m_X \equiv \sqrt{(k+p')^2} \leq 2.35$ GeV. The spectrum for $dB(b \to sg\eta')/dm_X$ is depicted in Fig. 1. The peak of the spectrum corresponds to $m_X \approx 2.4$ GeV. The destructive interference between F_1 and F_2 lowers down the branching ratio by about 14% which is quite different from the results obtained in Refs.[6,10] because our ΔF_1 is larger than theirs because the inclusion of the matrix element corrections.

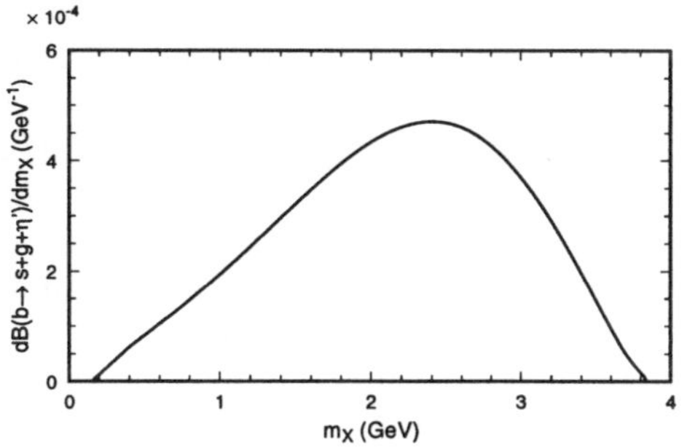

Figure 1: The distribution of $B(b \to s + g + \eta')$ as a function of the recoil mass m_X.

In the above calculation, $a_g(\mu)$ of the $\eta' - g - g$ vertex is treated as a constant independent of invariant-masses of the gluons, and μ is set to be 5 GeV. In practice, $a_g(\mu)$ should behave like a form-factor which becomes suppressed as the gluons attached to it go farther off-shell [7]. It is possible that the branching ratio obtained gets reduced significantly by the form-factor effect in $\eta' - g - g$ vertex. Should a large form-factor suppression occur, the additional contribution from $b \to \eta' s$ and $B \to \eta' s\bar{q}$ discussed earlier will become crucial. I however would like to stress that the estimate of $b \to sg\eta'$ with α_s evaluated at $\mu = 5$ GeV is conservative. To illustrate this, let me compare branching ratios for $b \to sg\eta'$ obtained at $\mu = 5$ GeV and $\mu = 2.5$ GeV respectively. The branching ratios at the above two scales with the kinematic al cut on m_X are 4.9×10^{-4} and 8.5×10^{-4} respectively. One can clearly see the significant scale-dependence! With the enhancement resulting from lowering the renormalization scale, there seems to be some room for the form-factor suppression in the attempt of explaining $B \to \eta' X_s$ by $b \to sg\eta'$.

It is clear that the Standard Model prediction for $B \to \eta' X_s$ is not in conflict with experimental data from CLEO.

2 $B \to J/\psi K^*$ and Test for CP Violation Beyond SM

The CLEO Collaboration has recently reported [5] the first full angular analysis of $B \to J/\psi K^{*0}$ decays. They find that the P wave component is small, $|P|^2 = |A_T|^2 = 0.16 \pm 0.08 \pm 0.04$, and final state interaction (FSI) phases $\phi(A_T) = -0.11 \pm 0.46 \pm 0.03\ rad$, and $\phi(A_{||}) = 3.00 \pm 0.37 \pm 0.04\ rad$ which are consistent with zero or π FSI phases in the convention where $\phi(A_0) = 0$. The small value for $|P|^2$ shows that $B \to J/\psi K^{*0} \to J/\psi K_S \pi^0$ decay is dominated by CP-even final states. This makes it practical to use $B \to J/\psi K^* \to J/\psi K_S \pi^0$ to observing mixing induced CP violation and to measure $\sin 2\beta$ without invoking an angular analysis [20]. The difference between $\sin 2\beta$ measured from $B \to J/\psi K_S$ ($\sin 2\beta_{J/\psi K_s}$) and $B \to J/\psi K^*$ ($\sin 2\beta_{J/\psi K_S \pi^0}$) can provide good test for CP violation beyond the SM [2]. With increased luminosities at CLEO, B-factories and other facilities, the charges of the B aand K^* can be identified, and therefore it is also possible to study direct CP violation in $B \to J/\psi K^*$ decays [3].

To proceed, let me show how $\sin 2\beta_{J/\psi K_S}$ and $\sin 2\beta_{J/\psi K_S \pi^0}$ can be used to test CP violation beyond the SM. The usual mixing induced CP violation measure is $\mathrm{Im}\,\xi = \mathrm{Im}\,\{(q/p)(A^* \bar{A}/|A|^2)\}$, where $q/p = e^{-2i\phi_B}$ is from B^0–$\bar{B}^0$ mixing, while A, $\bar{A}$ are B, $\bar{B}$ decay amplitudes into the same final state. For $B \to J/\psi K_S$, the final state is P-wave hence CP odd. Setting the weak phase to be σ_S, one has

$$\mathrm{Im}\,\xi(B \to J/\psi K_S) = -\sin(2\phi_B + 2\sigma_S) \equiv -\sin 2\beta_{J/\psi K_S}.$$

For $B \to J/\psi K^* \to J/\psi K_S \pi^0$, the final state has both P-amplitude (CP odd) and S- and D-amplitudes (CP even) (or their linear combinations $A_{||}$ and A_0). Let me denote their corresponding weak phases by σ_T, and $\sigma_{||}$ and σ_0, respectively. If $A_{||}$ and A_0 amplitudes have a common weak phase σ_*, one has

$$\begin{aligned}\mathrm{Im}\,\xi(B \to J/\psi K_S \pi^0) &= \mathrm{Im}\,\{e^{-2i\phi_B}[e^{-2i\sigma_T}|P|^2 - e^{-2i\sigma_*}(1-|P|^2)]\} \\ &\equiv -(1-2|P|^2)\sin 2\beta_{J/\psi K_S \pi^0}, \end{aligned} \tag{8}$$

In the SM one obtains the usual result of $\sin 2\beta_{J/\psi K_S} = \sin 2\beta_{J/\psi K_S \pi^0} = \sin 2\beta$. Clearly, both measurements provide true information about $\sin 2\beta$ within SM. However, this is no longer true if one goes beyond SM. Even for the case where the weak phases of $A_{||}$ and A_0 amplitudes are equal, if $\sigma_T \neq \sigma_*$ or if $\sigma_T, \sigma_* \neq \sigma_S$, then $\sin 2\beta_{J/\psi K_S} \neq \sin 2\beta_{J/\psi K_S \pi^0}$ follows.

To the first order in new weak phases, one has

$$\Delta_{KK^*} \approx (2\sigma_S - 2\sigma_* + \frac{2|P|^2}{1-2|P|^2}(\sigma_T - \sigma_*))\cos(2\phi_B). \tag{9}$$

If experiments will measure a non-zero Δ_{KK^*}. It is a signal for CP violation beyond the SM.

There are many ways where new physics may change the phases σ_i. To the lowest order they may arise from dimension 6 four quark operators and dimension 5 color dipole moment operators. New physics contributions from $C_L\bar{c}\gamma^\mu(1 \pm \gamma_5)c\bar{s}\gamma_\mu(1-\gamma_5)b$ type of four quark interactions are proportional to the dominant SM contribution in Eq.[1] which just generate a common weak phase for all the amplitudes and therefore Δ_{KK^*} is zero like the SM prediction. Δ_{KK^*} discussed here does not provide good test for new physics of this type. The interaction of the form $C_R\bar{c}\gamma^\mu(1\pm\gamma_5)c\bar{s}\gamma_\mu(1+\gamma_5)b$ will generate different phases for A_T and $A_{||,0}$. The reason is that the current $\bar{s}\gamma^\mu(1+\gamma_5)b$ contribution to A_T is proportional to $C_{SM}+C_R$, but to $A_{||,0}$ is proportional to $C_{SM}-C_R$. To a good approximation, $\delta_T = -\delta_{||,0}$. Δ_{KK^*} can be different from zero.

As an example, let us study R-parity violating supersymmetric models. In R-parity violating supersymmetric models, exchange of charged sleptons or down type squarks can generate non-zero C_R with an arbitrary phase δ_R. The allowed value for C_R is constrained from experimental data on $b \to s\gamma$. This still allow C_R term to contribute to $B \to J/\psi K^*$ at the amplitude level as large as 20% of the SM contribution. Stronger constraints can be obtained by assuming that $b \to c\bar{c}s$ is similar in strength to $b \to c\bar{u}s$ [21]. The upper bound of the weak phases are approximately given by

$$\delta_T = -\delta_* \approx 0.1 \sin\delta_R. \qquad (10)$$

Using Eq.(9), one can easily see that the difference Δ_{KK^*} can reach $-0.35\sin\delta_R\cos(2\phi_B)$. Such a large difference can be detected at B factories.

The contribution from dimension 5 color dipole operator has been estimated by assuming that color octet operators contribute the amount as determined in generalized factorization approximation. The magnitude of the color dipole coefficient as large as 10 times of the SM, if the chiral structure is the same as the SM one, is not ruled out and may in fact play some important role in the missing charm problem. The weak phases $\sigma_{T,||,0}$ can be as large as $0.08\sin\delta_c$, where δ_c is the weak phase of the color dipole phase [2]. The phases are approximately equal for σ_T and σ_*. The phase σ_S is suppressed by a factor of m^2_ψ/m^2_B due to helicity structure of the operator. The value for Δ_{KK^*} can be as large as $0.18\sin\delta_c\cos(2\phi_B)$. If it turns out that the color dipole interaction has opposite chirality, although there is an enhancement factor of $1/(1-2|P|^2)$, the coefficient C_R is more stringently constrained resulting in a smaller Δ_{KK^*}.

I now discuss direct CP violation in angular distribution for charge $B \to J/\psi K^*$. The full angular distribution for $B \to J/\psi K^*$ is given by

$$\frac{1}{\Gamma}\frac{d\Gamma}{d\cos\theta_{tr} d\cos\theta_{K^*} d\phi_{tr}} = \frac{9}{32\pi}\{2|A_0|^2\cos^2\theta_{K^*}(1-\sin^2\theta_{tr}\cos^2\phi_{tr})$$
$$+|A_{||}|^2\sin^2\theta_{K^*}(1-\sin^2\theta_{tr}\sin^2\phi_{tr}) + |A_T|^2\sin^2\theta_{K^*}\sin^2\phi_{tr}$$
$$-Im(A_{||}^*A_T)\sin^2\theta_{K^*}\sin 2\theta_{tr}\sin\phi_{tr}$$
$$+\frac{1}{\sqrt{2}}Re(A_0^*A_{||})\sin 2\theta_{K^*}\sin^2\theta_{tr}\sin 2\phi_{tr}$$
$$+\frac{1}{\sqrt{2}}Im(A_0^*A_T)\sin 2\theta_{K^*}\sin 2\theta_{tr}\cos\phi_{tr}\}, \quad (11)$$

where the transversity angles θ_{tr} and ϕ_{tr} are defined as polar and azimuth angles of the charge lepton in the J/ψ rest frame with x axis along the direction of K^*, x-y plane parallel to $K\pi$ plane. The angle θ_{K^*} is defined as that of the K in the rest frame of K^* relative to the negative of the J/ψ direction in that frame.

In the CLEO analysis, the FSI phase for A_0 was taken to be zero. For convenience I will use the convention that each amplitude A_i has both CP conserving FSI phase ϕ_i and CP violating phase σ_i as indicated before. With this convention, $A_j = |A_j|e^{i(\phi_j+\sigma_j)}$ while $\bar{A}_T = -|A_T|e^{i(\phi_T-\sigma_T)}$, $\bar{A}_{(||,0)} = |A_{(||,0)}|e^{i(\phi_{(||,0)}-\sigma_{(||,0)})}$.

It is clear that the coefficients $\alpha = -Im(A_{||}^*A_T)$, $\beta = Re(A_0^*A_{||})$, and $\gamma = Im(A_0^*A_T)$ of the last three terms in the angular distribution, and similarly $\bar{\alpha}$, $\bar{\beta}$ and $\bar{\gamma}$ for $\bar{B}$ decays, contain information about CP violation. Without separating B and $\bar{B}$ decays, however, which was the case for the CLEO analysis mentioned earlier, information on CP violation cannot be extracted. One must obtain the angular distributions for $B \to J/\psi K^*$ and $\bar{B} \to J/\psi \bar{K}^*$ decays separately and determine the coefficients for the interference terms in each case. The following three quantities then measure CP violation[3,22]

$$\begin{aligned} a_1 &= \alpha + \bar{\alpha} = 2|A_{||}||A_T|\cos(\phi_{||T})\sin(\sigma_{||T}), \\ a_2 &= \beta - \bar{\beta} = -2|A_{||}||A_0|\sin(\phi_{||0})\sin(\sigma_{||0}), \\ a_3 &= \gamma + \bar{\gamma} = -2|A_T||A_0|\cos(\phi_{T0})\sin(\sigma_{T0}), \end{aligned} \quad (12)$$

where $\phi_{ij} = \phi_i - \phi_j$ and $\sigma_{ij} = \sigma_i - \sigma_j$.

It is interesting to note that the CP violating observable $a_{1,3}$ do not require FSI phase differences and is especially sensitive to CP violating weak phases.

The present CLEO data on angular distributions which provide information about FSI phases are the charge averaged CP conserving quantities which are proportional to $\sin(\phi_{||T})\cos(\sigma_{||T})$, $\cos(\phi_{||0})\cos(\sigma_{||0})$, and $\sin(\phi_{T0})\cos(\sigma_{T0})$. At present these data do not exclude $a_{1,3}$ up to 50% and smaller for a_2

Again in the SM, the three CP violating observables are zero. They provide good tests for CP violation beyond the SM there it is possible to have non-zero values for a_i. For example in R-parity violating supersymmetric model considered earlier, for the central values for the magnitude of the amplitudes and assuming that the FSI are zero, the asymmetries can be as large as $a_1 = 0.10\sin\delta_R$ and $a_3 = -0.12\sin\delta_R$.

The color dipole interaction to these asymmetries are small because with the same chiral structure as the SM, the new phases are approximately equal as discussed before and therefore small asymmetries. For the case with opposite chiral structure as the SM the strength of the interaction is more stringently constrained, the phases are all small.

In all cases discussed above the weak phases for $\sigma_{||}$ and σ_0 are equal (or approximately equal). The asymmetry a_2 is approximately zero in all cases considered, and does not seem to be a good quantity to study for CP violation using this method.

The sensitivities for $a_{1,2}$ is similar to the sensitivity to the phase angles of the amplitudes. It is interesting to note that the systematic error in CLEO analysis is already as low as 0.03 [5]. With increases statistics, $a_{1,3}$ as large as 0.10 should be accessible at CLEO, at CDF, and at future B factories. Since the errors are determined through a fit, it is not clear how the statistical error scales with actual number of events. The question can only be answered by actual studies, but naive scaling implies that one would need 10^8 events to be able to distinguish the deviations given earlier. Nevertheless, the needed number of events may be less and measurement of the observables a_i will provide us with useful information about CP violation.

I thank W.-S. Hou and G.-L. Lin for collaborations on the work reported here. This work is supported in part by ROC National Science Council and by Australian Research Council.

1. Xiao-Gang He and Guey-Lin Lin, e-print hep-ph/9809204.
2. Xiao-Gang He and Wei-Shu Hou, e-print hep-ph/9804293.
3. Xiao-Gang He and Wei-Shu Hou, e-print hep-ph/9807507 (Phys. Rev. D in press).
4. CLEO Collaboration T. E. Browder *et al.*, Phys. Rev. Lett. 81, 1786(1998).
5. CLEO Collaboration C.P. Jessop et al., Phys. Rev. Lett. **79**, 4533 (1997).

6. W.S. Hou and B. Tseng, Phys. Rev. Lett. **80**, 434 (1998).
7. A. L. Kagan and A. Petrov, e-print hep-ph/9707354; A. L. Kagan, hep-ph/98-6266.
8. H. Fritzsch, Phys. Lett. **B 415**, 83 (1997); X.-G. He, W.-S. Hou and C.S. Huang, Phys. Lett. **B429**, 99 (1998).
9. CLEO collaboration, B.H. Behrens et al., Phys. Rev. Lett. **80**, 3710 (1998).
10. D. Atwood and A. Soni, Phys. Lett. **B 405**, 150 (1997).
11. A. Datta, X.-G. He, and S. Pakvasa, Phys. Lett. **B 419**, 369 (1998).
12. H.-Y. Cheng, and B. Tseng, e-print hep-ph/9803457; A. Ali, J. Chay, C. Greub and P. Ko, Phys. Lett. **B424**, 161 (1998); N. Deshpande, B. Dutta and S. Oh, Phys. Rev. **D57**, 5723(1998).
13. A. Ali, G. Kramer and C.-D. Lu, hep-ph/9804363.
14. I. Halperin and A. Zhitnitsky, Phys. Rev. Lett. **80**, 438 (1998); F. Araki, M. Musakonov and H. Toki, e-print hep-ph/9803356; D.S. Du, Y.-D. Yang and G.-H. Zhu, e-print hep-ph/9805451; M. Ahmady, E. Kou and A. Sugamoto, Phys. Rev. **D58**, 014015 (1998); D.S. Du, C.S. Kim and Y.-D. Yang, Phys. Lett. **B426**, 133 (1998); F. Yuan and K.-T. Chao, Phys. Rev. **D 56**, 2495 (1997); A. Dighe, M. Gronau and J. Rosner, Phys. Rev. Lett. **79**, 4333 (1997).
15. For a review see G. Buchalla, A. J. Buras and M. E. Lautenbacher, Review of Modern Physics, **68**, 1125 (1996).
16. N. G. Deshpande and X.-G. He, Phys. Lett. **B 336**, 471 (1994).
17. M. Ciuchini et al., Nucl. Phys. **B421**, 41(1994).
18. T. Feldmann, P. Kroll and B. Stech, hep-ph/9802409.
19. H. Simma and D. Wyler, Nucl. Phys. **B344**, 283(1990); A. Lenz, N. Nierste and G. Ostermier, Phys. Rev. **D56**, 7228(1997).
20. I. Dunietz et al., Phys. Rev. D**43**, 2193 (1991); A.S. Dighe et al., Phys. Lett. B**369**, 144 (1996).
21. Y. Grossman and M. Worah, Phys. Lett. B**395**, 241 (1997).
22. German Valencia, Phys. Rev. **D39**, 3339(1989).

ALTERNATIVE CP VIOLATION

David Bowser-Chao[(1)], Darwin Chang[(2)], and Wai-Yee Keung[(3)]
[(1)]*Physics Department, University of Texas at Austin, TX 78712-1081, USA*
[(2)]*NCST and Physics Department, National Tsing-Hua University Hsinchu 30043, Taiwan, R.O.C.*
[(3)]*Physics Department, University of Illinois at Chicago, IL 60607-7059, USA*

We briefly review some new classes of models of soft CP violation recently proposed by us and by Georgi and Glashow(GG). We emphasize their interesting consequences in the B physics. Both classes of models use softly broken CP symmetry to suppress tree level KM phase as well as the strong CP phase. The observed CP-violation in the kaon system is accounted for by the contribution due to a new heavy sector of scalars and vectorial fermions. The models can be roughly classified as the righted-handed in our case, or the left-handed in GG's case depending on how the chiral light quark sector relate to the new sector. The models can be milliweak or superweak in nature depending on the scale of the heavy sector.

Contribution to fourth Workshop on Particle Physics Phenomenology in Taiwan, presented by DC.

Introduction

While the standard model (SM) has been repeatedly confirmed by data from e^-e^+ and $p\bar{p}$ colliders, however, its neutrino sector as well as the CP violating sector (usually refered to as the Kobayashi-Maskawa[1] Model) have remained the least tested sectors of the theory. Recently, many new data on atmospheric as well as the solar neutrinos have given strong indication that there are indeed new physics beyond Standard Model. Therefore it is important that the many leptonic (BELLE, Babar, as well as CLEO) as well as hadronic (CDF, HERA-B) B factories probe the CP violating sector directly in search for new physics. It is also important, at the same time, for theorists to meditate on the potential alternative mechanism of CP violation and the corresponding new physics. With this perspective, models [2,3] of soft CP violation have been proposed recently as alternatives to the standard Kobayashi-Maskawa (KM) model [1]. The goal of these efforts is to mimic the successful CP phenomenology of the KM model and at the same time to solve the long-standing strong CP problem. Note that similar ideas in the same directions were presented by Barr[6] and his collaborators some time ago.

These models share the features of imposing soft (or spontaneous) CP violation in order to avoid tree level KM phase as well as the tree level strong CP θ phase. To account for the measured value of the CP-violating parameter

ϵ of the neutral kaon system, the models employ a new heavy sector of scalars and vectorial fermions. The first class of models, that we shall roughly classified as right-handed (RH) models [2], uses a heavy sector that couples only to right-handed down type quarks, which are $SU_L(2)$ singlets. The alternative left-handed (LH) models [3] use new particles that couple only to the ordinary left-handed quarks, which are $SU_L(2)$ doublets. This report is an expanded version of the short review of the phenomenology of these models in Ref.[4] with more discussions on B physics issues. An extensive presentation is under preparation [5].

The heavy sector of a typical model for our purposes requires two additional Higgs singlets, $h_\alpha(\alpha = 1,2)$ of charge q_h and a vectorial pair of heavy fermions, $Q_{L,R}$, of electromagnetic charge $-\frac{1}{3} + q_h$. One can also choose to have two pairs of fermions and only one heavy Higgs singlet. In addition, one can choose to assign the scalars or the fermions to be carrying the color such that together they will couple to the right-handed down type quarks, d_{Ri}. In the case of neutral, colorless fermions (the neutrinos), it is not even necessary to have vectorial pairs. One can also use the freedom in choosing charge q_h to avoids fractionally charged hadrons. Most of the phenomenology mention below are more or less independent of the choice of q_h and color. Relevant new terms in the Lagrangian are:

$$\begin{aligned}\mathcal{L}_{h_i} &= \left[(g\lambda_{i\alpha}\bar{Q}_L d_{iR} h_\alpha + M_Q \bar{Q}_L Q_R) + \text{h.c.}\right] \\ &\quad -(m^2)_{\alpha\beta} h_\alpha{}^\dagger h_\beta - \kappa_{\alpha\beta}(\phi^\dagger\phi - |\langle\phi\rangle|^2)\, h^\dagger_\alpha h_\beta \\ &\quad -\kappa'_{\alpha\beta\gamma\delta} h^\dagger_\alpha h_\beta h^\dagger_\gamma h_\delta\ , \end{aligned} \tag{1}$$

where ϕ is the Standard Model Higgs doublet. The soft breaking of CP symmetry implies a special basis where λ, κ, κ' and the SM Yukawa couplings are real. If fermions carry color, we also require (see below) that dim-3 couplings, M_Q, are real to avoid tree level contribution to θ. This leaves only a single CP violating parameter: $\text{Im}(m^2)_{12}$. We can diagonalize $(m^2)_{\alpha\beta}$ by a unitary matrix $U_{\alpha i}$ which in general is complex: $h_\alpha = U_{\alpha i} H_i$, with H_i the mass eigenstates. The quark-Higgs interaction in the mass eigenstate basis becomes

$$\mathcal{L}_{QqH} = g \sum_{q=d,s,b} \xi_{qj} (\bar{Q}_L q_R) H_j^- \ + \ \text{h.c.}\ , \tag{2}$$

where $H_i = U^*_{\alpha i} h_\alpha$ is the mass eigenstates, $\xi_{qj} \equiv \lambda_{q\alpha} U_{\alpha j}$. The Yukawa couplings ξ_{qj} are defined relative to the gauge coupling g of $SU_L(2)$. The rephasing-invariant measure of flavor-changing CP violation are

$$\mathcal{A}_{qq';ij} = \lambda_{q\alpha}\lambda_{q'\beta} U_{\beta i} U^*_{\alpha j} = \xi^*_{qi}\xi_{q'j} \tag{3}$$

with $(q, q' = d, s, b)$ and $i, j = 1, 2$. For flavor conserving amplitudes like electric dipole moments (EDM), we must also consider the coupling $\mathcal{B}_{ij} = \kappa_{\alpha\beta} U_{\beta i} U^*_{\alpha j}$. $\mathcal{A}_{qq;ij}$ as well as $\mathcal{B}_{ij}$ are Hermitian in indices $i.j$. Thus CP is broken only in the off-diagonal terms. As a result, they contribute to CP violation only when both the light and heavy charged Higgs are involved in a diagram at the lowest order. It is also possible to avoid the CP violation effect through coupling κ if one chooses to have only one heavy Higgs singlet and breaks CP symmetry in the dimension-3 heavy fermion mixing terms instead, as long as these heavy fermions are colorless. The contribution related to coupling κ' in the last term of (1) occurs only at the higher loop level and generally can be ignored.

Before continuing, we like to emphasize again that if the new fermions carry color it is necessary to impose CP symmetry on their bare masses also in order to avoid tree level strong CP problem. When there is more than one pair of vectorial fermions, one can potentially make the mass matrix hermitian, however that would require some additional symmetry to be implemented. Note also that while we employ soft CP breaking in our discussions, the CP breaking can easily be implemented spontaneously if one so desrires as shown in Ref.[2,4].

Constraint from ϵ

With CP broken only softly, the CKM matrix is real at tree level. Leading contribution to the CP violating parameter ϵ is due to the box diagram with only heavy particles in loop.

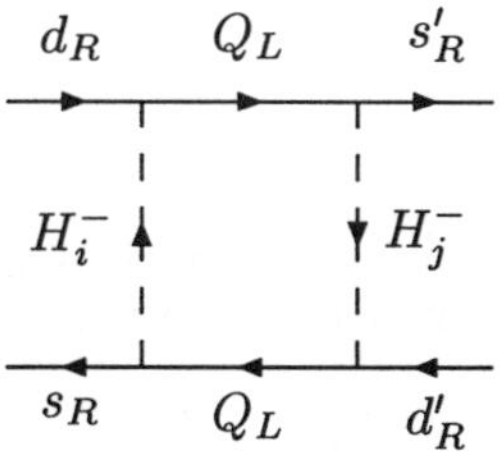

We evaluate all possible diagrams and match them with the low energy effective Hamiltonian which has the form

$$\mathcal{H}^{\Delta S=2} = \frac{G_F^2 m_W^2}{16\pi^2} \sum_{I=R,L} C^I_{\Delta S=2}(\mu) O^I_{\Delta S=2}(\mu) \ , \tag{4}$$

$$\text{with} \qquad O^{R,L}_{\Delta S=2} = \bar{s}\gamma_\mu(1 \pm \gamma_5) d \, \bar{s}\gamma^\mu(1 \pm \gamma_5) d \ . \tag{5}$$

The $W^{\pm}$ diagrams yield a purely real Wilson coefficient $C^{L}_{\Delta S=2}(\mu)$; CP violation is due solely to the operator $O^{R}_{\Delta S=2}$ rather than $O^{L}_{\Delta S=2}$, in contrast to the KM model, because the complex coefficient $C^{R}_{\Delta S=2}(\mu)$ is generated by the charged Higgs in RH models. At the scale $\mu = M_Q$, we have

$$C^{R}_{\Delta S=2}(M_Q) = 2\xi_{d1}\xi^{*}_{s1}\xi_{d2}\xi^{*}_{s2}\frac{m_W^2}{M_Q^2}\frac{f(x_2)-f(x_1)}{x_2-x_1} + \sum_{i=1,2}(\xi_{di}\xi^{*}_{si})^2\frac{m_W^2}{M_Q^2}\frac{df}{dx}(x_i)\ , \tag{6}$$

with $f(h) = (1-h)^{-2}(1+2h+h^2+h^2\ln h)$.

The real part of the diagram can contribute to part of the $K_L - K_s$ mass difference while the imaginary gives rise to ϵ. This is analyzed in detail in Ref. [2]. For illustration here, we can take the (decoupling) limit $m_2 \gg m_1$ and assume $m_1 = M_Q$ for simplicity. Demanding that the imaginary part of the box diagram gives enough contribution to ϵ while the corresponding real part gives just a fraction $\mathcal{F}$ of the mass difference Δm_K, we obtain[2] the constraints,

$$\mathrm{Im}\left(\mathcal{A}_{sd}/(0.058)^2\right)^2 R_Q^2 = 1\ , \tag{7}$$

$$\mathrm{Re}\left(\mathcal{A}_{sd}/(0.058)^2\right)^2 R_Q^2 = 156\mathcal{F}\ ; \tag{8}$$

where $R_Q = 300\ \mathrm{GeV}/M_Q$. The reasonable constraint $|\mathcal{F}| < 1$ can be easily satisfied. It is important to emphasize that, in RH models, the heavy particle box diagrams induce a right-handed four fermion operator, contrary to the LH models and the KM model in which the leading CP violating operators are left-handed.

Constraints from (ϵ'/ϵ)

The leading contribution to the direct CP violating parameter ϵ' is due to the gluonic penguin diagrams with only the heavy particles in the loop.

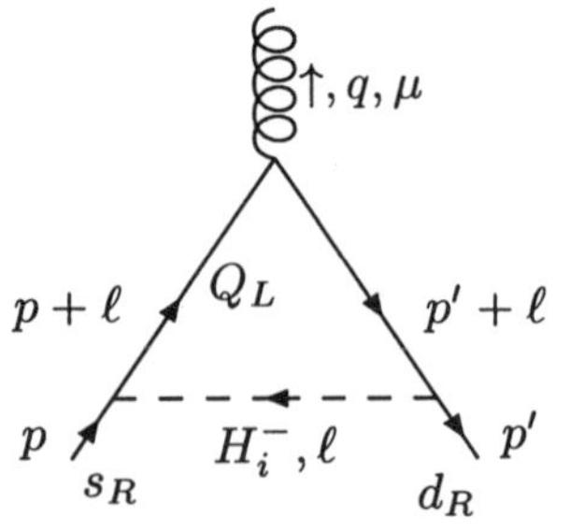

The CP violating piece of the effective Hamiltonian is parametrized as

$$\mathcal{H}^{\Delta S=1} = (G_F/\sqrt{2})\tilde{C}(\bar{s}T^a\gamma_\mu(1+\gamma_5)d) \times \sum_q (\bar{q}T^a\gamma^\mu q) \, .$$

At the electroweak scale, the Wilson coefficient is

$$\tilde{C} = -\alpha_s \sum_i \frac{\xi_{di}\xi^*_{si}}{6\pi} \frac{m_W^2}{M_Q^2} F\left(\frac{m_{H_i}}{M_Q^2}\right) ,$$

$$F(h) = \left[\frac{(2h-3)h^2 \ln h}{(1-h)^4} + \frac{7-29h+16h^2}{6(1-h)^3}\right] .$$

The electromagnetic penguin and long distance effects can contribute to the imaginary part of the $\Delta I = 3/2$ amplitude and give a small contribution. This is analyzed in detail in Ref. [2]. In our decoupling limit, the result is

$$\epsilon'/\epsilon = -1.9 \times 10^{-5} \mathrm{Im}(\mathcal{A}_{sd}/(0.058)^2) R_Q^2 \tag{9}$$

$$= \pm 1.9 \times 10^{-5} \left(\sqrt{(156\mathcal{F})^2+1} - 156\mathcal{F}\right)^{\frac{1}{2}} R_Q/\sqrt{2} \, ,$$

using the constraints in Eq.(8). For $R_Q = 1$ and $\mathcal{F} \approx 0$ (or -0.3), $\epsilon'/\epsilon = 1.4 \times 10^{-5}$ (or 1.3×10^{-4}), which is roughly the same order of magnitude as the KM model. One can of course makes the model more superweak [7] by setting the scale higher (and R_Q smaller) and $\mathcal{A}_{sd}$ larger. For $\mathcal{A}_{sd}$ of order one, M_Q is roughly 100 TeV.

B^0–$\bar{B}^0$ Mixing

Another (much weaker) constraint to be considered is that from the $B^0_{s,d}$ mass splitting [2]. Using the usual estimate of strong form factors involved and the experimental value for ΔM_{B^0}, we have

$$\delta(\Delta M_{B^0})/\Delta M_{B^0} = 1.1 \times 10^{-3} \, R_Q^2 \, \left|\mathcal{A}_{bd}/0.058^2\right|^2 \, ,$$

so even taking $\mathcal{A}_{bd} = (0.15)^2$, the fractional contribution is only about 5% for $M_Q = 300$ GeV.

Nevertheless, the relevant parameter $\mathcal{A}_{bd}$ is not subjected to constraints from ϵ or ϵ'. If it is of the same size as $\mathcal{A}_{sd}$, the box amplitude from the exotic sector C_{RR}^{RH} is much smaller than that from SM C_{LL}^{SM}. The future B factory would observe only a collapsed unitarity KM angle [9]. However, if $\mathcal{A}_{bd} \gg \mathcal{A}_{sd}$, the triangle can looks substantially different from that predicted by standard KM model.

In SM, constraint on the quark mixing based on the measurement of the B_d^0–$\bar{B}_d^0$ oscillation is given by

$$|1 - \rho - i\eta| = 1.01 \pm 0.21 \ ,$$

in terms of Wolfenstein[8] parameterization. This constraint is translated into an relation for the RH model as

$$|(1 - \rho)^2 + \mathcal{A}_{bd}^2/N| = (1.01 \pm 0.21)^2 \ ,$$

with

$$\frac{1}{N} \equiv \frac{C_{RR}^{RH}/\mathcal{A}_{bd}^2}{C_{LL}^{SM}} \ ,$$

which is the ratio of box amplitudes of the RH model and the SM with the coupling $\mathcal{A}_{bd}^2$ removed. Numerically, $N \approx (0.3)^4$. This requires $\xi_{1b}, \xi_{1d} \approx 0.3$ in order that the new interaction plays a role in the B physics. On the other hand, the $|V_{ub}/V_{cb}|$ constraint in the SM gives

$$\rho^2 + \eta^2 = (0.41 \pm 0.07)^2 \ .$$

In softly broken CP models, as η is zero, the ρ parameter is restricted in two possible regions

$$\rho = \pm(0.41 \pm 0.07) \ .$$

Combining all these bounds in the RH model, we derive the favorable regions of $\mathcal{A}_{bd}^2/N$ in the complex plane,

For $\rho > 0$,

$$|(1 - 0.41 \pm 0.07)^2 + \mathcal{A}_{bd}^2/N| = (1.01 \pm 0.21)^2 \ . \tag{10}$$

For $\rho < 0$,

$$|(1 + 0.41 \pm 0.07)^2 + \mathcal{A}_{bd}^2/N| = (1.01 \pm 0.21)^2 \ . \tag{11}$$

They are rings at the centers $-(0.59 \pm 0.07)^2$ and $-(1.41 \pm 0.07)^2$ respectively with radius $(1.10 \pm 0.21)^2$.

The point $\mathcal{A}_{bd} = 0$ is not favorable at the 1-σ fit, but certainly not excluded at the 3-σ level. In the figure, we illustrate a possible location of $\mathcal{A}_{bd}^2$. The angle 2β corresponds to the phase between B_d-$\bar{B}_d$ oscillation and the (W) decay amplitude.

If the penguin diagram from the exotic sector is not important, all CP properties of B physics is coded in this phase 2β, unlike the SM where there are three angles for various channels of the B decay.

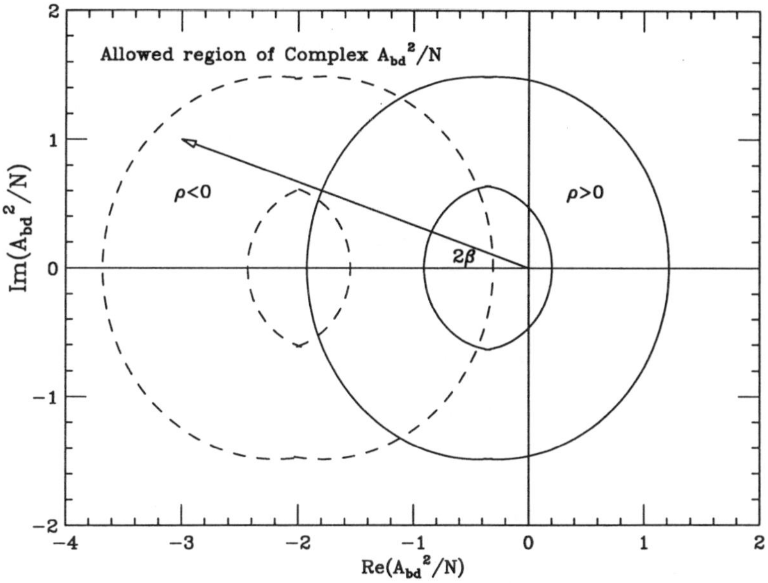

$b \to s\gamma$ and Other Constraints

In RH models, the operator induced by the exotic sector has helicity opposite to that of the SM contribution in the process of $b \to s\gamma$. These two kinds of amplitudes do not interfere in the rate. In the decoupling limit ($m_2 \gg m_1$ and $m_1 = M_Q$), we calculate the ratio $\mathcal{F}_{b\to s\gamma} \equiv \delta B(b \to s\gamma)/B(b \to s\gamma)_{\mathrm{SM}}$,

$$\mathcal{F}_{b\to s\gamma} = 6.4 \times 10^{-6} \left| R_Q^2 \cdot \frac{0.0389}{V_{tb}V_{ts}^*} \cdot \frac{\mathcal{A}_{bs}}{(0.058)^2} \right|^2 .$$

Even for a very large $\mathcal{A}_{bs} \approx (0.3)^2$, $\mathcal{F}_{b\to s\gamma}$ reaches an insignificant value of 0.04. It is worthwhile to point out that, in LH models, the exotic contribution gives rise to operator that will interefere with that of SM and therefore the model is more severely constrained [5].

Constraints from Strong CP θ_{QCD} and the induced KM phase

In both RH and LH models, θ_{QCD} is only induced starting at the two-loop level, via generation of complex down-flavor quark masses as long as the coupling

κ exists. A typical diagram is shown in Fig. 1 in Ref. [2]; this effect does not require more than one flavor of down-quark. However, it does require both charged Higgs to be involved.

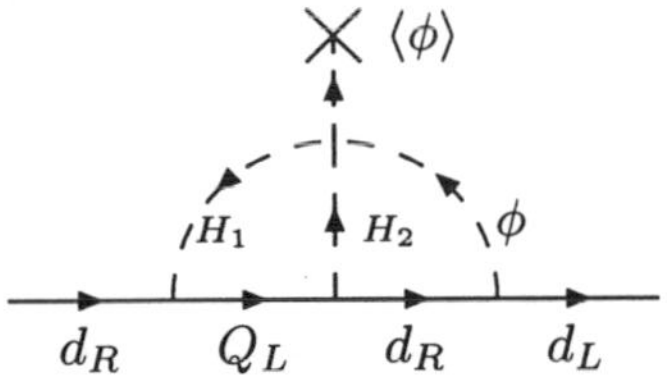

Roughly,

$$\theta_{\rm QCD} \simeq g^2 I \ {\rm Im}(\mathcal{A}_{dd;12}\ \mathcal{B}_{12})/(16\pi^2)^2 \ . \tag{12}$$

The factor I, of order one, is given by the integral

$$I = \int_0^1 \frac{dz}{(1-z)} \int_0^{1-z} dx \ \times \tag{13}$$

$$\left[\frac{zm_{H_2}^2 + xM_Q^2 + ym_{H_1}^2}{zm_{H_2}^2 + xM_Q^2 + ym_{H_1}^2 - z(1-z)M_{\phi^0}^2} \cdot \right.$$

$$\left. \log \frac{z(1-z)M_{H^0}^2}{zm_{H_2}^2 + xM_Q^2 + ym_{H_1}^2} - (m_{H_1} \leftrightarrow m_{H_2}) \right] ,$$

with the Feynman parameters $y = 1 - x - z$. The integral I vanishes at the degenerate case $m_{H_2} = m_{H_1}$, but its size approaches [5] 1 as $m_{H_2} \to \infty$. This non-decoupling phenomena is not surprising because, in the large m_{H_2} limit, the CP is a broken symmetry. Numerically,

$$\theta_{\rm QCD} \sim 6 \times 10^{-10} {\rm Im} \left(\frac{\mathcal{A}_{dd;12}}{0.058^2} \frac{\mathcal{B}_{12}}{0.01} \right) \ . \tag{14}$$

The present constraint, $\theta_{\rm QCD} < 10^{-9}$, can easily be accommodated. In the scenario of a large $\mathcal{A}_{bd;11} \sim 0.3$ for interesting B physics, the relevant parameter $\mathcal{A}_{bb;12}$, which appears in the formula of $\theta_{\rm QCD}$ due to the b quark contribution, can still be small because it involves the another generation charged Higgs. Even $\mathcal{A}_{bb;12} \sim 0.3$, we can reduce the κ such that $\mathcal{B}_{12} \sim 0.0004$ in order not to give too large $\theta_{\rm QCD}$.

In addition, there are also three loop diagrams due to the gluonic contribution. A typical graph is shown below.

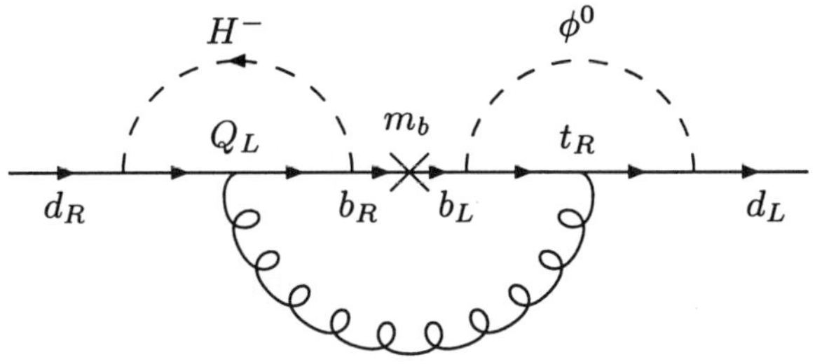

This contribution is independent of the coupling κ. However, unless κ happens to be very small, it may not be competitive with the two loop contributions because of the KM angle suppression and the additional loop factor [5]. Both the 2-loop and the 3-loop contributions also exist generically in LH models. However, they are typically numerically smaller in that case. The two loop contributions disappear (in both RH and LH models) of course when only one complex scalar boson is used as mentioned earlier.

Since CP is broken at higher energy, a non-vanishing KM phase η (defined in the Wolfenstein parametrization) [8] can in general be loop induced. It can originate either from the loop-induced complex mass matrix or from that of the complex kinetic energy terms. The contribution from complex mass matrix is similar to the analysis of θ and is therefore small. The contribution from complex kinetic terms is induced at the one loop level in LH models as analyszed in Ref. [3]. It is in general also suppressed by some small KM mixing angles and small mass ratios and therefore numerically tiny enough to be ignored phenomenologically. In contrast, in RH models, since the KM phase is related only to the rotation of the left-handed quarks, such contribution will not arise until at the two loop level as given in the following figure. Therefore induced η is even smaller [5].

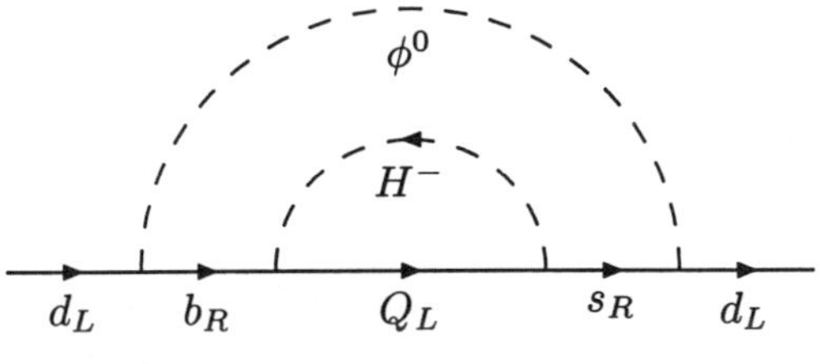

Constraints from electric dipole moments

A down-flavor quark EDM, however, is generated at the two-loop (or three-loop) level, in parallel with the generation of complex down quark masses discussed above. The typical contribution is given by diagrams similar to those for θ, except with an external photon attached to internal charged lines. An estimate of the two loop contribution gives EDM which is consistent with the current experimental bound. The contribution from chromo-electric dipole moment of gluon (the Weinberg operator) won't arise until three loop level (even with κ coupling) and therefore expected to be small. The electron couples only indirectly with the CP violating sector, so its EDM vanishes at two loops and the three loop contributions are insignificantly small.

Decays of New Particles

In the generic RH model, h and Q can be assigned a new conserved quantum number which guarantees a stable lightest exotic particle, either H_1 or Q. Stable heavy hadron in the mass range of 100 GeV or higher is not ruled out cosmologically (similar to the stable gluino analysis in Ref.[10]. However, when $q_h = -1$, an interaction, $h_\alpha L_i L_j$, is allowed, which can lead to H^- (on-shell or off-shell) decays into $l^-\nu$. Even so, lepton number is still conserved, just as in the SM, since Q and H will naturally carry lepton number ($L = \pm 2$). Another way for H to decay is to introduce a second Higgs doublet and let H couple to two different Higgs doublets. In that case H can decay into a neutral Higgs, plus a charged Higgs which in turn decays into ordinary quarks and leptons.

Conclusion

With strong evidence for new physics in the neutrino sector based on super-Kamiokande data, it is very likely that new physics may also appear in the other least tested sector of the standard model: *i.e.* CP violation. In addition, the well known strong CP problem as well as the electroweak baryo-genesis problem associated the KM mechanism seem to give hints that new physics can surface in the B factories which will operate in the near future.

We have reviewed two new classes of models of CP violation which are aiming at providing alternative interpretation of the CP violating data. The model naturally prevents the strong CP problem. Both classes of models are surprisingly similar to the KM model in the sense that the CP-breaking mechanism is seemingly milliweak(if the exotic particle scale is chosen to be as low as possible), while its phenomenology (as studied here) is quite superweak-like. The phenomenological distinction between the KM model and these soft breaking models will likely be made clear in experiments planned for the B factories. A detailed analysis of such issues will be necessary [5].

W.-Y. K. is supported by a grant from the DOE of USA, and D. C. by the NSC of R.O.C. We thank H. Georgi, S. Glashow, P. Frampton, R. Mohapatra, and L. Wolfenstein for very useful discussions.

References

1. M. Kobayashi and T. Maskawa, Prog. Theor. Phys. **49**, 652 (1973).
2. D. Bowser-Chao, D. Chang, and W.-Y. Keung, Phys. Rev. Lett. **81**, 2028 (1998).
3. H. Georgi and S. Glashow, hep-ph/9807399. See also, P. Frampton and M. Harada, hep-ph/9809402.
4. D. Bowser-Chao, D. Chang, and W.-Y. Keung, hep-ph/9811258, contribution to International Conference on High Energy Physics 1998, Triumf, Canada.
5. D. Bowser-Chao, D. Chang, and W.-Y. Keung, in preparation. For a short review, see Ref.[4].
6. S. Barr and A. Zee, Phys. Rev. Lett. **55**, 2253 (1985); J. Nieves, Nucl. Phys. **B189**, 189 (1981); Phys. Lett **164B**, 85 (1985); S. Barr, Phys. Rev. **D34**, 1567 (1986); S. Barr and E. M. Freire Phys. Rev. **D41**, 2129 (1990).
7. L. Wolfenstein, Phys. Rev. Lett. **13**, 562 (1964);
8. L. Wolfenstein, Phys. Rev. Lett. **51**, 1945 (1983); L.–L. Chau and W.–Y. Keung, Phys. Rev. Lett. **53**, 1802 (1984).
9. S. Mele, hep-ph/9810333; see also R. Barbieri, L. Hall, A. Stocchi, and N. Weiner, Phys. Lett. **B425** 119 (1998) or hep-ph/9712252.
10. H. Baer, K. Cheung and J. F. Gunion, hep-ph/9806361.

Solving the $\rho-\pi$ puzzle by higher charmonium Fock states

Yu-Qi Chen
Physics Department, Ohio State University, Columbus OH 43210
E-mail: ychen@pacific.mps.ohio-state.edu

A new explanation for the longstanding puzzle of the tiny branching fraction of $\psi' \to \rho\pi$ relative to that for $J/\psi \to \rho\pi$ was proposed. In the case of J/ψ, we argue that this decay is dominated by a higher Fock state in which the $c\bar{c}$ pair is in a color-octet $^3S_1$ state and annihilates via the process $c\bar{c} \to q\bar{q}$. In the case of the ψ', we argue that the probability for the $c\bar{c}$ pair in this higher Fock state to be close enough to annihilate is suppressed by a dynamical effect related to the small energy gap between the mass of the ψ' and the $D\bar{D}$ threshold.

1 Introduction

Usually, a well-defined puzzle of physics provides us with a gateway to look at something new. We get insight into the physics behind the puzzle through seeking its solution. There is such a long-standing puzzle in charmonium physics which is called as the "$\rho - \pi$ puzzle" of J/ψ and ψ' decays. The puzzle is that, J/ψ decays to $\rho\pi$ but ψ' does not, which contradicts our understanding of the decay processes. The discrepancy between the conventional theoretical expectation and the experiments is greater than 65. This extremely large number clearly tells us that there is something unusual in charmonium decays. We hope to learn something from this 15 year mystery. Recently, Chen and Braaten proposed a new explanation for it [1]. That's what I am going to talk about.

J/ψ and ψ' are nonrelativistic bound states of a charm quark and its antiquark. Their decays into light hadrons are believed to be dominated by the annihilation of the $c\bar{c}$ pair into three gluons. In order to annihilate, the c and $\bar{c}$ must have a separation of order $1/m_c$, which is much smaller than the size of the charmonium state. Thus the annihilation amplitude for an S-wave state like J/ψ or ψ' must be proportional to the wavefunction at the origin, $\psi(\mathbf{r}=0)$. The width for decay into any specific final state h consisting of light hadrons is therefore proportional to $|\psi(0)|^2$. The width for decay into e^+e^- is also proportional to $|\psi(0)|^2$. This leads to the simple prediction that the ratio of the branching fractions for ψ' and J/ψ is given by the "15% rule":

$$Q_h \equiv \frac{B(\psi' \to h)}{B(J/\psi \to h)} = Q_{ee} = (14.7 \pm 2.3)\% . \tag{1}$$

This rule follows simply from the fact that ψ' can make a transition to J/ψ

and χ_c by emitting π's and γ. This decreases the branching ratio of the $\psi' \to h$ annihilation decay. This rule is obeyed by most of the J/ψ and ψ' annihilation decay modes. The ρ–π puzzle is that the prediction (1) is severely violated in the $\rho\pi$ and several other decay channels. The Mark II collaboration found that $Q_{\rho\pi} < 0.6\%$ and $Q_{K^*K} < 2\%$ [2]. Recent data from the BES collaboration has made the puzzle even sharper. They obtained $Q_{\rho\pi} < 0.23\%$, $Q_{K^{*+}K^-} < 0.64\%$, and $Q_{K^{*0}\bar{K}^0} = (1.7 \pm 0.6)\%$ [3]. Thus the suppression of Q relative to the 15% rule is about 10 for $K^{*0}\bar{K}^0$ and greater than 65 for $\rho\pi$.

2 Previous explanations

The $\rho - \pi$ puzzle challenges our conventional theory. It has attracted broad interest. In literature, several different explanations have been proposed. A summary of those explanations has recently been given by Chao [4]. Hou and Soni [5] suggested that $J/\psi \to \rho\pi$ is enhanced by a mixing of the J/ψ with a glueball $\mathcal{O}$ that decays to $\rho\pi$. Brodsky, Lepage, and Tuan [6] emphasized that $J/\psi \to \rho\pi$ violates the helicity selection rule of perturbative QCD, and argued that the data requires $\mathcal{O}$ to be narrow and nearly degenerate with the J/ψ. Present data from BES constrains the mass and width of the glueball to the ranges $|m_{\mathcal{O}} - m_{J/\psi}| < 80$ MeV and 4 MeV $< \Gamma_{\mathcal{O}} < 50$ MeV [7]. This mass is about 700 MeV lower than the lightest $J^{PC} = 1^{--}$ glueball observed in lattice simulations of QCD without dynamical quarks [8]. There are explanations of the ρ–π puzzle that involve the dependence of the decay amplitude on the energy of the charmonium state. Karl and Roberts [9] suggested that the decay proceeds through $c\bar{c} \to q\bar{q}$ followed by the fragmentation of the $q\bar{q}$ into $\rho\pi$. They argued that the fragmentation probability is an oscillatory function of the energy which could have a minimum near the mass of the ψ'. Chaichian and Tornqvist [10] pointed out that the suppression of ψ' decays could be explained if the form factors for two-body decays fall exponentially with the energy as in the nonrelativistic quark model. There are other explanations of the ρ–π puzzle that rely on the fact that there is a node in the radial wavefunction for ψ', but not for J/ψ. Pinsky [11] suggested that this node makes $\psi' \to \rho\pi$ a "hindered M1 transition" like $J/\psi \to \eta_c\gamma$. Brodsky and Karliner [12] suggested that the decay into $\rho\pi$ proceeds through intrinsic charm components of the ρ and π wavefunctions. They argued that the $c\bar{c}$ pair in the $|u\bar{d}c\bar{c}\rangle$ Fock state of the ρ^+ or π^+ has a nodeless radial wavefunction which gives it a larger overlap with J/ψ than ψ'. Finally, Li, Bugg, and Zou [13] have suggested that final-state interactions involving the rescattering of $a_1\rho$ and $a_2\rho$ into $\rho\pi$ could be important and might interfere destructively in the case of the ψ'.

3 A new explanation

Recently Braaten and I proposed a new explanation of the ρ–π puzzle [1]. We argue that the decay $J/\psi \to \rho\pi$ is dominated by a higher Fock state of the J/ψ in which the $c\bar{c}$ is in a color-octet ${}^3S_1$ state. The $c\bar{c}$ pair in this Fock state can annihilate via $c\bar{c} \to g^* \to q\bar{q}$. The amplitude for forming $\rho\pi$ is dominated by the endpoint of the meson wavefunctions, where a single q or $\bar{q}$ carries most of the momentum of the meson. The suppression of $\psi' \to \rho\pi$ is attributed to a suppression of the $c\bar{c}$ wavefunction at the origin for the higher Fock state of the ψ'. Such a suppression can arise from a dynamical effect associated with the small energy gap between the mass of the ψ' and the $D\bar{D}$ threshold.

3.1 J/ψ and ψ' inclusive annihilation decay rate

It is convenient to analyze charmonium decays using nonrelativistic QCD (NRQCD) [14], the effective field theory obtained by integrating out the energy scale of the charm quark mass m_c. In Coulomb gauge, a charmonium state has a Fock state decomposition in which the probability of each Fock state scales as a definite power of v, the typical relative velocity of the charm quark. The dominant Fock state of the J/ψ and the ψ' is $|c\bar{c}_1({}^3S_1)\rangle$, whose probability $\mathcal{P}$ is of order 1. We denote the color state of the $c\bar{c}$ pair by a subscript (1 for color-singlet, 8 for color-octet) and we put the angular momentum quantum numbers in parentheses. The Fock states $|c\bar{c}_8({}^3P_J) + S\rangle$, where S represents dynamical gluons or light quark-antiquark pairs with energies of order $m_c v$ or less, have probability $\mathcal{P} \sim v^2$. The next most important Fock states include $|c\bar{c}_8({}^1S_0) + S\rangle$ with $\mathcal{P} \sim v^3$ (or $\mathcal{P} \sim v^4$ if perturbation theory is sufficiently accurate at the scale $m_c v$) and $|c\bar{c}_8({}^3S_1) + S\rangle$ with $\mathcal{P} \sim v^4$.

The decay of the J/ψ into light hadrons proceeds via the annihilation of the c and $\bar{c}$, which can occur through any of the Fock states. This is expressed in the NRQCD factorization formula for the inclusive decay rate [14], which includes the terms

$$\Gamma_{J/\psi \to l.h.} = \left(\frac{20(\pi^2-9)\alpha_s^3}{243 m_c^2} + \frac{16\pi\alpha^2}{27 m_c^2} \right) \langle \mathcal{O}_1({}^3S_1) \rangle_{J/\psi} + \frac{5\pi\alpha_s^2}{6m_c^2} \langle \mathcal{O}_8({}^1S_0) \rangle_{J/\psi}$$
$$+ \frac{19\pi\alpha_s^2}{6m_c^4} \langle \mathcal{O}_8({}^3P_0) \rangle_{J/\psi} + \frac{\pi\alpha_s^2}{m_c^2} \langle \mathcal{O}_8({}^3S_1) \rangle_{J/\psi} + \ldots . \quad (2)$$

where $\alpha_s = \alpha_s(m_c)$. The matrix elements are expectation values in the J/ψ of local gauge-invariant NRQCD operators that measure the inclusive probability of finding a $c\bar{c}$ in the J/ψ at the same point and in the color and angular-momentum state specified. The matrix element of the $c\bar{c}_1({}^3S_1)$ term in (2) is proportional to the square of the wavefunction at the origin and scales as v^3. Its coefficient includes a term of order α_s^3 from $c\bar{c} \to ggg$ and a term of order α^2 from the electromagnetic annihilation process $c\bar{c} \to \gamma^* \to q\bar{q}$. The color-octet terms in (2) represent contributions from higher Fock states. Their matrix elements scale like v^6, v^7 and v^7, respectively. Their coefficients are all of order α_s^2 and come from $c\bar{c} \to gg$ for the $c\bar{c}_8(^1S_0)$ and $c\bar{c}_8(^3P_0)$ terms and from $c\bar{c} \to g^* \to q\bar{q}$ for the $c\bar{c}_8(^3S_1)$ term. Note that the coefficients of the color-octet matrix elements are two orders of magnitude larger than that of $\langle \mathcal{O}_1(^3S_1) \rangle_{J/\psi}$, which suggests that the higher Fock states may play a more important role in annihilation decays than is commonly believed.

3.2 Color-octet contribution to $J/\psi \to \rho\pi$

According to a general QCD factorization formula, the amplitude for a two-body annihilation decay can be expressed in terms of hard-scattering factors $\widehat{\mathcal{T}}$ that involve only the scale m_c, initial-state factors that involve scales of order $m_c v$ and lower, and final-state factors that involve only the scale $\Lambda_{\rm QCD}$. If m_c was asymptotically large, the dominant terms in the factorization formula would have the minimal number of partons involved in the hard scattering. Terms involving additional soft partons in the initial state are suppressed by powers of v. Terms involving additional hard partons in the final state are suppressed by powers of $\Lambda_{\rm QCD}/m_c$. As pointed out by Brodsky and Lepage[15], the leading term in the asymptotic factorization formula is strongly constrained by the vector character of the QCD interaction between quarks and gluons. It vanishes unless the sum of the helicities of the mesons is zero. Rotational symmetry then requires the angular distribution to be $1 - \cos^2\theta$. This helicity selection rule is violated by the decay $J/\psi \to \rho\pi$. Parity and rotational symmetry require the helicity of the ρ to be ± 1 and the angular distribution to be $1 + \cos^2\theta$. Since the helicity of the pion is 0, the helicity selection rule is violated. Thus the amplitude for $J/\psi \to \rho\pi$ is suppressed by a factor of $\Lambda_{\rm QCD}/m_c$ relative to that for generic mesons. The leading contribution is a term with a hard scattering factor of order α_s^2 from the process $c\bar{c}_1 \to u\bar{d}d\bar{u}g$.

Since the charm quark mass m_c is less than an order of magnitude larger than $\Lambda_{\rm QCD}$, there can be large corrections to the asymptotic decay amplitude. In particular, there can be significant regions of phase space in which some of the gluons involved in the hard scattering are relatively soft. It might therefore be more appropriate to absorb them into the initial-state or final-state factors. In this case, not all of the soft partons in S need be involved in the hard scattering, and not all the partons that form the ρ and π need be produced by the hard scattering. For example, there can be a contribution from the Fock state $|c\bar{c}_8(^3S_1) + S\rangle$ that involves the hard-scattering process $c\bar{c} \to q\bar{q}$, where $q = u$ or d. This produces a state $|q\bar{q} + S\rangle$ that consists of the soft partons S together with a q and a $\bar{q}$ that are back-to-back and whose momenta are approximately m_c. Such a state has a nonzero overlap with the final state $|\rho\pi\rangle$. The overlap comes from the endpoints of the meson wavefunctions, in which most of the momenta of the ρ and π are carried by the q and $\bar{q}$. When S is a $q\bar{q}$ pair, the ρ and π mesons are formed from the 2-quark antiquark pairs, with the hard quark pair carries most of the momenta of the ρ and π.

This contribution to the T-matrix element can be written schematically in the form

$$\mathcal{T}_{J/\psi\to\rho\pi} = \sum_{q\bar{q}} \widehat{\mathcal{T}}(c\bar{c}_8(^3S_1) \to q\bar{q}) \sum_{S} \langle\rho\pi|q\bar{q} + S\rangle \, \langle c\bar{c}_8(^3S_1) + S|J/\psi\rangle. \quad (3)$$

Note that the initial-state and final-state factors on the right side of (3) cannot be separated, because they are connected by the sum over soft modes S. The T-matrix element (3) would contribute to the $c\bar{c}_8(^3S_1)$ term in the factorization formula (2) for inclusive decays.

The endpoint contribution in (3) leads to a definite angular distribution. Since the q and $\bar{q}$ carry most of the momenta of the mesons, the angular distribution of the mesons will follow that of the q and $\bar{q}$, which is $1 + \cos^2\theta$. Thus (3) will contribute most strongly to form factors which allow the angular distribution $1 + \cos^2\theta$. It will also contribute most strongly to decays into mesons like ρ and π for which most of the momentum can be carried by a single q or $\bar{q}$. There are also endpoint contributions involving the hard-scattering process $c\bar{c}_8(^3P_J) \to gg$, which produces a pair of hard gluons with the angular distribution $2 + \cos^2\theta$, and $c\bar{c}_8(^1S_0) \to gg$, for which the angular distribution is isotropic. Their contributions to $J/\psi \to \rho\pi$ are suppressed by the small probabilities for most of the momentum of the ρ or π to be carried by a single gluon and by the mismatch between the angular distribution of the gluons and the $1 + \cos^2\theta$ distribution of the $\rho\pi$.

We argue that the color-octet term in (3) may actually dominate the decay rate for $J/\psi \to \rho\pi$. We compare the various factors in that term with those in the asymptotic amplitude. The hard-scattering factor $\widehat{\mathcal{T}}$ in the color-octet term in (3) is only of order α_s, compared to α_s^2 for the asymptotic amplitude. The suppression from the initial-state factor in the color-octet term in (3), including the sum over S, might be as little as a factor of v^2 relative to the asymptotic amplitude. This follows from the fact that the color-octet amplitude contributes in quadrature to the $\langle\mathcal{O}_8(^3S_1)\rangle$ term in (2), which is suppressed by v^4. As for the final-state factors, (3) is suppressed by the endpoints of the meson wavefunctions, while the asymptotic amplitude is suppressed by

Λ_{QCD}/m_c from the violation of the helicity selection rule. Considering the various suppression factors, it is certainly plausible that the $c\bar{c}_8(^3S_1)$ term in (3) could dominate over the leading contribution from the $c\bar{c}_1(^3S_1)$ Fock state.

3.3 Suppression of $\psi' \to \rho\pi$

We have argued that the decay $J/\psi \to \rho\pi$ may be dominated by the annihilation of the $c\bar{c}$ pair in the $|c\bar{c}_8(^3S_1) + S\rangle$ Fock state via $c\bar{c} \to q\bar{q}$. If this is true, then the ρ–π puzzle can be explained by a suppression of this decay mechanism in the case of the ψ'. This suppression can arise from the initial-state factor $\langle c\bar{c}_8(^3S_1) + S|J/\psi\rangle$ if the $c\bar{c}$ wavefunction for the $|c\bar{c}_8(^3S_1) + S\rangle$ Fock state is suppressed in the region in which the separation of the $c\bar{c}$ is less than or of order $1/m_c$. Note that it does not require a suppression of the probability for the higher Fock state, but just a shift in the probability away from the region in which the c and $\bar{c}$ are close enough to annihilate. A possible mechanism for this suppression is a dynamical effect related to the small energy gap between the mass of the ψ' and the $D\bar{D}$ threshold. In the Born-Oppenheimer approximation[16], the $c\bar{c}$ pair in the dominant $|c\bar{c}_1\rangle$ Fock state moves adiabatically in response to a potential $V_1(R)$ given by the minimal energy of QCD in the presence of a color-singlet $c\bar{c}$ pair with fixed separation R. Similarly, the $c\bar{c}$ pair in the $|c\bar{c}_8(^3S_1) + S\rangle$ Fock state moves adiabatically in response to a potential $V_8(R)$ given by the minimal energy of the soft modes S in the presence of a color-octet $c\bar{c}$ pair with fixed separation R. At short distances, this potential approaches a repulsive Coulomb potential $\alpha_s/(6R)$. At long distances, the minimal energy state consists of D and $\bar{D}$ mesons separated by a distance R, and $V_8(R)$ therefore approaches a constant $2(M_D - m_c)$ equal to the energy of the $D\bar{D}$ threshold. A charmonium state spends most of the time on the color-singlet adiabatic surface, but it occasionally makes a transition to the color-octet adiabatic surface. Since the J/ψ is 640 MeV below the $D\bar{D}$ threshold, a $c\bar{c}$ pair on the color-octet adiabatic surface is far off the energy shell. The time spent by the J/ψ on this surface is too short for the $c\bar{c}$ pair to respond to the repulsive short-distance potential. Since the wavefunction for the $|c\bar{c}_1\rangle$ Fock state peaks at the origin, the $c\bar{c}$ wavefunction for the $|c\bar{c}_8 + S\rangle$ Fock state should have significant support near the origin. However the mass of the ψ' is only 43 MeV below D^+D^- threshold and 53 MeV below $D^0\bar{D}^0$ threshold. A $c\bar{c}$ pair on the color-octet adiabatic surface can be very close to the energy shell. The ψ' can therefore spend a sufficiently long time on this surface for the $c\bar{c}$ pair to respond to the repulsive short-distance potential. This response can lead to a significant suppression of the $c\bar{c}$ wavefunction at the origin for the $|c\bar{c}_8 + S\rangle$ Fock state.

This suppression effect can be illustrated further as follows. The repulsive interactions between c and $\bar{c}$ in the color-octet higher Fock state $|c\bar{c}_8 + S\rangle$ leads to the increase of the distance between c and $\bar{c}$. This suppresses the annihilation probability of the $c\bar{c}$ pair. Although this suppression effect is complicated we can expect that it is sensitive to the lifetime of the color-octet higher Fock state in J/ψ and ψ'. The longer the lifetime is, the larger the suppression effects are. In the case of J/ψ, the lifetime of this state is shorter because it is far off-shell. An extreme case is the decay of the J/ψ or Υ to 3 hard gluons. After one or two hard gluons are emitted, the c or $\bar{c}$ quark is far off-shell and the lifetime of the intermediate state is very short. The effect of the repulsive interaction can be accounted for as part of the radiative correction to the short-distance coefficients. Consequently, we need only one parameter $\psi(0)$ to describe the annihilation of the $c\bar{c}$ pair in this process. In the case of ψ', the lifetime of the intermediate higher Fock state is much longer because it's very close to the $D\bar{D}$ threshold. The repulsive interactions separate c and $\bar{c}$ to a very large distance until it is screened by a $D\bar{D}$ state, in which the annihilation rate is very tiny. In this case, the amplitude will be enhanced by a factor $1/\Delta E$, the energy difference between the mass of the ψ' and the $D\bar{D}$ state, which arises from the time integral from 0 to ∞ in perturbative calculation. However, the smaller phase space may compensate this enhancement. This is similar to the electromagnetic transitions between two different energies in atoms, where transition rates are smaller for those with smaller energy difference.

If the initial-state factor $\langle c\bar{c}_8(^3S_1) + S|J/\psi\rangle$ in (3) is suppressed for all soft partons S, the suppression can be expressed in the form of a relation between the NRQCD matrix elements in (2):

$$\frac{\langle \mathcal{O}_8(^3S_1)\rangle_{\psi'}}{\langle \mathcal{O}_1(^3S_1)\rangle_{\psi'}} \ll \frac{\langle \mathcal{O}_8(^3S_1)\rangle_{J/\psi}}{\langle \mathcal{O}_1(^3S_1)\rangle_{J/\psi}} . \tag{4}$$

This inequality can be tested by calculating the matrix elements using Monte Carlo simulations of lattice NRQCD. Since the ψ' is so close to the $D\bar{D}$ threshold, it would be essential to include dynamical light quarks in the simulations.

4 Signatures

Our proposal implies interesting signatures. It leads to predictions for the flavor-dependence of the suppression of the decays of ψ' into vector/pseudoscalar final states. It can explain the J^{PC} dependence of the suppression of the two-body decays of ψ'. Also it gives prediction for the angular distribution of these two-body decays.

4.1 Flavor dependence

One of the signatures of our explanation is the prediction for the flavor-dependence of the suppression of the decays of ψ' into vector/pseudoscalar final states. Bramon, Escribano, and Scadron [17] have analyzed the decays $J/\psi \to VP$ assuming that the decay amplitude is the sum of a flavor-connected amplitude g, a flavor-disconnected amplitude rg, and an isospin-violating amplitude e. The expressions for the amplitudes are given in Ref. [17]. They allowed for violations of $SU(3)$ flavor symmetry through parameters s and $x-1$. The authors give two sets of parameters that fit the existing data, one with $x=1$ and the other with $x=0.64$. Both sets have e comparable in magnitude to rg and an order of magnitude smaller than g. If the decay $J/\psi \to \rho\pi$ is dominated by endpoint contributions, we can identify g and e with the two terms in (3). While rg may also have endpoint contributions from $c\bar{c} \to gg$, we assume for simplicity that it is dominated by subasymptotic contributions from the $c\bar{c}_1(^3S_1)$ Fock state. The amplitudes for the decays $\psi' \to VP$ can be expressed in a similar way in terms of amplitudes g', e', and $(rg)'$. Our explanation of the ρ–π puzzle implies that $|g'|$ is much smaller than $|g|$, and that e' and $(rg)'$ differ from e and rg only by the factor required by the 15% rule. The unknown amplitude g' is constrained by the BES data on $\psi' \to \rho\pi$ and $\psi' \to K^{*0}\bar{K}^0$. The upper bound on $B(\psi' \to \rho\pi)$ gives an upper bound on $|g'+e'|^2$, which implies that g' lies in a circle in the complex g'-plane. The BES measurement of $B(\psi' \to K^{*0}\bar{K}^0)$ gives an allowed range for $|(1-s)g'-(1+x)e'|^2$, which constrains g' to an annulus. The intersection of the interior of the circle with the annulus is the allowed region for g'. By varying g' over that region and taking into account the uncertainties in the parameters of Ref. [17], we obtain the predictions for Q_{VP} in Table 1. Measurements of the ψ' branching fractions consistent with these predictions would imply that the suppression of the vector/pseudoscalar decays is due to the suppression of g'. This would lend support to our explanation of the ρ–π puzzle.

There is a possibility that the isospin-violating amplitude e is dominated by the contribution from the $c\bar{c}$ color-singlet higher Fock state $|c\bar{c}_1(^3S_1)+S\rangle$ of the J/ψ and the ψ' through an annihilation of the $c\bar{c} \to \gamma^* \to q\bar{q}$. If this is the case, then we expect there to be no suppression of the 15% rule for the ψ' decay from this amplitude, since it arises from the $c\bar{c}$ color-singlet state. The ratio of the cross sections of $e^+e^- \to \omega\pi$ and $e^+e^- \to \mu^+\mu^-$ near the J/ψ and the ψ' resonances should then be significantly smaller than the corresponding branching ratios at the resonances. This possibility can also be tested by observing the decay rates of J/ψ, $\psi' \to e^+e^-\pi\pi$ or J/ψ, $\psi' \to \mu^+\mu^-\pi\pi$ with soft π's, since these processes receive contribution from the $|c\bar{c}_1(^3S_1)+S\rangle$

Table I. Predictions for Q_{VP} in units of 1% for all the vector/pseudoscalar final states. The values for $\rho\pi$ and $K^{*0}\bar{K}^0 + c.c.$ were used as input. The columns labelled $x = 1$ and $x = 0.64$ correspond to the two parameter sets of Ref. [17].

VP	$x = 1$	$x = 0.64$
$\rho\pi$	$0 - 0.25$	$0 - 0.25$
$K^{*0}\bar{K}^0 + c.c.$	$1.2 - 3.0$	$1.2 - 3.0$
$K^{*+}K^- + c.c.$	$0 - 0.36$	$0 - 0.52$
$\omega\eta$	$0 - 1.6$	$0 - 1.6$
$\omega\eta'$	$10 - 51$	$12 - 55$
$\phi\eta$	$0.8 - 3.6$	$0.4 - 3.0$
$\phi\eta'$	$0.7 - 2.5$	$0.5 - 2.2$
$\rho\eta$	$14 - 22$	$14 - 22$
$\rho\eta'$	$12 - 20$	$13 - 21$
$\omega\pi$	$11 - 17$	$11 - 17$

higher Fock states.

4.2 J^{PC} dependence

A solution to the ρ–π puzzle should also be able to explain the pattern of suppression for various J^{PC} states with the same flavors. A preliminary measurement of the axial-vector/pseudoscalar decay mode $\psi' \to b_1\pi$ by the BES collaboration[18] gives $Q_{b_1\pi} = (24 \pm 7)\%$, consistent with no suppression relative to the 15% rule. A preliminary measurement of the vector/tensor decay mode $\psi' \to \rho a_2$ [18] gives $Q_{\rho a_2} = (2.9 \pm 1.6)\%$, which, though suppressed relative to the 15% rule, is an order of magnitude larger than $Q_{\rho\pi}$. This pattern can be explained by also taking into account the orbital-angular-momentum selection rule for exclusive amplitudes in perturbative QCD [19]. The decay modes $b_1\pi$ and ρa_2 both have form factors that are allowed by the helicity selection rule. They also both have form factors that violate the helicity selection rule, but are compatible with an endpoint contribution from $c\bar{c} \to q\bar{q}$. However, in the case of $b_1\pi$, the endpoint contribution is further suppressed by the violation of the orbital-angular-momentum selection rule. Thus we should expect no suppression of $\psi' \to b_1\pi$ and only a partial suppression of $\psi' \to \rho a_2$.

We also expect that the decays of the J/ψ and the ψ' into the isospin violating axial-vector/pseudoscalar modes $b_1\eta$ and $b_1\eta'$ etc. should be seriously suppressed relative to $b_1\pi$ decay modes, because they are dominated by a contribution from the leading Fock state which is suppressed by $(\alpha_{\rm em}/\alpha_s)^2$. The decays of the ψ' into isospin violating vector/tensor modes such as $a_1\omega$ and $f_1\rho$ should not be suppressed relative to the J/ψ decays, because they are dominated by the contribution from the color-singlet Fock state. A thorough analysis of J/ψ and ψ' decays into axial-vector/pseudoscalar and vector/tensor final states will be presented elsewhere.

4.3 Angular distribution

Our proposal also has implications for the angular distributions of two-body decay modes. In general, the angular distribution must have the form $1 + \alpha \cos^2 \theta$, with $-1 < \alpha < +1$. Our solution to the ρ–π puzzle is based on the suppression of a contribution to ψ' decays that gives the angular distribution $1 + \cos^2 \theta$. In the case of $J/\psi \to \rho\pi$, the angular momentum selection rule implies that $\alpha = +1$. Since $J/\psi \to b_1\pi$ and $\psi' \to b_1\pi$ are dominated by the contribution from the $c\bar{c}$ leading Fock state, we expect that α is close to -1. The parameter α for the $\psi' \to \rho a_2$ should be less than α for the corresponding J/ψ decay.

5 Conclusion

In conclusion, we have proposed a new explanation of the ρ–π puzzle. We suggest that the decay $J/\psi \to \rho\pi$ is dominated by a Fock state in which the $c\bar{c}$ is in a color-octet $^3S_1$ state which decays via $c\bar{c} \to q\bar{q}$. The suppression of this decay mode for the ψ' is attributed to a dynamical effect that suppresses the $c\bar{c}$ wavefunction at the origin for Fock states that contain a color-octet $c\bar{c}$ pair. Our explanation for the ρ–π puzzle can be tested by studying the flavor dependence of the two-body decay modes of the J/ψ and ψ', their angular distributions, and their dependence on the J^{PC} quantum numbers of the final-state mesons.

Acknowledgments

This work had been done in a close collaboration with E. Braaten. I would like to thank E. Braaten for a careful reading of the manuscript. It was supported in part by the U.S. Department of Energy, Division of High Energy Physics, under Grant DE-FG02-91-ER40690. We thank Steve Olsen for discussions of the data on charmonium decays from the BES collaboration.

References

1. Y.-Q. Chen and E. Braaten, Phys. Rev. Lett. **80**, 5060 (1998).
2. M.E.B. Franklin et al., Phys. Rev. Lett. **51**, 963 (1983).
3. Y.S. Zhu, in **Proceedings of the 28th International Conference on High Energy Physics**, ed. Z. Adjuk and A.K. Wroblewski (World Scientific, 1997).

4. K.-T. Chao, in **17th International Symposium on Lepton-Photon Interactions**, ed. Z.-P. Zheng and H.-S. Chen (World-Scientific, 1996).
5. W.S. Hou and A. Soni, Phys. Rev. Lett. **50**, 569 (1983).
6. S.J. Brodsky, G.P. Lepage, and S.F. Tuan, Phys. Rev. Lett. **59**, 621 (1987).
7. W.-S. Hou, Phys. Rev. D **55**, 6952 (1997).
8. See, for example, M. Peardon, hep-lat/9710029.
9. G. Karl and W. Roberts, Phys. Lett. **144B**, 263 (1984).
10. M. Chaichian and N.A. Tornqvist, Nucl. Phys. **B323**, 75 (1989).
11. S.S. Pinsky, Phys. Lett. **B236**, 479 (1990).
12. S.J. Brodsky and M. Karliner, Phys. Rev. Lett. **78**, 4682 (1997).
13. X.-Q. Li, D.V. Bugg, and B.-S. Zou, Phys. Rev. D **55**, 1421 (1997).
14. G.T. Bodwin, E. Braaten, and G.P. Lepage, Phys. Rev. D **51**, 1125 (1995).
15. S.J. Brodsky and G.P. Lepage, Phys. Rev. D **24**, 2848 (1984).
16. K.J. Juge, J. Kuti, and C.J. Morningstar, hep-lat/9709131; hep-ph/9711451.
17. A. Bramon, R. Escribano, and M.D. Scadron, Phys. Lett. **B403**, 339 (1997).
18. S.L. Olsen (BES collaboration), Int. J. Mod. Phys. **A12**, 4069 (1997).
19. V.L. Chernyak and A.R. Zhitnitsky, Phys. Rep. **112**, 173 (1984).

Constraints on Electron-quark Contact Interactions

Kingman Cheung
Department of Physics, University of California,
Davis CA 95616
E-mail: cheung@gluon.ucdavis.edu

In this talk, I summarize a global analysis on electron-quark contact interactions using the updated NC DIS data at HERA, Drell-yan production at the Tevatron, total hadronic cross sections at LEP, atomic parity violation measurement, low energy e-nucleon scattering data, and the ν-nucleon scattering data. The global data do not show any evidence for contact interactions. Thus, we obtain limits of 8–15 TeV on the compositeness scale, which are significantly better than those published by each individual experiment.

1 Introduction

Four-fermion contact interaction is not something new, but was proposed decades ago by Fermi to account for the nuclear beta decay. The interaction is represented by

$$\mathcal{L} \sim G_F \left(\bar{e}\gamma^\mu(1-\gamma^5)\nu\right) \; \left(\bar{u}\gamma_\mu(1-\gamma^5)d\right) \tag{1}$$

where G_F is the fermi constant with dimension $(\text{mass})^{-2}$. This interaction is not renormalizable because amplitudes grow indefinitely with the energy scale if G_F is kept constant. It was only until 60's that the electroweak theory was proposed. The four-fermion contact interaction was then replaced by exchange of weak gauge bosons and G_F replaced by the W boson propagator: $G_F \to 1/(p^2 - m_W^2)$. The weak gauge bosons were only discovered later when the energy scale reached the hundred GeV level. In the above history we learn a couple of lessons: (i) the existence of four-fermion contact interactions is a signal of new physics beyond the existing standard theory; (ii) the exact nature of new physics is unknown at the low energy scale. Only when the energy scale is high enough can the nature of new physics be probed.

The purpose of this analysis [1] is to examine the data from current accelerator experiments to see if there is any sign of contact interactions. If so it is a signal of new physics; if not we put limits on the compositeness scale Λ. Four-fermion contact interactions have been searched in recent high energy experiments: (i) the $qqqq$ contact interaction studied at the Tevatron by CDF [2]; (ii) $eeqq$ interactions at LEP [3], HERA [4], Drell-yan production at the Tevatron [5,6], and low energy e-Nucleon scattering experiments [7]. We concentrate on $eeqq$ contact interactions. In particular, the neutral current (NC) deep-inelastic scattering (DIS) data collected by H1 [8] and ZEUS [9] at HERA between 1994–96 showed a significant excess in cross section in the high-Q^2 region, which aroused an enormous amount of phenomenological activities. One of the explanations is the *eeqq* contact interaction at the scale of 3 TeV. However, the data collected in 1997 alone agreed very well with the SM and, therefore, the logical explanation for the excess in 1996 was statistical fluctuation. The overall result of the combined 94-97 data is as follows [10]: (i) data by ZEUS agreed with the SM expectation up to $Q^2 \sim 30000\,\mathrm{GeV}^2$ while there are 2 events at $Q^2 > 35000\,\mathrm{GeV}^2$, where only 0.29 ± 0.02 is expected; (ii) data by H1 only showed a slight deviation above the SM for $Q^2 > 15000\,\mathrm{GeV}^2$ and the excess events in the mass window of 200 GeV are now much less significant. Although the outcome is somewhat discouraging for searching for new physics, we can, however, use the data to constrain new physics. The objective here is to constrain the *eeqq* contact interactions using the global data, which include: (i) NC DIS data at HERA [10], (ii) Drell-yan production at the Tevatron [5], (iii) total hadronic cross sections at LEP and the left-right asymmetry at SLD [3], (iv) atomic physics parity violation measurement [11] on ^{133}Cs, (v) low energy e-N scattering experiments [7], and (vi) low energy ν-N scattering experiment by CCFR [12]. We shall obtain fits of parameters of *eeqq* contact interactions and finally the limits on the compositeness scale Λ.

In this write-up, we shall summarize the analysis in Ref. [1]. The results presented here, however, use the more updated data since Ref. [1].

2 Parameterization

The conventional effective Lagrangian of *eeqq* contact interactions has the form

$$L_{NC} = \sum_q \Big[\eta_{LL} \left(\overline{e_L}\gamma_\mu e_L\right)\left(\overline{q_L}\gamma^\mu q_L\right) + \eta_{RR}\left(\overline{e_R}\gamma_\mu e_R\right)\left(\overline{q_R}\gamma^\mu q_R\right)$$
$$+\eta_{LR}\left(\overline{e_L}\gamma_\mu e_L\right)\left(\overline{q_R}\gamma^\mu q_R\right) + \eta_{RL}\left(\overline{e_R}\gamma_\mu e_R\right)\left(\overline{q_L}\gamma^\mu q_L\right)\Big] , \qquad (2)$$

where eight independent coefficients $\eta^{eu}_{\alpha\beta}$ and $\eta^{ed}_{\alpha\beta}$ have dimension $(\mathrm{TeV})^{-2}$ and are conventionally expressed as $\eta^{eq}_{\alpha\beta} = \epsilon g^2/\Lambda^2_{eq}$, with a fixed $g^2 = 4\pi$. The sign factor $\epsilon = \pm 1$ allows for either constructive or destructive interference with the SM γ and Z exchange amplitudes and Λ_{eq} represents the mass scale of the exchanged new particles, with coupling strength $g^2/4\pi = 1$. A coupling of this order is expected in substructure models and therefore Λ_{eq} is sometimes called the "compositeness scale".

Left-handed electrons and quarks belong to SU(2) doublets $L = (\nu_L, e_L)$ and $Q = (u_L, d_L)$ and thus from SU(2) symmetry one expects relations between contact terms involving left-handed u or d quarks; similarly, contact terms for left-handed electrons and neutrinos should be related. We start with the most general $SU(2) \times U(1)$ invariant contact term Lagrangian,

$$\begin{aligned}\mathcal{L}_{SU(2)} = \eta_1 \left(\overline{L}\gamma^\mu L\right)\left(\overline{Q}\gamma_\mu Q\right) + \eta_2 \left(\overline{L}\gamma^\mu T^a L\right)\left(\overline{Q}\gamma_\mu T^a Q\right) \\ +\eta_3 \left(\overline{L}\gamma^\mu L\right)\left(\overline{u_R}\gamma_\mu u_R\right) + \eta_4 \left(\overline{L}\gamma^\mu L\right)\left(\overline{d_R}\gamma_\mu d_R\right) \\ +\eta_5 \left(\overline{e_R}\gamma^\mu e_R\right)\left(\overline{Q}\gamma_\mu Q\right) + \eta_6 \left(\overline{e_R}\gamma^\mu e_R\right)\left(\overline{u_R}\gamma_\mu u_R\right) \\ +\eta_7 \left(\overline{e_R}\gamma^\mu e_R\right)\left(\overline{d_R}\gamma_\mu d_R\right) . \end{aligned} \tag{3}$$

By expanding the η_5 term we have

$$\eta_{RL}^{eu} = \eta_5 = \eta_{RL}^{ed} . \tag{4}$$

In addition, the four neutrino and the lepton couplings are related by SU(2),

$$\eta_{LL}^{\nu u} = \eta_{LL}^{ed} , \eta_{LL}^{\nu d} = \eta_{LL}^{eu} , \eta_{LR}^{\nu u} = \eta_{LR}^{eu} , \eta_{LR}^{\nu d} = \eta_{LR}^{ed} . \tag{5}$$

In our analysis, the relations of Eqs. (4) and (5) are only used when neutrino scattering data are included in the analysis. Even though we expect that $SU(2) \times U(1)$ will be a symmetry of the renormalizable interactions which ultimately manifest themselves as the contact terms of Eq. (2), electroweak symmetry breaking may break the degeneracy of SU(2) multiplets of new, heavy quanta whose exchanges give rise to (2). This would result in a violation of the relations of Eqs. (4) and (5). One example is the exchange of the stop $\tilde{t}_1$, $\tilde{t}_2$, and the sbottom $\tilde{b}_L, \tilde{b}_R$ in R-parity violating SUSY models. The large top-quark mass may lead to substantial splitting of the masses of these squarks which could easily lead to violations by up to a factor of two of SU(2) relations such as $\eta_{LR}^{\nu d} = \eta_{LR}^{ed}$.

Because of severe experimental constraints on intergenerational transitions like $K \to \mu e$ we restrict our discussion to first generation contact terms. Only where required by particular data (e.g. the muon sample of Drell-yan production at the Tevatron) will we assume universality of contact terms between e and μ.

3 Global Data

The global data used in this analysis have been described in Ref. [1]. Here we only list those that have been updated since then.

Table 1: The measured number of events as a function of $Q^2_{\rm min}$ at HERA.

ZEUS ($\mathcal{L} = 46.60\,\mathrm{pb}^{-1}$)			H1 ($\mathcal{L} = 37.04\,\mathrm{pb}^{-1}$)		
$Q^2_{\rm min}$ (GeV^2)	$N_{\rm obs}$	N_{exp}	$Q^2_{\rm min}$ (GeV^2)	$N_{\rm obs}$	N_{exp}
2500	1817	1792±93	2500	1297	1276±98
5000	440	396±24	5000	322	336±29.6
10000	66	60±4	10000	51	55.0±6.42
15000	20	17±2	15000	22	14.8±2.13
35000	2	0.29±0.02	20000	10	4.39±0.73
			25000	6	1.58±0.29

3.1 HERA data

The 1997 data alone by H1 and ZEUS agreed very well with the SM expectation, though the combined 1994–1997 data still showed an excess of cross section at high Q^2. The significance of excess is far less severe now. We use the Q^2 distribution presented in the 1998 spring conferences[10]. Note that using Q^2 distribution will not reduce appreciably the sensitivity to contact interactions than using the x-y distribution, because x distribution is not sensitive to contact interactions (unlike the narrow-width leptoquark model) but y is somewhat sensitive to it. The updated data are tabulated in Table 1.

3.2 Drell-yan Production

Since our previous analysis, in which we used the plotted data on a CDF graph, CDF[5] has published the observed number of events in bins of invariant mass of the lepton pair. The data are given in Table 2. In this CDF paper, they obtained limits on the compositeness scale using only the CDF data, in the order of a few TeV [a]. At the end, we shall obtain limits significantly better than these limits.

3.3 LEP

The LEP collaborations have published new measurements of total hadronic cross sections at $\sqrt{s} = 130, 172$, and 183 GeV. The data that we used are shown in Table 3.

Table 2: The electron and muon samples of Drell-yan production in each mass bin by CDF.

	e^+e^-		$\mu^+\mu^-$	
$M_{\ell\ell}$	$N_{\rm obs}$	$N_{\rm exp}$	$N_{\rm obs}$	$N_{\rm exp}$
50–150	2581	2581	2533	2533
150–200	8	10.8	9	9.7
200–250	5	3.5	4	3.2
250–300	2	1.4	2	1.3
300–400	1	0.97	1	0.94
400–500	1	0.25	0	0.27
500–600	0	0.069	0	0.087

4 Fits and Limits

The fits of contact parameters are obtained by minimizing the χ^2 of the data sets. In order to see how each data set affects the fit, we obtain the fit with each data set added one at a time. The fits with various combinations of data sets are shown in Table 4. Two important observations are offered as follows. (i) When the Drell-yan data are added to "HERA+APV+eN", the fitted parameters change dramatically and so are the χ^2's of each data set. This can be understood as follows. The HERA data actually favor non-zero contact parameters (especially the last entries of ZEUS and H1 data): see Fig. 1(a). However, this fit of contact parameters very much contradicts the Drell-yan data: see Fig. 1(b). Therefore, when DY data are taken into account, the fit changes drastically. The curves of the fit with all data sets are also shown in Fig. 1. (ii) The goodness of the fits is indicated by the χ^2 per degree of freedom (d.o.f.). The χ^2/d.o.f. ($\chi^2_{\rm cont.}$/d.o.f.=1.003) of contact interactions is very close to that of the SM ($\chi^2_{\rm SM}$/d.o.f.=1.067) for the last column in Table 4. For the second last column in Table 4, $\chi^2_{\rm cont.}$/d.o.f.= 1.045 and $\chi^2_{\rm SM}$/d.o.f.= 1.089.

In view of these, we conclude that the global data do not show any sign of contact interactions. Thus, we can derive 95% CL limits on the compositeness scale, below which the contact interaction is ruled out. The limits on $\Lambda_\pm$ are listed in Table 5–7. In Table 5, for each chirality coupling considered the others are put to zero. The limits on Λ obtained range from 8–15 TeV, which improve significantly from each individual experiment [3,4,5,6]. We also calculate the limits on the compositeness scale when some symmetries on contact terms are considered, as shown in Tables 6 and 7. VV stands for vector-vector:

[a] D0 Collaboration [6] has also recently published a paper on "Search for Quark-lepton Compositeness using the Drell-Yan process at D0".

Table 3: Total hadronic cross sections σ_{had} measured by the LEP collaborations.

$\sqrt{s}$ (GeV)	σ_{had}	σ_{SM}
	ALEPH	
130	79.5 $\pm$ 4.14	77.16
136	64.5 $\pm$ 3.85	62.52
183	23.6 $\pm$ 0.73	23.05
	DELPHI	
130.2	82.2 $\pm$ 5.2	83.1
136.2	65.9 $\pm$ 4.7	67.0
161.3	40.2 $\pm$ 2.1	34.8
172.1	30.6 $\pm$ 2.0	28.9
	L3	
130.3	81.8 $\pm$ 6.4	78
136.3	70.5 $\pm$ 6.2	63
140.2	67 $\pm$ 47	56
161.3	37.3 $\pm$ 2.2	34.9
170.3	39.5 $\pm$ 7.5	29.8
172.3	28.2 $\pm$ 2.2	28.9
	OPAL	
130.25	64.3 $\pm$ 5.1	77.6
136.22	63.8 $\pm$ 5.2	62.9
161.34	35.5 $\pm$ 2.2	33.7
172.12	27.0 $\pm$ 1.9	27.6

$\eta_{LL} = \eta_{LR} = \eta_{RL} = \eta_{RR} = \eta_{VV}$, while AA stands for axial-vector-axial-vector: $\eta_{LL} = -\eta_{LR} = -\eta_{RL} = \eta_{RR} = \eta_{AA}$. These limits, in general, are not as strong as those in the previous table because the additional symmetry automatically satisfies the parity violation experiments: APV and e-N.

In conclusion, the global data have been examined and do not support the existence of $eeqq$ contact interactions with the compositeness scale upto 8–15 TeV. Although the 1994–97 the NC DIS data at HERA favor a slightly non-zero contact interaction, the other data, especially the Drell-yan data at the Tevatron and the atomic parity violation measurement, severely constrain it. Finally, the limits on the compositeness scale obtained in this analysis are significantly better than the results published by each individual experiment. We urge others to use the limits of Λ obtained in this analysis. The above analysis can also be applied in a straight-forward fashion to other new physics such as Z' and leptoquark models.

I would like to thank Vernon Barger, Karou Hagiwara, and Dieter Zeppenfeld for collaboration.

Table 4: The best estimate of the $\eta^{eq}_{\alpha\beta}$ parameters when various data sets are added successively. In the last column when the ν-N data are included the $\eta^{\nu q}_{L\beta}$ are given in terms of $\eta^{eq}_{L\beta}$ by Eq. (5) and we assume $\eta^{eu}_{RL} = \eta^{ed}_{RL}$ in the last column.

	HERA only	HERA+APV +eN	HERA+APV +eN+DY	HERA+APV +eN+DY+LEP	HERA+DY+APV +eN+LEP+νN
η^{eu}_{LL}	$2.04^{+3.97}_{-5.26}$	$2.25^{+2.29}_{-3.63}$	$0.22^{+0.67}_{-0.57}$	$0.049^{+0.63}_{-0.43}$	$-0.046^{+0.62}_{-0.38}$
η^{eu}_{LR}	$-4.30^{+4.30}_{-0.78}$	$-2.77^{+3.20}_{-1.70}$	$0.60^{+0.51}_{-0.66}$	$0.76^{+0.38}_{-0.63}$	$0.76^{+0.35}_{-0.68}$
η^{eu}_{RL}	$-1.75^{+3.75}_{-2.59}$	$-3.53^{+2.91}_{-0.90}$	$-0.004^{+0.72}_{-0.72}$	$0.042^{+0.73}_{-0.75}$	$0.13^{+0.73}_{-0.77}$
η^{eu}_{RR}	$2.62^{+4.28}_{-5.36}$	$2.23^{+1.77}_{-3.41}$	$0.040^{+0.66}_{-0.62}$	$-0.051^{+0.68}_{-0.57}$	$-0.091^{+0.71}_{-0.56}$
η^{ed}_{LL}	$-1.71^{+7.87}_{-6.77}$	$-2.22^{+5.62}_{-4.55}$	$0.25^{+1.64}_{-1.72}$	$0.39^{+0.76}_{-0.92}$	$0.11^{+0.82}_{-0.53}$
η^{ed}_{LR}	$-0.011^{+4.85}_{-4.48}$	$-0.95^{+3.76}_{-3.47}$	$1.65^{+1.39}_{-2.79}$	$0.79^{+1.33}_{-2.08}$	$0.36^{+1.16}_{-1.82}$
η^{ed}_{RL}	$-1.86^{+4.86}_{-4.38}$	$-0.95^{+3.87}_{-3.23}$	$1.97^{+1.30}_{-2.74}$	$1.11^{+1.27}_{-2.02}$	$= \eta^{eu}_{RL}$
η^{ed}_{RR}	$-2.28^{+7.87}_{-7.22}$	$-1.61^{+5.59}_{-4.85}$	$0.55^{+1.60}_{-1.73}$	$0.73^{+0.85}_{-1.03}$	$0.86^{+0.60}_{-1.16}$
HERA	7.57	7.86	12.10	12.34	12.73
APV		0.00	0.00	0.001	0.001
eN		0.46	0.47	0.46	0.51
DY			4.40	4.38	4.39
LEP				23.57	23.49
νN					0.00
Total χ^2	7.57	8.32	16.97	40.75	41.12
SM χ^2	17.27	20.27	24.55	51.20	51.21
SM d.o.f.	11	16	28	47	48

References

1. V. Barger, K. Cheung, K. Hagiwara, and D. Zeppenfeld, *Phys. Rev.* D **57**, 391 (1998).
2. F. Abe *et al.* (CDF Coll.), *Phys. Rev. Lett.* **77**, 5336 (1996); erratum ibid. **78**, 4307 (1997).
3. ALEPH Coll., conference contribution ALEPH 98-021; DELPHI Coll., *submission to HEP'97 Jerusalem conference*, DELPHI 97-132; L3 Coll., *Phys. Lett.* B **407**, 361 (1997); OPAL Coll. *Eur. Phys. J.* C **2**, 441 (1998).
4. S. Aid *et al.* (H1 Coll.), *Phys. Lett.* B **353**, 578 (1995); talk by D. Williams (ZEUS Coll.) at *DIS98*, Brussels, April 1998.
5. F. Abe *et al.* (CDF Coll.), *Phys. Rev. Lett.* **79**, 2198 (1997).
6. D0 Coll., submission to the *ICHEP98*, Vancouver, Canada, July 1998
7. C.Y. Prescott *et al.*, *Phys. Lett.* B **84**, 524 (1979); W. Heil *et al.*, *Nucl.*

Table 5: The best estimate on $\eta^{eq}_{\alpha\beta}$ and the 95% CL limits on the compositeness scale $\Lambda^{eq}_{\alpha\beta}$, where $\eta^{eq}_{\alpha\beta} = 4\pi\epsilon/(\Lambda^{eq}_{\alpha\beta\epsilon})^2$. When one of the η's is considered the others are set to zero. SU(2) relations are assumed and νN data are included.

		95% CL Limits	
Chirality (q)	η (TeV^{-2})	Λ_+ (TeV)	Λ_- (TeV)
LL(u)	0.026 ± 0.056	9.9	11.6
LR(u)	0.11 ± 0.079	7.3	11.1
RL(u)	-0.043 ± 0.038	15.5	10.8
RR(u)	-0.12 ± 0.078	11.5	7.0
LL(d)	0.072 ± 0.060	8.5	12.5
LR(d)	0.079 ± 0.072	7.8	11.2
RR(d)	-0.064 ± 0.072	10.9	8.1

Phys. B **327**, 1 (1989); P.A. Souder *et al.*, *Phys. Rev. Lett.* **65**, 694 (1990).
8. C. Adloff *et al.* (H1 Coll.), *Z. Phys.* C **74**, 191 (1997).
9. J. Breitweg *et al.* (ZEUS Coll.), *Z. Phys.* C **74**, 207 (1997).
10. See, e.g., DESY seminar by G. Bernardi (H1 Coll.) at DESY, Hamburg, March 13th, 1998, and the talk presented by S. Dasu (ZEUS Coll.) at the *Phenmomenology Workshop*, University of Wisconsin-Madison, March 1998.
11. C.S. Wood *et al.*, *Science* **275**, 1759 (1997).
12. CCFR Coll., K.S. McFarland *et al.*, *Eur. Phys. J.* **C1**, 509 (1998).

Table 6: The best estimate on η^{eq} for the minimal setting, VV, AA, and SU(12), and the corresponding 95% CL limits on the compositeness scale Λ, where $\eta = 4\pi\epsilon/(\Lambda_\epsilon)^2$. When one of the η's is considered the others are set to zero. Here we do not use SU(2) relations nor do we include the νN data.

Chirality (q)	η (TeV^{-2})	95% CL Limits Λ_+ (TeV)	Λ_- (TeV)
$\eta^{eu}_{LR} = \eta^{eu}_{RL}$	$0.41^{+0.23}_{-0.27}$	4.0	6.1
$\eta^{ed}_{LR} = \eta^{ed}_{RL}$	$-1.19^{+0.37}_{-0.30}$	2.3	2.8
η^{eu}_{VV}	$-0.064^{+0.090}_{-0.089}$	9.3	7.6
η^{ed}_{VV}	$0.38^{+0.18}_{-0.21}$	4.4	4.9
η^{eu}_{AA}	$-0.30^{+0.12}_{-0.11}$	9.4	5.1
η^{ed}_{AA}	$0.31^{+0.15}_{-0.16}$	4.8	7.5
$\eta^{eu}_{LL} = -\eta^{eu}_{LR}$	$-0.47^{+0.19}_{-0.18}$	6.5	4.1
$\eta^{eu}_{RL} = -\eta^{eu}_{RR}$	$0.54^{+0.19}_{-0.21}$	3.9	5.8
$\eta^{ed}_{LL} = -\eta^{ed}_{LR}$	$0.54^{+0.24}_{-0.27}$	3.7	4.8
$\eta^{ed}_{RL} = -\eta^{ed}_{RR}$	$-0.62^{+0.30}_{-0.26}$	3.6	3.5

Table 7: Same as the last Table but with a further condition: $\eta^{eu} = \eta^{ed}$. Here $q = u = d$.

Chirality (q)	η (TeV^{-2})	95% CL Limits Λ_+ (TeV)	Λ_- (TeV)
$\eta^{eq}_{LR} = \eta^{eq}_{RL}$	$0.46^{+0.25}_{-0.33}$	3.9	5.5
η^{eq}_{VV}	$-0.026^{+0.15}_{-0.13}$	5.3	6.9
η^{eq}_{AA}	$-0.40^{+0.17}_{-0.15}$	4.3	4.5
$\eta^{eq}_{LL} = -\eta^{eq}_{LR}$	$-0.61^{+0.25}_{-0.21}$	3.4	3.7
$\eta^{eq}_{RL} = -\eta^{eq}_{RR}$	$0.65^{+0.21}_{-0.25}$	3.6	3.3

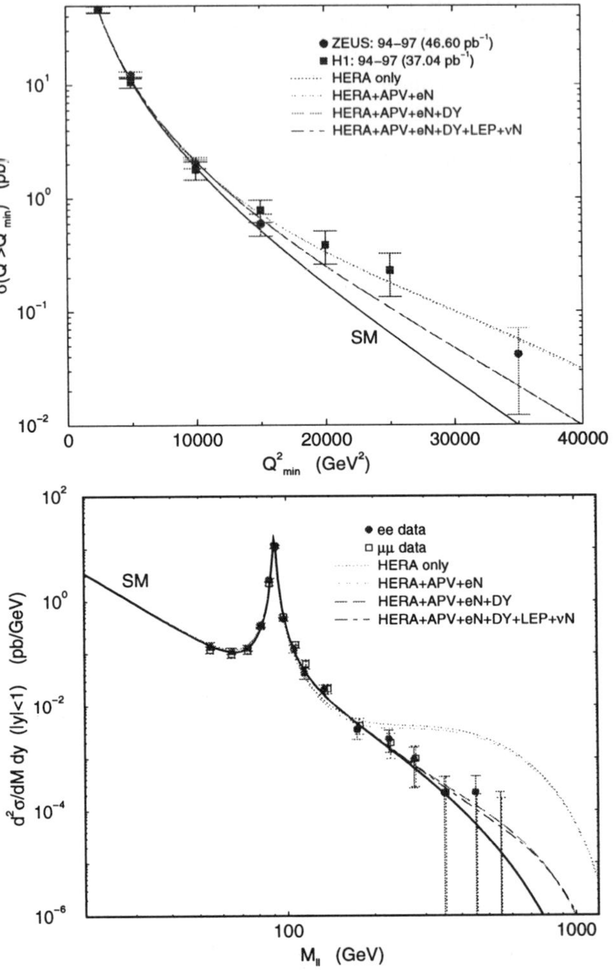

Figure 1: (a) The cumulated cross section $\sigma(Q^2 > Q^2_{\mathrm{min}})$ at HERA as a function of Q^2_{min}. The 94–97 H1 and ZEUS data, and curves of fits to various data sets are shown, (b) the differential cross section $d^2\sigma/dMdy$ for Drell-yan production at the Tevatron.

QCD RESUMMATION IN SINGLE AND DOUBLE GAUGE BOSON PRODUCTIONS AT HADRON COLLIDERS

C.-P. YUAN
Department of Physics and Astronomy, Michigan State University,
East Lansing, MI 48824, USA

After a brief review on the technique of resumming the large logarithmic terms $\alpha_s^n \ln^m(Q^2/Q_T^2)$ due to the effects of multiple gluon emission predicted by the QCD theory, I discuss its application in the production of $W^\pm$, Z, $\gamma\gamma$ and ZZ in hadron collisions.

1 Introduction

It is the prediction of the QCD theory that at hadron colliders the production of Drell-Yan pairs or weak gauge bosons ($W^\pm$ and Z) are often accompanied by gluon radiation. Therefore, to test the QCD theory or to probe the electroweak properties in the vector boson productions, it is necessary to include the effects of multiple gluon emission. At the Tevatron (a $p\bar{p}$ collider), we expect about 2×10^6 $W^\pm$ and 6×10^5 Z bosons produced at $\sqrt{S} = 1.8\,\text{TeV}$, per $100\,\text{pb}^{-1}$ of luminosity. This large sample of data is useful for (i) QCD studies (with either single or multiple scales), (ii) precision measurement of the W boson mass and width, and (iii) probing for new physics (e.g., Z'), *etc.* To achieve the above physics goals requires detailed information on the distributions of the rapidity and the transverse momentum of $W^\pm/Z$ bosons and of their decay products.

Consider the production process $h_1 h_2 \to VX$. Denote Q_T and Q to be the transverse momentum and the invariant mass of the vector boson V, respectively. When $Q_T \sim Q$, there is only one hard scale in this problem, and the fixed-order perturbation calculation is reliable. When $Q_T \ll Q$, this becomes a two scale problem, and the convergence of the conventional perturbative expansion becomes impaired. Hence, it is necessary to apply the technique of QCD resummation to resum the *singular* terms:

$$\begin{aligned}\frac{d\sigma}{dQ_T^2} \sim \frac{1}{Q_T^2}\{\, &\alpha_S(L+1) + \alpha_S^2(L^3+L^2) + \alpha_S^3(L^5+L^4) + \alpha_S^4(L^7+L^6) + \ldots \\ &+\alpha_S^2(L+1) + \alpha_S^3(L^3+L^2) + \alpha_S^4(L^5+L^4) + \ldots \\ &+\alpha_S^3(L+1) + \alpha_S^4(L^3+L^2) + \ldots \,\},\end{aligned}$$

where L denotes $\ln(Q^2/Q_T^2)$ and the explicit coefficients multiplying the logs are suppressed.

Resummation of large logarithms yields a Sudakov form factor and cures divergence at $Q_T \to 0$. It was pioneered by Dokshitzer, D'yakonov and Troyan (DDT) who performed the analysis in Q_T-space which lead to the leading-log resummation formalism [1]. Later, Parisi-Petronzio showed [2] that for large Q the $Q_T \to 0$ region can be calculated perturbatively by imposing the condition of the transverse momentum conservation,

$$\delta^{(2)}\left(\sum_{i=1}^{n} \vec{k}_{T_i} - \vec{q}_T\right) = \int \frac{d^2 b}{4\pi^2}\, e^{i\vec{Q}_T\cdot\vec{b}} \prod_{i=1}^{n} e^{i\vec{k}_{T_i}\cdot\vec{b}},$$

in the b-space (the impact parameter, which is the Fourier conjugate to Q_T). Their improved formalism sums also some subleading-logs. They showed that as $Q \to \infty$, events at $Q_T \sim 0$ may be obtained asymptotically by the emission of at least two gluons whose transverse momenta are not small and add to zero. The intercept at $Q_T = 0$ is predicted to be [2]

$$\left.\frac{d\sigma}{dQ_T^2}\right|_{Q_T\to 0} \sim \sigma_0 \left(\frac{\Lambda^2}{Q^2}\right)^{\eta_0},$$

where $\eta_0 = A \ln\left[1 + \frac{1}{A}\right]$ with $A = 12C_F/(33 - 2n_f)$, and $\eta_0 \simeq 0.6$ for $n_f = 4$ and $C_F = 4/3$. Collins-Soper-Sterman (CSS) extended [3] the work by Parisi-Petronzio in b-space and applied the properties of the renormalization group invariance to set up a formalism that resums all the large log terms to all orders in α_s. This is the formalism that we will concentrate on in this work.

2 Collins-Soper-Sterman (CSS) resummation formalism

In the resummation formalism by Collins, Soper and Sterman [3], the cross section is written in the form

$$\frac{d\sigma(h_1 h_2 \to VX)}{dQ^2\, dQ_T^2 dy} = \frac{1}{(2\pi)^2}\int d^2 b\, e^{i\vec{Q}_T\cdot\vec{b}}\widetilde{W}(b, Q, x_1, x_2) + Y(Q_T, Q, x_1, x_2),$$

where Y is the *regular* piece which can be obtained by subtracting the *singular* terms from the exact fixed-order result. $\widetilde{W}$ satisfies a renormalization group equation. Its solution is of the form

$$\widetilde{W}(Q, b, x_1, x_2) = e^{-\mathcal{S}(Q,b,C_1,C_2)}\widetilde{W}\left(\frac{C_1}{C_2 b}, b, x_1, x_2\right),$$

where the Sudakov exponent is defined as

$$\mathcal{S}(Q, b, C_1, C_2) = \int_{C_1^2/b^2}^{C_2^2 Q^2} \frac{d\overline{\mu}^2}{\overline{\mu}^2}\left[A\left(\alpha_s(\overline{\mu}), C_1\right) \ln\left(\frac{C_2^2 Q^2}{\overline{\mu}^2}\right) + B\left(\alpha_s(\overline{\mu}), C_1, C_2\right)\right],$$

and the x_1 and x_2 dependence of $\widetilde{W}$ factorizes as

$$\widetilde{W}\left(\frac{C_1}{C_2 b}, b, x_1, x_2\right) = \sum_j e_j^2 \, \mathcal{C}_{jh_1}\left(\frac{C_1}{C_2 b}, b, x_1\right) \, \mathcal{C}_{jh_2}\left(\frac{C_1}{C_2 b}, b, x_2\right),$$

where $\mathcal{C}_{jh}$ is a convolution of the parton distribution with a calculable Wilson coefficient, called C_{ja} function. They are defined through

$$\mathcal{C}_{jh}(Q, b, x) = \sum_a \int_x^1 \frac{d\xi}{\xi} \, C_{ja}\left(\frac{x}{\xi}, b, \mu = \frac{C_3}{b}, Q\right) \, f_{a/h}\left(\xi, \mu = \frac{C_3}{b}\right),$$

where a sums over incoming partons, and j denotes the quark flavors with (electroweak) charge e_j. A few comments about this formalism is listed below:

- The A, B and C functions can be calculated order-by-order in α_s.
- A special choice of the renormalization constants C_i can be made so that the singular terms obtained from the expansion of the CSS resummed calculation agrees with that from the fixed-order calculation. This is the canonical choice. It has $C_1 = C_3 = 2e^{-\gamma_E} \equiv b_0$ and $C_2 = C_1/b_0 = 1$, where $\gamma_E = 0.577\ldots$ is Euler's constant.
- b is integrated from 0 to ∞. For $b \gg 1/\Lambda_{QCD}$, the perturbative calculation is no longer reliable. Hence, a non-perturbative function is needed in this formalism to compare theory prediction with experimental data.

We refer the readers to Ref. [4] for a more detailed discussion on how to apply the CSS resummation formalism to phenomenological physics at hadron colliders.

2.1 Non-perturbative function

As noted in the previous section, it is necessary to include a non-perturbative function in the CSS resummation formalism to incorporate some long distance physics that is not accounted by the perturbative derivation. Collins-Soper postulated [3]

$$\widetilde{W}_{j\bar{k}}(b) = \widetilde{W}_{j\bar{k}}(b_*) \widetilde{W}_{j\bar{k}}^{NP}(b),$$

with

$$b_* = \frac{b}{\sqrt{1 + (b/b_{max})^2}},$$

so that b never exceeds b_{max} and $\widetilde{W}_{j\bar{k}}(b_*)$ can be reliably calculated perturbatively. (In a numerical calculation, b_{max} is typically set to be of the order of GeV^{-1}.) Based upon a renormalization group analysis, they found that the non-perturbative function can be generally written as

$$\widetilde{W}_{j\bar{k}}^{NP}(b,Q,Q_0,x_1,x_2) = \exp\left[-F_1(b)\ln\left(\frac{Q^2}{Q_0^2}\right) - F_{j/h_1}(x_1,b) - F_{\bar{k}/h_2}(x_2,b)\right],$$

where F_1, F_{j/h_1} and $F_{\bar{k}/h_2}$ must be extracted from data with the constraint

$$\widetilde{W}_{j\bar{k}}^{NP}(b=0) = 1.$$

Furthermore, F_1 only depends on Q, while F_{j/h_1} and $F_{\bar{k}/h_2}$ in general depend on x_1 or x_2, and their values can depend on the flavor of the initial state partons (j and $\bar{k}$ in this case). Later, in Ref. [5], it was shown that the $F_1(b)\ln\left(\frac{Q^2}{Q_0^2}\right)$ dependence is also suggested by analyzing the infrared renormalon contribution in Q_T distribution.

2.2 *Testing the universality of* $\widetilde{W}_{j\bar{k}}^{NP}$

The CSS resummation formalism suggests the non-perturbative function to be universal. Its role is similar to the parton distribution function (PDF) in any fixed order perturbative calculation, and its value needs to be determined by data. The first attempt to determine such a universal non-perturbative function was done by Davies, Webber and Stirling (DWS) [6] in 1985 using Duke and Owens parton distribution function. In 1994, Ladinsky and Yuan (LY) [7] observed that the prediction of the DWS set of $\widetilde{W}_{j\bar{k}}^{NP}$ largely deviates from the R209 data ($p+p \to \mu^+\mu^- + X$ at $\sqrt{S} = 62\,\mathrm{GeV}$) using the CTEQ2M PDF. To incorporate possible $\ln(\tau)$ dependence, LY postulated

$$\widetilde{W}_{j\bar{k}}^{NP}(b,Q,Q_0,x_1,x_2) = \exp\left[-g_1 b^2 - g_2 b^2 \ln\left(\frac{Q}{2Q_0}\right) - g_1 g_3 b \ln(100 x_1 x_2)\right],$$

where $x_1 x_2 = \tau$. A "two-stage fit" of the R209, CDF-Z ($4\,\mathrm{pb}^{-1}$ data) and E288 ($p + Cu$) data gave $g_1 = 0.11^{+0.04}_{-0.03}$ GeV^2, $g_2 = 0.58^{+0.1}_{-0.2}$ GeV^2, and $g_3 = -1.5^{+0.1}_{-0.1}$ GeV^{-1}, for $Q_0 = 1.6$ GeV and $b_{max} = 0.5$ GeV^{-1}. Unfortunately, a fortran code error in calculating the parton densities inside the neutron lead to a wrong g_3 value. Currently, Brock, Ladinsky, Landry, and Yuan [8] are revisiting this problem using the R209 ($p+p$), CDF-Z ($p+\bar{p}$ with $4\,\mathrm{pb}^{-1}$ data), E288 ($p + Cu$), and E605 ($p + Cu$) data, with CTEQ3M PDF. The preliminary results show that, for $Q_0 = 1.6\,\mathrm{GeV}$ and $b_{max} = 1/(2\,\mathrm{GeV})$, two forms for $\widetilde{W}_{jk}^{NP}(b, Q, Q_0, x_1, x_2)$ give good fits ($\chi^2/dof \simeq 1.4$): (i) $g_1 = 0.24$, $g_2 = 0.34$ and $g_3 = 0.0$ (DWS form, pure Gaussian form in b-space, without x dependence), and (ii) $g_1 = 0.15$, $g_2 = 0.48$ and $g_3 = -0.58$ (LY form, with a linear b term and x dependence). We are in the process of determining the uncertainties of these fitted parameters g_i.

3 W/Z Production and Decay at the Tevatron

A detailed study on the distributions of the decay leptons from the W/Z boson produced at the Tevatron has been given in Ref. [4]. Here, we shall concentrate on the effects of the non-perturbative function in various distributions predicted by the CSS resummation formalism.

3.1 Transverse Momentum Q_T of the W/Z Boson

First of all, let us examine the distribution of the transverse momentum of the W/Z boson produced at the Tevatron. By construction of the theory, the CSS formalism will give a smooth Q_T distribution if A, B, C and Y functions were known to all orders in α_s. However, in practice we only calculate those functions to some fixed order in α_s, therefore, some "switching" procedure (to switch from the total resummed result to the fixed-order perturbative result) should be applied to have a continuous distribution that would match the fixed-order result in the large transverse momentum Q_T (of the order Q) where $\ln(Q^2/Q_T^2)$ is small and the resummed result is less accurate. Our matching prescription is to switch from the resummed prediction to the fixed-order perturbative calculation as they cross around $Q_T \sim Q$. This switching procedure is done for any given Q (invariant mass) and y (rapidity) of the W/Z boson.[a] As shown in the figures of Ref. [4], including a higher order Y term, as expected, improves the "matching" between the resummed and the fixed-order results. At $O(\alpha_s)$, that crossing occurs at about 50 GeV, while at $O(\alpha_s^2)$, it occurs at about 70 GeV for W boson production. Different non-perturbative functions do not change the distributions of Q_T for $Q_T > 10$ GeV, though they can give very different prediction in the small Q_T (near peak) region.

3.2 Total cross section

It was shown by an analytical calculation in Ref. [4] that if "matching" (switching) is chosen to be at $Q_T = Q$, then the CSS resummed rate is the same

[a]This procedure is different from that proposed in Ref. [9].

as the next-to-leading order (NLO) rate. However, in that case, the Q_T distribution is not smoothly continuous for Q_T close to Q. Since the region of $Q_T > 50\,\text{GeV}$ only contributes to the total rate by less than 2%, and the CSS resummed formalism includes also some even higher order contributions beyond NLO, we regard the difference between the resummed rate and the NLO rate as a gauge of the uncalculated higher order contribution which is shown to be small at the Tevatron. We found that the total rate is not sensitive to either non-perturbative function or matching prescription.

3.3 Lepton transverse momentum

At the 1.8 TeV Tevatron, the best method for measuring the mass of the W boson is to examine the transverse mass distribution of its decay leptons, which has been known to be not sensitive to the non-perturbative function. At the Run-2 of the Tevatron (at 2 TeV, with a $2\,\text{fb}^{-1}$ luminosity), the measurement of the transverse mass of the W boson becomes less accurate. It is useful to also measure the inclusive lepton $P_T^{e^+}$ spectrum. In Fig. 1, we show that the difference between the resummed and the NLO results is much larger than the dependence on the non-perturbative function in the resummed calculation (although its dependence is not negligible).

3.4 Lepton charge asymmetry

The distribution of the rapidity of the charged lepton from $W^\pm$ boson decay provides useful information to determine the ratio of the up and down quark parton distribution functions. It was shown in Ref. [4] that without imposing kinematic cuts, the prediction of the resummed calculation is the same as that of the NLO. This is obvious because an integration over the entire Q_T space in the CSS formalism reproduces the NLO rate as discussed above. However, with kinematic cuts, the rapidity distribution of the charged lepton or the W boson is different between the resummed and the NLO calculations. Hence, they yield different predictions on the lepton charge asymmetry in the large rapidity region[4]. We have also checked that the dependence of this distribution on the non-perturbative function is negligible.

The resummed calculation also predicts different correlations between the two decay leptons from Z boson. A couple of examples are the distribution showing the balance in transverse momentum $\Delta p_T = |\vec{p}_T^{\,\ell_1}| - |\vec{p}_T^{\,\bar{\ell}_2}|$, and the angular correlation $z = -\vec{p}_T^{\,\ell_1} \cdot \vec{p}_T^{\,\bar{\ell}_2}/[\max(p_T^{\ell_1}, p_T^{\bar{\ell}_2})]^2$. We refer the reader to Ref. [4] for more details. It is expected that the effect of the non-perturbative function is maximal near $\Delta p_T = 0$ and $z = 1$.

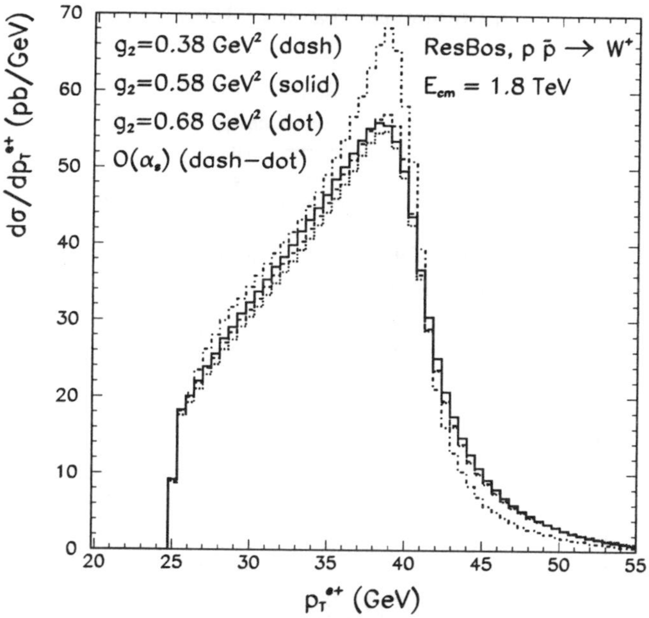

Figure 1: NLO (dashed) and resummed $\mathcal{O}(\alpha_S)$ (solid) calculations Using the LY fit with cuts ($Q_T < 30$ GeV, $p_T^{e^+,\nu} > 25$ GeV).

4 $\gamma\gamma$ and ZZ Productions

4.1 At Colliders

It is straightforward to extend the above CSS formalism to calculate the distribution of photons from $h_1 h_2 \to \gamma\gamma X$. In Ref. [10], we have included the full content of NLO contributions from $q\bar{q}$ and qg subprocesses. The accuracy of the resummed result is similar to that for the Drell-Yan and W/Z productions. In addition, we also included part of the higher order contribution from the gg process whose lowest order contribution comes from one-loop box diagrams. This was done by including the $A^{(2)}$ term in the Sudakov factor to resum higher order large logs due to initial state radiation, and the NLO contribution, of $\mathcal{O}(\alpha_{em}^2 \alpha_s^3)$, to the hard scattering subprocess is added in an approximate fashion.

We found that the $gg \to \gamma\gamma X$ rate is not small at the Tevatron. Including

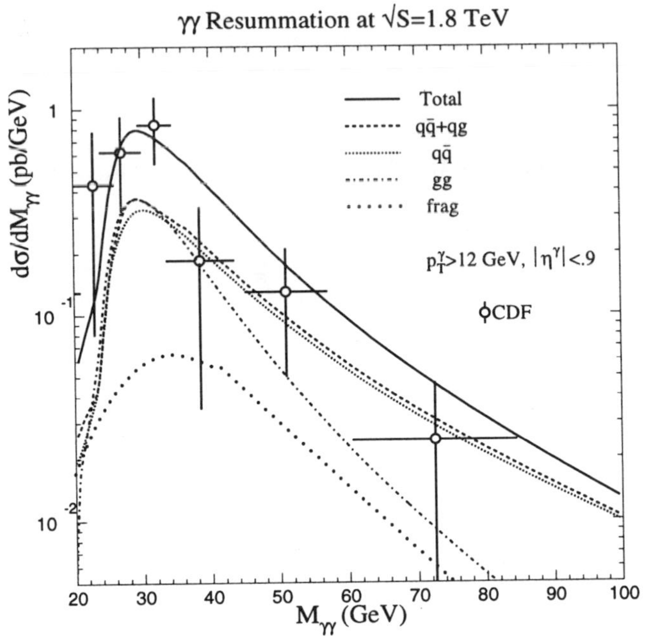

Figure 2: The predicted distribution for the invariant mass of the photon pair $M_{\gamma\gamma}$ from the resummed calculation compared to the CDF data, with the CDF cuts imposed in the calculation.

part of the NLO $gg \to \gamma\gamma$ approximately doubles the LO gg box contribution to the cross section. This is clearly seen in the distribution of the invariant mass of the photon pair, as shown in Fig. 2. The other distributions, such as the distribution of the transverse momentum of the photon pair Q_T, compared with CDF and DØ data, can be found in Ref. [10]. In general, the theory predictions agree well with the data. We note that to accurately predict the production rates of the di-photon pairs at the CERN Large Hadron Collider (LHC) requires a better calculation of the gg fusion rate. Namely, a full NLO result is needed to predict the distribution of the di-photon pair with large Q_T or small $\Delta\phi$ [11].

A similar study for the ZZ pair production at the Tevatron and the LHC can be found in [11]. In Fig. 3, we show the predicted transverse momentum distribution of Z^0 pairs prodcued at the LHC.

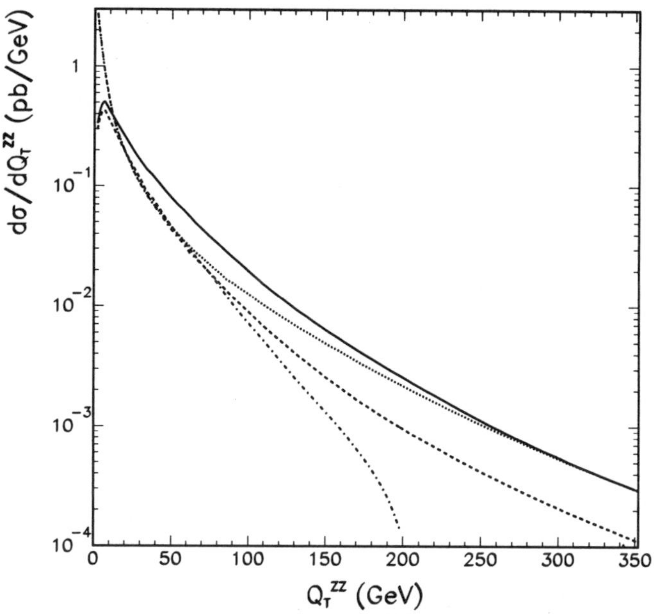

Figure 3: Transverse momentum distribution of Z^0 pairs from $q\bar{q}+qg$ partonic initial states at the LHC. The $\mathcal{O}(\alpha_s)$ (dotted) and the asymptotic (dash-dotted) pieces are coincide and diverge as $Q_T \to 0$. The resummed (solid) curve matches the $\mathcal{O}(\alpha_s)$ curve at about $Q_T = 320$ GeV. The resummed $q\bar{q}$ contribution (excluding the qg contribution) is shown as dashed line.

4.2 At Fixed Target Experiments

In Ref. [10], we also presented our results for the di-photon production rate and distributions at the fixed-target experiment (E706 at Fermilab): $p + Be \to \gamma\gamma + X$ at $\sqrt{S}$=31.5 GeV.

In this case, we found that in general the gg rate is small, and the distribution of the transverse momentum of the photon pair Q_T is dominated by the non-perturbative contribution from the partonic $q\bar{q}$ process. Therefore, when this set of data becomes available, it should also be included in a global fit to determine the non-perturbative function associated with the $q\bar{q}$ initiated processes in the CSS formalism. As shown in Ref. [10], in some region of phase space (*e.g.*, di-photon pair with relatively large Q_T), the resummed gg contribution can become important to provide some useful information about the non-perturbative function associated with the partonic gg process in the CSS formalism. We refer the readers to Ref. [10] for more details.

5 Conclusion

The effects of QCD gluon resummation are important in many precision measurements. A Monte Carlo package ResBos ((Resummed Boson Production and Decay)) is available[12] for studying the effects of gluon resummation (with NLO hard part corrections) in hadron collisions.

Acknowledgments

I thank the organizers for the warm hospitality. It is always a pleasure to come back to my home country and discuss physics with friends here. I would also like to thank my collaborators: C. Balázs, E. Berger, R. Brock, G. Ladinsky, F. Landry, S. Mrenna, and J.W. Qiu for the fruitful collaborations. From them and my CTEQ colleagues, I have learned a great deal of QCD physics. This work was supported in part by the U.S. NSF under grant PHY-9802564

References

1. Y.I. Dokshitser, D.I. D'Yakonov, S.I. Troyan, Phys. Lett. **B79** (1978) 269.
2. G. Parisi, R. Petronzio, Nucl. Phys. **B154** (1979) 427.
3. J. Collins, D. Soper, Nucl. Phys. **B193** (1981) 381; Erratum **B213** (1983) 545; **B197** (1982) 446;
   J. Collins, D. Soper, G. Sterman, Nucl. Phys. **B250** (1985) 199.
4. C. Balázs and C.–P. Yuan, Phys. Rev. **D56** (1997) 5558, and the references therein.
5. G.P. Korchemsky and G. Sterman, Nucl. Phys. **B437** (1995) 415.
6. C. Davies, B. Webber, W. Stirling, Nucl. Phys. **B256** (1985) 413.
7. G.A. Ladinsky, C.–P. Yuan, Phys. Rev. **D50** (1994) 4239.
8. R. Brock, G. Ladinsky, F. Landry, and C.–P. Yuan, in preparation.
9. P.B. Arnold, R.P. Kauffman, Nucl. Phys. **B349** (1991) 381.
10. C. Balázs, E.L. Berger, S. Mrenna, and C.–P. Yuan, Phys. Rev. **D57** (1998) 6934, and the references therein.
11. C. Balázs, C.–P. Yuan, hep-ph/9810319.
12. It can be obtained from http://www.pa.msu.edu/~balazs/ResBos/.

SUSY COSMOLOGY

Kin-Wang Ng
Institute of Physics, Academia Sinica, Taipei, Taiwan 11529, R.O.C.
E-mail: nkw@phys.sinica.edu.tw

A brief account of the playing role of supersymmetric theories in cosmology is given. In particular, we will discuss in more details the Affleck-Dine mechanism for baryogenesis, and the lightest supersymmetric particle and the dark matter problem.

1 Supersymmetry and Cosmology

The hot big-bang creates extreme conditions such as high temperature and density, long length and time scales, thus opening possibilites to test particle physics that needs energies or other conditions beyond the reach of ground-based laboratories. For examples, inflation driven by a inflaton potential can resolve the flatness, homogeneity, and monopole problems, although a convincing origin for the flatness of the potential near the metastable state is lacking. The dominance of matter over antimatter in the universe may originate from out-of-equilibrium baryon-violating processes such as decays of heavy mesons in grand unified theories (GUTs), or electroweak phase transition via sphaleron configuration. Big-bang nucleosynthesis can explain very well the abundances of primordial elements. So, it has been a great success to connect the micro world to the macro world. Even more, the detection of relic photons which decoupled from the cosmic matter at the era of hydrogen recombination, and recent observations of the cosmic microwave background anisotropy has already brought new excitement and development in cosmology.[1,2]

Supersymmetry (SUSY) is a very important concept in particle physics that relate boson to fermion. One of the main goals of next-generation accelerators is to look for SUSY partners of ordinary particles as predicted by SUSY models[3]. It is worthwhile to discuss SUSY in the context of cosmology. In fact, there are some hints that SUSY might play an important role in cosmology.

First of all, cosmological and astrophysical observations all converge to a conclusion that almost 90% of the matter in the universe is unseen. This dark component is highly probably non-baryonic, and preferably cold (means that they are weakly-interacting, massive, and non-relativistic at the time of structure formation). Although the nature of this cold dark matter is unknown, SUSY particle such as neutralino is a natural and excellent candidate for it[4].

As is well known, there is a fine-tuning problem in conventional inflation models, where one has to adjust the self-coupling constant of the inflaton field incredibly small in order to have a flat plateau for slow-rollover of the field near the origin. In the context of SUSY, working and viable models of inflation can be easily constructed, though none of these models are too compelling. However, SUSY can allow that the scale associated with symmetry breaking is greater than the scale associated with the vacuum energy density driving inflation (the so-called primordial inflation), and it would also be a powerful tool in alleviating fine-tuning problems encountered in inflationary models (due to the fact that SUSY can stabilize the mass scales in radiative corrections because of exact cancellations between boson and fermion loops) [5].

In the inflationary universe the conventional mechanisms for baryogenesis by the out-of-equilibrium decays of superheavy gauge or Higgs bosons faces a number of difficulties. It has become increasingly clear that especially in superstring-inspired and other supergravity models many problems related to the baryon asymmetry can be solved if baryogenesis took place much below the reheating temperature due to inflaton decays [6]. In SUSY models there exists a convenient and natural mechanism first proposed by Affleck and Dine [7]; where scalar quark and lepton fields obtain large vacuum expectation values along flat directions of the scalar potential. These coherent fields start to decay at temperature $\sim 10^4$ GeV when SUSY-breaking effects start to become important, and a baryon asymmetry will be produced provided C and CP are also violated.

In this talk, we will review the Affleck-Dine mechanism for baryogenesis, and SUSY cold dark matter. We will omit most of the basic theories of SUSY which are covered in other talks of this conference. We will not talk about inflation and SUSY models for inflation. Interested readers may refer to the excellent review [5].

2 Affleck-Dine Mechanism for Baryogenesis

2.1 Flat Directions and Baryon Asymmetry

It is well-known that the effective potential in globally supersymmetric theories typically have degenerate minima. For example, in a minimal $N = 1$ globally supersymmetric $SU(5)$ with superpotential,

$$W = \lambda\left(\frac{1}{2}M\mathrm{tr}\Sigma^2 + \frac{1}{3}\mathrm{tr}\Sigma^3\right) \tag{1}$$

for the adjoint Higgs field Σ, the effective potential $V(\Sigma)$ has at least three degenerate minima: $SU(5)$, $SU(4) \times U(1)$, and $SU(3) \times SU(2) \times U(1)$. The degenerate vacuua are separated by barriers of height of order $\lambda^2 M^4$. When coupling to $N = 1$ supergravity, the effective potential becomes

$$V = \exp\left(K^2 \sum_i \phi^{i*}\phi_i\right)\left[\sum_i \left|\frac{\partial W}{\partial \phi_i} + \frac{K^2}{2}\phi^{i*}W\right|^2 - 3K^2 |W|^2\right], \qquad (2)$$

where ϕ_i are the fields in the theory and $K^2 = 8\pi/M_P^2$. The degeneracy of the vacuua is removed by the leading term, $-3K^2 |W|^2$, but in a rather unfortunate way: the lowest energy state becomes the $SU(5)$ invariant vacuum [8].

In fact, the effective potential of a globally supersymmetric theory can have supersymmetric vacuum (i.e., $V = |F|^2 + D^2/2 = 0$) in the sector of matter fields which are generated by combining non-zero vevs of sfermion fields in the theory. The vacuum usually depends on one or more free parameters of vevs so making itself flat in the potential. For example, in minimal $SU(5)$ with Yukawa couplings

$$W = \beta^a T_a \bar{F}_a \bar{H} + \gamma^{ab} T_a T_b H, \qquad (3)$$

where T and $\bar{F}$ are the $\underline{10}$ and $\underline{\bar{5}}$ lepton-quark fields. H and $\bar{H}$ are $\underline{5}$ and $\underline{\bar{5}}$ Higgs fields. (Note that β has been diagonalized in the generation space.) For three generations of matter, a F- and D- flat direction is given by

$$\langle d_1^c \rangle = \langle s_2^c \rangle = \langle t_3^c \rangle = v, \qquad (4)$$

where the subscripts are color indices. Similar flat directions are given by permutating color indices or generations. This kind of supersymmetric vacuum (flat direction) will not be affected by the modified F-terms in supergravity (2) since W is always vanishing.

Affleck and Dine [7] have shown that in the supersymmetric standard model there is a flat direction in the scalar potential for the following set of vacuum states:

$$\langle u_3^c \rangle = a, \quad \langle u_1 \rangle = v; \qquad (5)$$

$$\langle s_2^c \rangle = a, \quad \langle \mu \rangle = v; \qquad (6)$$

$$\langle b_1^c \rangle = e^{i\phi}\left(|v|^2 + |a|^2\right)^{\frac{1}{2}}, \qquad (7)$$

and all other fields vanishing; a and v are arbitrary complex numbers. A large baryon asymmetry could be generated by decays of these vevs via baryon-number violating dimension-four scalar couplings which appear after supersymmetry has been broken.

As envisaged by Affleck and Dine, the whole problem can be reduced to solving the evolution of a single complex scalar field ϕ in an expanding universe with the effective potential,

$$V(\phi) = m_\phi^2 \phi^* \phi + i\lambda \left(\phi^4 - \phi^{*4}\right), \tag{8}$$

where the ϕ^4 term is associated with the baryon-number violating couplings with $\lambda \sim \epsilon m_\phi^2/\Lambda_{GUT}^2$ in which ϵ is a parameter characterizing CP-violation, and Λ_{GUT} is the GUT scale. For small ϕ, this theory has an approximately conserved current $j_\mu = i(\phi^* \partial_\mu \phi - \phi \partial_\mu \phi^*)$, which will be referred to as baryon number. Classically, in an expanding universe, ϕ obeys the equation of motion,

$$\frac{d^2\phi}{dt^2} + 3H\frac{d\phi}{dt} + m_\phi^2 \phi = 4i\lambda\phi^{*3}, \tag{9}$$

where H is the Hubble parameter. With an initial condition at $t = 0$,

$$\phi = i\phi_0, \quad \text{and} \quad \frac{d\phi}{dt} = 0, \tag{10}$$

it is found that the baryon number per particle at large times ($t >> m_\phi^{-1}$) in a matter or radiation-dominated universe is given by

$$r \simeq \lambda \frac{\phi_0^2}{m_\phi^2}. \tag{11}$$

Thus, with a large initial vev ϕ_0, the baryon ratio r can be of order of one, therefore resulting in a baryon asymmetry which is several order of magnitudes larger than observed.

2.2 Dilution Mechanisms

Now the problem of the AD mechanism is on how to dilute the large baryon asymmetry to an acceptable value. It has been shown [9] that the largest contribution to the entropy density of the universe is due to inflation decays. In addition the scalar fermion vevs are determined by scalar field quantum fluctuations produced during inflationary epoch resulting in a baryon number to entropy density of order 10^{-10}.

Another dilution mechanism involves the presence of non-renormalizable terms in the effective potential [10,11]. We expect that all fundamental forces are unified in a grand theory. The standard model is the low-energy effective theory of a GUT after integrating out the heavy fields which are contained in the representations of the gauge group of the GUT. Therefore, the existence of a flat direction in the supersymmetric standard model does not mean that it is flat in the full theory.

For example, assume that the early universe is in a symmetric state of $SU(5)$ with three generations of matter fields coupling to at least two pairs of Higgs 5-plets. After the universe has undergone a phase transition to $SU(3) \times SU(2) \times U(1)$, some dimension-six non-renormalizable terms arise from integrating out the colored Higgs triplets. They can be represented as ϕ^6/Λ_{GUT}^2 in the low-energy effective potential. Then, the effective potential (8) becomes [10]

$$V(\phi) = m_\phi^2 \phi^* \phi + i\lambda \left(\phi^4 - \phi^{*4}\right) + \frac{\phi^{*3}\phi^3}{\Lambda_{GUT}^2}. \tag{12}$$

This added term will affect the evolution of the ϕ field and eventually damps the baryon ratio. To work out the efficiency of this dilution mechanism, we have to study systematically the existence of flat directions and higher-dimensional non-renormalizable terms in the low-energy effective potential. Since these non-renormalizable terms are usually related to dimension-five baryon-violating processes in grand unified theories, it is very interesting to investigate the relation between the structure of the grand gauge group, its representations and cosmological consequences.

2.3 Cold Universe

As mentioned above, if the initial vevs for the scalar fermion fields are quite large, the resulting baryon-photon number density ratio could be as large as 10^3 just after the decays of the scalar fields, thus leading to a cold universe filled with very dense fermions.

The development of a cold universe as a non-standard scenario for the evolution of the universe has been explored previously [12,13]. They [13] considered a possibility that a cold universe filled with very dense fermions undergoes a first-order quark-hadron phase transition to a reheated universe with an acceptable baryon asymmetry. In addition, density perturbations generated during the phase transition can form massive stars which could conceivably play a role as seeds for explosive galaxy formation models, and their remnants may form the dark matter in galactic halos and clusters. Since a cold universe as an alternative to the standard hot big-bang model may be a distinct feature of the supersymmetric theory, it is worhwhile to investigate in detail the baryon production and its subsequent evolution which leads to a cold universe. Furthermore, the density perturbations associated with the phase transition to a reheated universe from a cold dense solid of fermions are very interesting.

2.4 Non-equilibrium Production of Baryon Asymmetry

The non-equilibrium evolution of the vev of a scalar field has been studied [14]. They found that the particles produced by parametric amplification give rise to dissipative behavior for this mode. They showed explicitly that the mechanism of dissipation via particle production cannot be explained with a simple derivative term in the equation of motion. In the case of the AD mechanism, this method can be applied to study the evolution of a complex scalar field with an self-interacting potential. This is equivalent to treating two interacting real scalar fields. To account for the decays of the scalar fields, we can introduce Yukawa-type couplings of the scalar fields to light fields. Then, using the non-equilibrium particle production we can calculate the time evolution of the number operators of the complex scalar field, from which we can construct the baryon ratio. Due to the nonlinear coupling of the complex scalar and a might-be large initial vevs, this non-equilibrium baryon ratio can be significantly different from that obtained in naive classical calculations. The work along this line is in progress [15].

3 Dark Matter

The abundances of primordial light elements in hot big-bang nucleosynthesis constraints the density of baryonic matter to lie within the range $0.01 \leq \Omega_B \leq 0.25$, where $\Omega_B \equiv \rho_B/\rho_c$ is the ratio of the density of baryonic matter to the critical density. Inflation predicts total $\Omega_0 = 1$. On the other hand, observations of luminous matter in the cosmos indicate that $\Omega_{lum} \leq 0.02$, and velocity measurements of stellar objects in clusters reveal that $0.1 < \Omega_0 < 0.4 - 4$ [16].

So, both theoretical prejudices and observational data prefer the existence of dark matter. More convincing evidences of the existence of dark matter are from the observations of the rotational curves of spiral galaxies. For a star near the edge of a galaxy, its rotational velocity should vary as the inverse square root of its distance from the galactic center. But observations show that the rotational curve becomes asymptotically flat at large distances, indicating that a large spherical halo of mass encloses the galaxy [4].

Among cosmological dark matter models, cold dark matter models can almost fit all available cosmological and astrophysical data such as large-scale-structure formation, cosmic microwave background anisotropy, and gravitational lensing [2]. Although recently Super-Kamiokande experiment has announced evidence for massive neutrinos in the study of atmospheric neutrino oscillation [17], and several cosmological measurements have indicated the existence of significant vacuum energy [2], cold dark matter should still constitutes a major component of the unseen matter. Below we discuss the SUSY particles as the cold dark matter.

3.1 Relic Density of Neutralinos

In the early universe, all species of SUSY particles are present and at thermal equilibrium. As the universe expands, the temperature falls and unstable heavy particles begin to decay into light particles. Eventually, only the lightest SUSY particles, protected by R parity, are left over. They can disappear only through pair annihilations. It has been proposed that the LSP is probably a neutralino (a mixture of photino, zino, and higgsino) [18,19,20]. Assuming that photino is the LSP, we can calculate its present relic density as follows. The number density $n(t)$ of photino at time t is governed by the Boltzman equation,

$$\frac{dn}{dt} = -3\frac{\dot{R}}{R}n \ - \ \langle\sigma v_{rel}\rangle(n^2 - n_0^2), \tag{13}$$

where n_0 is the number density at thermal equilibrium, σ is the annihilation cross section of two photinos into a light fermion pair via a scalar fermion exchange, and v_{rel} is the relative velocity of the photinos. By solving the time evolution of n, we find that photino annihilations will stop at a freezing temperature about one-twentieth of its rest mass, and that the left-over can easily have the closure density [18].

3.2 Direct and Indirect Search of Galactic Halo Dark Matter

Dynamical studies show that the galactic halo may be composed of weakly interacting massive particles (WIMPS) such as neutralinos. Direct detections of WIMPs by cryogenic detectors [21] as well as indirect observations of their decay products (such as high-energy neutrinos from the sun [22], cosmic γ-rays, antiprotons, and positions [23]) by, for instances, Super-Kamiokande underground neutrino detector and Alpha Magnetic Spectrometer [24] in space, are being carried out.

A cyrogenic detector contains many small superconducting metal grains frozen at very low temperature. When a dark matter candidate collides with the metal, it may be stopped. Its kinetic energy is then converted into thermal energy, which will heat up the grain and turn it into normal state. The transition then triggers an electromagnetic signal. There are two kinds of interaction between dark matter and the nuclei in the metal grains: vector-vector type (e.g. if it is heavy Dirac neutrino) and spin-spin type (axial-axial) (e.g. if it is neutralino) [21].

For vector-vector interaction, the effective lagrangian is

$$L_{eff} = \frac{1}{\sqrt{2}} G_F \bar{\psi}_\chi \gamma_\mu \psi_\chi \bar{\psi}_N \gamma^\mu \psi_N, \tag{14}$$

then the cross section for a Dirac fermion scattering off a nucleus is

$$\sigma_v = \frac{1}{2\pi} G_F^2 \frac{m_\chi^2 m_N^2}{(m_\chi + m_N)^2} \bar{Y}^2 \bar{N}^2, \tag{15}$$

where $\bar{Y} = \frac{1}{2}(Y_L + Y_R)$ ($Y_{L,R}$ are the hypercharges of the left and right hand fermion respectively), and $\bar{N} = N - (1 - 4sin^2\theta_w)Z$ (N and Z are respectively the neutron and atomic number of the nucleus, and θ_w is Weinberg angle). For Dirac neutrino, $Y_{L,R} = -1, 0$. If the target is made of Ge, $N = 41$ and $Z = 32$. Then $\sigma_v \approx 10^{-34}\ cm^2$ for $m_\nu = 10\ GeV$. Note that Majorana neutrino has no vector-vector interaction because its $Y_{L,R}$ are equal but opposite in sign.

For spin-spin interaction, the effective lagrangian is

$$L_{eff} = \frac{1}{\sqrt{2}} G_F \bar{\psi}_\chi \gamma_\mu \gamma_5 \psi_\chi \bar{\psi}_N \gamma^\mu \gamma_5 \psi_N. \tag{16}$$

In non-relativistic limit, $\bar{\psi}_\chi \gamma_\mu \gamma_5 \psi_\chi$ can be identified with the spin of χ, whereas $\bar{\psi}_N \gamma^\mu \gamma_5 \psi_N$ is proportional to the spin of the nucleus. Then, the cross section for a neutralino scattering off a nucleus with total spin J_N and proportionality constant λ is

$$\sigma_s = \frac{32}{\pi} G_F^2 \frac{m_\chi^2 m_N^2}{(m_\chi + m_N)^2} \lambda^2 J_N(J_N + 1), \tag{17}$$

which is much smaller than σ_v (we assume $\lambda^2 J_N(J_N + 1) \approx 1$).

So far, all experimental results are negative. Let us assume that the halo of our galaxy is composed of cold dark matter. Collision rate per nucleus in the cyrogenic detector is then given by $n_\chi v \sigma$, where $n_\chi = 0.3/m_\chi(GeV)\ cm^{-3}$ is the number density of χ and $v = 300\ km\ s^{-1}$ is their average velocity. The data thus give constraints to the masses of halo cold dark matter with vector-vector interactions. Dirac and scalar neutrinos with masses $12\ GeV < m_\chi < 1.4\ TeV$ have been excluded at 95 percent confidence level [25].

Now, let us turn to indirect detection of dark matter. Here, we mention only the solar high-energy neutrinos [22], which are induced by decays of cold dark matter inside the sun. Cold dark matter particles in the halo are being gravitationally attracted by the sun. If a particle loses enough energy when it hits the sun, it will be trapped inside the sun. Their evaporation rate from the sun will be negligible if their mass is above 4 GeV. As soon as they accumulate in the sun core, they annihilate and release a flux of high-energy neutrinos. These neutrinos can escape from the sun and can be detected by underground proton-decay detectors. So far, no excess high-energy neutrinos from the sun above the atmospheric neutrino background (which is induced by cosmic ray collisions) has been observed, thus ruling out scalar electron and muon neutrinos (they are different from other candidates, emitting a large flux of monochromatic neutrinos) as halo dark matter [26]. Recent Kamiokande experiment has ruled out neutralino with mass less than 1 TeV as halo dark matter [27].

Acknowledgments

The author would like to thank the organizers for their assistance. This work was supported in part by the R.O.C. NSC Grant No. NSC88-2112-M-001-042.

References

1. For a review, see *Particle and Nuclear Astrophysics and Cosmology in the Next Millennium*, ed. E. W. Kolb and R. D. Peccei (World Scientific, 1995).
2. For a review, see *Critical Dialogues in Cosmology*, ed. N. Turok (World Scientific, 1997).
3. Y. Okada, in this Proceedings.
4. For a review, see *Particle Physics and Cosmology: Dark Matter*, ed. M. Srednicki (North-Holland, 1990).
5. K. A. Olive, *Phys. Rep.* **190**, 307 (1990).
6. B. A. Campbell, J. Ellis, D. V. Nanopoulos, and K. A. Olive, *Mod. Phys. Lett.* A **1**, 389 (1986).
7. I. Affleck and M. Dine, *Nucl. Phys.* B **249**, 361 (1985).
8. S. Weinberg, *Phys. Rev. Lett.* **48**, 1303 (1982).
9. J. Ellis, K. Enqvist, D. V. Nanopoulos, and K. A. Olive, *Phys. Lett.* B **191**, 343 (1987).
10. K.-W. Ng, *Nucl. Phys.* B **321**, 528 (1989).
11. M. Dine, L. Randall, and S. Thomas, *Nucl. Phys.* B **458**, 291 (1996).
12. A. D. Linde, *Rep. Prog. Phys.* **42**, 390 (1979); G. Lasher, *Phys. Rev. Lett.* **42**, 1646 (1979); C. J. Hogan, *Astrophys. J.* **252**, 418 (1982).
13. K.-W. Ng and W.-K. Sze, *Phys. Rev.* D. **43**, 3813 (1991).
14. D. Boyanovsky, H.J. de Vega, R. Holman, D.-S. Lee, and A. Singh, *Phys. Rev.* D **51**, 4419 (1995).
15. D.-S. Lee and K.-W. Ng, in preparation.
16. Particle Data Group, *Phys. Rev.* D **50**, 1173 (1994).

17. Y. Fukuda *et al.*, *Phys. Rev. Lett.* **81**, 1562 (1998).
18. J. Ellis *et al.*, *Nucl. Phys.* B **238**, 453 (1984).
19. G. Jungman, M. Kamionkowski, and K. Griest, *Phys. Rep.* **267**, 195 (1996).
20. C. Kao, in this Proceedings.
21. M. W. Goodman and E. Witten, *Phys. Rev.* D **31**, 3059 (1985).
22. J. Silk, K. A. Olive, and M. Srednicki, *Phys. Rev. Lett.* **55**, 257 (1985); T. Gaisser, G. Steigman, and S. Tilav, *Phys. Rev.* D **34**, 2206 (1986); J. Hagelin, K.-W. Ng, and K. A. Olive, *Phys. Lett.* B **180**, 375 (1987).
23. J. Silk and M. Srednicki, *Phys. Rev. Lett.* **53**, 624 (1984).
24. See the AMS homepage: http://hpl3tri1.cern.ch/.
25. D. O. Caldwell *et al.*, *Phys. Rev. Lett.* **61**, 510 (1988).
26. K.-W. Ng, K. A. Olive, and M. Srednicki, *Phys. Lett.* B **188**, 138 (1987).
27. M. Mori *et al.*, *Phys. Rev.* D **48**, 5505 (1993).

SUSY PARTICLE MASS SPECTRA AND PHENOMENOLOGY IN THE MINIMAL SUPERGRAVITY MODEL

V. Barger and Chung Kao
Department of Physics, University of Wisconsin
Madison, WI 53706, USA

Recent developments of two special topics in the phenomenology of the minimal supergravity model are discussed: (i) the detection of trilepton signature from chargino-neutralino associated production at the upgraded Tevatron, and (ii) the search for neutral Higgs bosons via their decays into muon pairs at the LHC.

1 Introduction

A supersymmetry (SUSY) between fermions and bosons provides a natural explanation for the Higgs mechanism of electroweak symmetry breaking (EWSB) in the framework of a grand unified theory (GUT). The evolution of renormalization group equations (RGEs) [1] with particle content of the minimal supersymmetric model (MSSM) [2] is consistent with a grand unified scale ($M_{\rm GUT}$) at about 2×10^{16} GeV and an effective SUSY mass scale in the range $M_Z < M_{\rm SUSY} \lesssim 1$ TeV [3]. Radiative corrections with a large top quark Yukawa coupling to Higgs bosons at $M_{\rm GUT}$, can drive the corresponding Higgs boson mass squared parameter negative, spontaneously breaking the electroweak symmetry and naturally explaining the electroweak scale.

The MSSM requires at least two Higgs doublets ϕ_1 and ϕ_2 [2,4] that couple to the $t_3 = -1/2$ and $t_3 = +1/2$ fermions, respectively. After spontaneous symmetry breaking, there remain five physical Higgs bosons: a pair of singly charged Higgs bosons $H^{\pm}$, two neutral CP-even scalars H^0 (heavier) and h^0 (lighter), and a neutral CP-odd pseudoscalar A^0. At the tree level, all Higgs boson masses and couplings are determined by just two independent parameters, which are commonly chosen to be the pseudoscalar mass (m_A) and the ratio of vacuum expectation values (VEVs) of Higgs fields $\tan\beta \equiv v_2/v_1$.

In the minimal supergravity model (mSUGRA) [5], it is assumed that SUSY is broken in a hidden sector with SUSY breaking communicated to the observable sector through gravitational interactions, leading naturally to a common scalar mass (m_0), a common gaugino mass ($m_{1/2}$), a common trilinear coupling (A_0) and a common bilinear coupling (B_0) at $M_{\rm GUT}$. Through minimization of the Higgs potential, the B parameter and magnitude of the Higgs mixing parameter μ are related to $\tan\beta$ and M_Z. The weak-scale SUSY mass spectra and couplings in the mSUGRA can be predicted by the evolution of RGEs from the $M_{\rm GUT}$ with four parameters: m_0, $m_{1/2}$, A_0 and $\tan\beta$; and the sign of the Higgs mixing parameter μ [6,7,8]. Since A_0 mainly affects the masses of third generation sfermions, it is taken to be zero in our analysis. In SUSY theories with a conserved R-parity[b], the lightest supersymmetric particle (LSP) cannot decay into normal particles and the LSP is an attractive candidate for dark matter [9]. In most of the SUGRA parameter space, the lightest neutralino (χ_1^0) is the LSP [6,7,10].

We calculate masses and couplings in the Higgs sector with one loop corrections from the top and the bottom Yukawa interactions in the one-loop effective potential [11] at the scale $Q = \sqrt{m_{\tilde{t}_L} m_{\tilde{t}_R}}$ [8,10]. At this high scale, the RGE improved one-loop corrections approximately reproduce the dominant two loop corrections [12] to the mass of the lighter scalar (m_h).

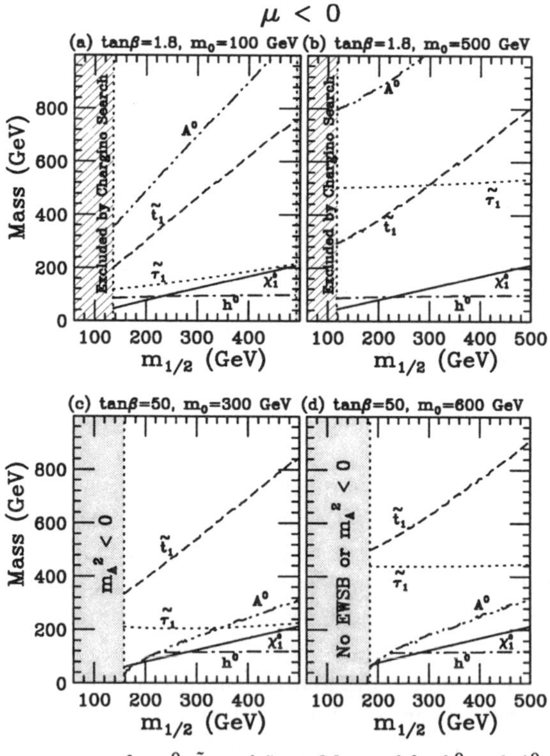

Figure 1: Masses versus $m_{1/2}$ for χ_1^0, $\tilde{t}_1$ and $\tilde{\tau}_1$ at M_Z, and for h^0 and A^0 at $Q = \sqrt{m_{\tilde{t}_L} m_{\tilde{t}_R}}$.

---

[b] $R = +1$ for Standard Model particles and Higgs bosons and $R = -1$ for their superpartners.

In Figure 1, we present masses, in the case of $\mu < 0$, for the lightest neutralino (χ_1^0), the lighter top squark ($\tilde{t}_1$), the lighter tau slepton ($\tilde{\tau}_1$), and two neutral Higgs bosons: the lighter CP-even (h^0) and the CP-odd (A^0). Also shown are the regions that do not satisfy the following theoretical requirements: electroweak symmetry breaking (EWSB), tachyon free, and the lightest neutralino as the LSP. The region excluded by the $m_{\chi_1^\pm} < 90$ GeV limit from the chargino search [13] at LEP 2 is indicated.

Recent measurements of the $b \to s\gamma$ decay rate by the CLEO [14] and LEP collaborations [15] place constraints on the parameter space of the minimal supergravity model [16]. It was found that $b \to s\gamma$ excludes most of the mSUGRA parameter space when $\tan\beta$ is large and $\mu > 0$ [16]. Therefore, we consider only $\mu < 0$ in our analysis when $\tan\beta \geq 10$. In our convention, $-\mu$ appears in the chargino mass matrix and $+\mu$ appears in the neutralino mass matrix [6].

2 Trilepton Signature at the Tevatron

In this section, we discuss the prospects for discovery of the trilepton signal with missing transverse energy ($3\ell + \not{E}_T$) [17,18,19,20] in the minimal supergravity model at the upgraded Tevatron with 2 TeV center of mass energy and an integrated luminosity (L) of 2 fb^{-1} (RUN II) to 30 fb^{-1} (TeV33). The source of this signal is associated production of the lightest chargino ($\chi_1^\pm$) and the second lightest neutralino (χ_2^0) with decays to leptons [17,18,19,20].

In the mSUGRA model with gauge coupling unification, the sleptons ($\tilde{\ell}$), the lighter chargino ($\chi_1^\pm$) and the lighter neutralinos (χ_1^0, χ_2^0) are less massive than the gluinos and squarks over most of the parameter space. Because of this, the $3\ell + \not{E}_T$ signal from associated production and decays of $\chi_1^\pm \chi_2^0$ is one of the most promising channels for supersymmetric (SUSY) particle searches at the Tevatron. The background to this signal from processes in the Standard Model (SM) can be greatly reduced with suitable cuts.

The Yukawa couplings of the bottom quark (b) and the tau lepton (τ) are proportional to $\tan\beta$ and are greatly enhanced when $\tan\beta$ is large. In SUSY GUTs, the masses of the third generation sfermions are consequently very sensitive to the value of $\tan\beta$. As $\tan\beta$ increases, the lighter tau slepton ($\tilde{\tau}_1$) and the lighter bottom squark ($\tilde{b}_1$) become lighter than charginos and neutralinos while other sleptons and squarks are heavy. Then, $\chi_1^\pm$ and χ_2^0 can dominantly decay into final states with tau leptons via real or virtual $\tilde{\tau}_1$. While these decays reduce the trilepton signal with hard cuts to suppress the backgrounds, they also open new discovery channels via τ leptons [20].

The relevance of τ leptons in the production and decays of $\chi_1^\pm \chi_2^0$, are illustrated in Figure 2, where the product of the cross section $\sigma(p\bar{p} \to \chi_1^\pm \chi_2^0 +$

X) and the branching fraction $B(\chi_1^\pm \chi_2^0 \to 3\ leptons + \not{E}_T)$ versus $\tan\beta$ is presented with $\mu < 0$, $m_{1/2} = 200$ GeV, $m_0 = 100$ and 200 GeV, for four final states (a) $\tau\tau\tau$, (b) $\tau\tau\ell$, (c) $\tau\tau\tau$ and (d) $\ell\ell\ell$, where $\ell = e$ or μ. For $m_0 \lesssim 200$ GeV and/or $\tan\beta \gtrsim 40$, channels with at least one τ lepton are dominant.

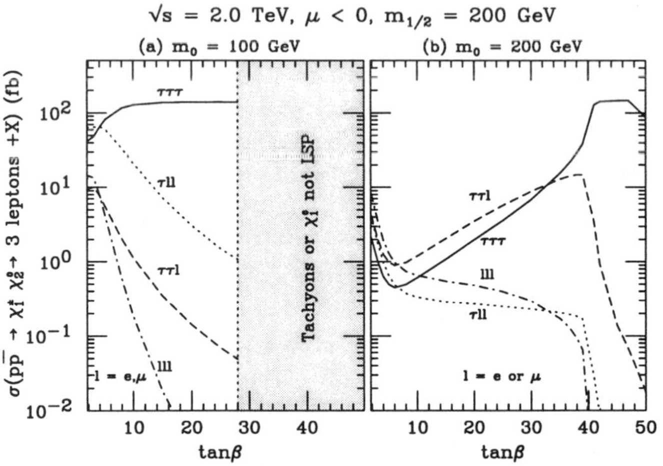

Figure 2: Cross section of $p\bar{p} \to \chi_1^\pm \chi_2^0 \to 3$ leptons $+ X$ without cuts at $\sqrt{s} = 2$ TeV versus $\tan\beta$, with $\mu < 0$, $m_{1/2} = 200$ GeV, $m_0 = 100$ GeV for 4 final states: (a) $\tau\tau\tau$ (solid), (b) $\tau\tau\ell$ (dot-dash), (c) $\tau\ell\ell$ (dash) and (d) $\ell\ell\ell$ (dot), where $\ell = e$ or μ.

One way to detect τ leptons is through their one prong and three prong hadronic decays. The CDF and the DØ collaborations are currently investigating the efficiencies for detecting these modes and for implementing a τ trigger. It has been suggested that the τ leptons in the final state may be a promising way to search for $\chi_1^\pm \chi_2^0$ production at the Tevatron if excellent τ identification becomes feasible [21,22].

Another way of exploiting the τ signals, is to detect the soft electrons and muons from leptonic τ decays by employing softer but realistic p_T cuts on the leptons [20] than conventionally used [18,21]. We find that this can improve the trilepton signal from $\chi_1^\pm \chi_2^0$ production by at least a factor of two [20].

In hadron collisions, associated production of chargino and neutralino occurs via quark-antiquark annihilation in the s-channel through a virtual W boson ($q\bar{q}' \to W^{\pm *} \to \chi_1^\pm \chi_2^0$) and in the t and u-channels through squark ($\tilde{q}$) exchanges. If the squarks are light, a destructive interference between the W boson and the squark exchange amplitudes can suppress the cross section by as much as 40% compared to the s-channel contribution alone. For squarks much heavier than the gauginos, the s-channel W-resonance amplitude dominates.

For $\tan\beta \lesssim 5$, the branching fractions are sensitive to the sign of μ. For

$\mu < 0$ and $\tan\beta \sim 3$, we observe that the dominant decays are: (i) for $m_0 \lesssim 170$ GeV, $\chi_1^\pm\chi_2^0 \to 3\ell + \not{E}_T$ via real or virtual $\tilde{\ell}_R$ and $\tilde{\tau}_1$; (ii) for $m_0 \gtrsim 180$ GeV, $\chi_1^\pm\chi_2^0 \to q\bar{q}'\chi_1^0$. For $\mu > 0$ and $\tan\beta \sim 3$, the dominant decays are: (i) for $m_0 \lesssim 100$ GeV, $\chi_2^0 \to \tilde{\nu}_L\nu$ and $\chi_1^\pm \to \tilde{\nu}_L\ell$. (ii) for 110 GeV $\lesssim m_0 \lesssim 170$ GeV, $\chi_1^\pm\chi_2^0 \to 3\ell + \not{E}_T$ via real or virtual sleptons; (iii) for $m_0 \gtrsim 180$ GeV, $\chi_1^\pm\chi_2^0 \to q\bar{q}'\chi_1^0$.

For $\mu < 0$ and $m_0 \sim 200$ GeV, χ_2^0 dominantly decays (i) into $\tau\bar{\tau}\chi_1^0$ for $25 \lesssim \tan\beta \lesssim 40$, (ii) into $\tau\tilde{\tau}_1$ for $\tan\beta \gtrsim 40$. For $m_0 \lesssim 300$ GeV and large $\tan\beta \gtrsim 35$, both $\tilde{\tau}_1$ and $\tilde{b}_1$ can be lighter than other sfermions, and $\chi_1^\pm$ and χ_2^0 can decay dominantly into final states with τ leptons or b quarks via virtual or real $\tilde{\tau}_1$ and $\tilde{b}_1$.

The ISAJET 7.37 event generator program [23] is employed to calculate the $3\ell + \not{E}_T$ signal from $\chi_1^\pm\chi_2^0$ and the dominant backgrounds from $t\bar{t}$ and WZ at the upgraded Tevatron. An energy resolution of $\frac{0.7}{\sqrt{E}}$ for the hadronic calorimeter and $\frac{0.15}{\sqrt{E}}$ for the electromagnetic calorimeter is assumed. Jets are defined to be hadron clusters with $E_T > 15$ GeV in a cone with $\Delta R = \sqrt{\Delta\eta^2 + \Delta\phi^2} = 0.7$. Leptons with $p_T > 5$ GeV and within $|\eta_\ell| < 2.5$ are considered to be isolated if the hadronic scalar E_T in a cone with $\Delta R = 0.4$ about the lepton is smaller than 2 GeV.

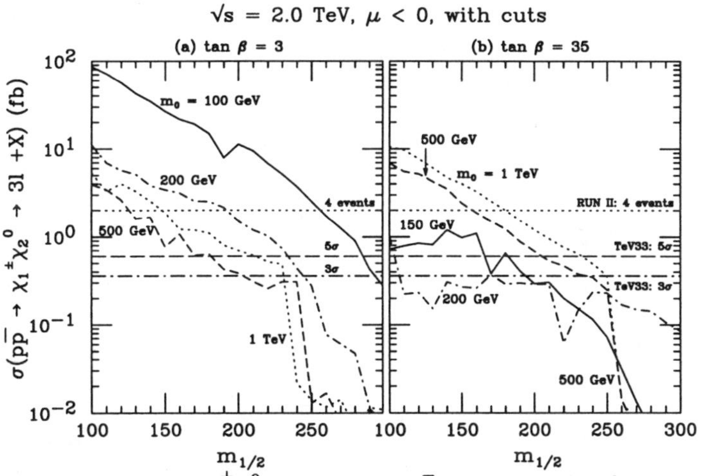

Figure 3: Cross section of $p\bar{p} \to \chi_1^\pm\chi_2^0 \to 3\ell + X$ at $\sqrt{s} = 2$ TeV with soft acceptance cuts, versus $m_{1/2}$, with (a) $\tan\beta = 3$ and (b) $\tan\beta = 35$, for $m_0 = 100$ GeV (solid), 200 GeV (dot-dash), 500 GeV (dash) and 1000 GeV (dot). Also presented are cross sections for (i) 4 signal events with $L = 2$ fb^{-1} (dot) and (ii) 5 σ (dash) signal for $L = 30$ fb^{-1}.

Our cuts are similar to those recently employed by the CDF [19]:

(i) We require 3 isolated leptons in each event, with

$$p_T(\ell_1) > 11 \text{ GeV}, \;\; p_T(\ell_2) > 7 \text{ GeV}, p_T(\ell_3) > 5 \text{ GeV}, \;\; \text{and}$$
$$|\eta(\ell_1)| < 1.0, \;\; |\eta(\ell_2, \ell_3)| < 2.0. \tag{1}$$

(ii) We require $\not{E}_T > 25$ GeV. This cut removes backgrounds from SM processes such as Drell-Yan dilepton production, where an accompanying jet may fake a lepton.
(iii) To reduce the background from WZ production, we require that $|m(\ell\bar{\ell}) - M_Z| \geq 10$ GeV. This cut reduces most of the WZ background.
The surviving total background cross section from WZ and $t\bar{t}$ is about 0.46 fb. The background is mainly due to WZ events, where $Z \to \tau\bar{\tau}$, with subsequent τ lepton decays.

To assess the discovery potential for the upgraded Tevatron, we present cross sections of the trilepton signal and the background after cuts versus $m_{1/2}$ in Fig. 3, for $\mu < 0$, $\tan\beta = 3$ and 35, and several vaules of m_0. At the Tevatron with the Main Injector and $L = 2$ fb^{-1}, we expect about one background event from the cross section of 0.46 fb. The signal cross section is required to yield at least four events for discovery. The Poisson probability for the SM background to fluctuate to this level is less than 0.4%. For TeV33 with $L = 30$ fb^{-1}, we expect about 14 background events and a 5σ signal would be 15 events, which corresponds to a signal cross section of 0.62 fb. The discovery contours for $\tan\beta = 2$ are presented in Figure 4.

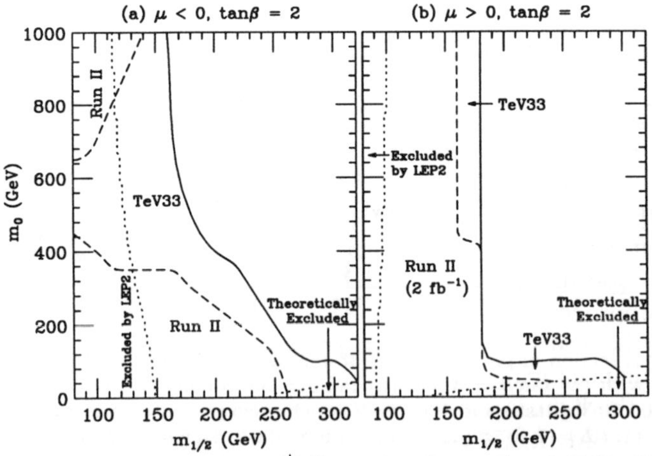

Figure 4: Discovery contours of $p\bar{p} \to \chi_1^{\pm}\chi_2^0 \to 3\ell + X$ at $\sqrt{s} = 2$ TeV with cuts in the parameter plane of $(m_{1/2}, m_0)$, with $\tan\beta = 2$ for (a) $\mu < 0$ and (b) $\mu > 0$.

In some regions of the mSUGRA parameter space, the $\chi_1^\pm$ and the χ_2^0 decay dominantly into τ leptons $+ \chi_1^0$. With soft but realistic lepton p_T cuts, the $\tau \to \ell$ contributions enhance the trilepton signal by at least a factor of two. The Tevatron trilepton searches are most sensitive to the region of mSUGRA parameter space with $m_0 \lesssim 100$ GeV and $\tan\beta \lesssim 10$. For $m_0 \sim 100$ GeV and $\tan\beta \sim 2$, the trilepton signal should be detectable (i) at the RUN II if $m_{1/2} \lesssim 260$ GeV, and (ii) at the TeV33 if $m_{1/2} \lesssim 290$ GeV. For $\tan\beta \sim 35$, the trilepton signal should be detectable at the TeV33, if $m_0 \sim 150$ GeV and $m_{1/2} \lesssim 170$ GeV, or if $m_0 \gtrsim 500$ GeV and $m_{1/2} \lesssim 200$ GeV. It is difficult to search for trileptons at the Tevatron for 180 GeV $\lesssim m_0 \lesssim 400$ GeV and $10 \lesssim \tan\beta \lesssim 35$, because $\chi_1^\pm$ and χ_2^0 dominantly decay into $q\bar{q}'\chi_1^0$.

3 Discovering the Higgs Bosons with Muons

In this section, we discuss the prospects for the discovery of neutral Higgs bosons via decays into muon pairs at the CERN LHC. Promising results are found for the CP-odd pseudoscalar (A^0) and the heavier CP-even scalar (H^0) Higgs bosons at large $\tan\beta \equiv v_2/v_1$. For $\tan\beta \gtrsim 10$ and 100 GeV $\lesssim m_A, m_H \lesssim 550$ GeV, the A^0 and the H^0 masses could be precisely reconstructed from the dimuon invariant mass.

In our analysis, the cross section of $pp \to \phi \to \mu\bar{\mu} + X$ ($\phi = A^0, H^0$, or h^0) is evaluated from the Higgs boson cross section $\sigma(pp \to \phi + X)$ multiplied with the branching fraction of the Higgs decay into muon pairs $B(\phi \to \mu\bar{\mu})$ with two dominant subprocesses: $gg \to \phi$ and $gg \to \phi b\bar{b}$. The parton distribution functions of CTEQ3L[24] are chosen to evaluate the cross section of $pp \to \phi + X$. We take $M_Z = 91.19$ GeV, $\sin^2\theta_W = 0.2319$, $M_W = M_Z \cos\theta_W$, $m_b = 4.8$ GeV, and $m_t = 175$ GeV.

At the LHC energy, the SM Higgs boson (h^0_{SM}) is produced dominantly from gluon fusion, and from vector boson fusion if the h^0_{SM} is heavy. In the MSSM, gluon fusion ($gg \to \phi$) is the major source of neutral Higgs bosons for $\tan\beta \lesssim 4$. If $\tan\beta$ is larger than about 10, neutral Higgs bosons in the MSSM are dominantly produced from b-quark fusion ($b\bar{b} \to \phi$) [25]. We have evaluated the cross section of Higgs bosons in pp collisions $\sigma(pp \to \phi + X)$, with two dominant subprocesses: $gg \to \phi$ and $gg \to \phi b\bar{b}$. Since the Yukawa couplings of $\phi b\bar{b}$ are enhanced by $1/\cos\beta$, the production rate of neutral Higgs bosons is usually enhanced with large $\tan\beta$. For m_A larger than about 150 GeV, the couplings of the lighter scalar h^0 to gauge bosons and fermions become close to those of the SM Higgs boson, therefore, gluon fusion is the major source of the h^0 even if $\tan\beta$ is large.

The QCD radiative corrections to the subprocess $gg \to \phi$ was found to be large [26], the same corrections to $gg \to \phi b\bar{b}$ are still to be evaluated. To be conservative, we take a K-factor of 1.5 for the contribution from $gg \to \phi$ and a K-factor of 1.0 for the contribution from $gg \to \phi b\bar{b}$ to evaluate the cross section of $pp \to \phi + X$. For the dominant Drell-Yan background [27,28,29], we have adopted the well known K-factor from referee [30].

If the $b\bar{b}$ mode dominates Higgs decays, the branching fraction of $\phi \to \mu\bar{\mu}$ is about $m_\mu^2/3m_b^2$, where 3 is the color factor of the quarks. It has been found that QCD radiative corrections greatly reduce the decay width of $\phi \to b\bar{b}$. Therefore, the branching fraction of $\phi \to \mu\bar{\mu}$ is about 2×10^{-4} when the $b\bar{b}$ mode dominates. For $\tan\beta \gtrsim 10$, the branching fraction of $B(\phi \to \mu\bar{\mu})$ ($\phi = A^0, H^0$, or h^0) in most of the parameter space is about 2×10^{-4}, even when the A^0 and the H^0 can decay into SUSY particles. For m_A less than about 80 GeV, the H^0 decays dominantly into h^0h^0, A^0A^0 and ZA^0.

We define the signal to be observable if the 99% confidence level upper limit on the background is smaller than the corresponding lower limit on the signal plus background [31], namely,

$$L(\sigma_s + \sigma_b) - N\sqrt{L(\sigma_s + \sigma_b)} > L\sigma_b + N\sqrt{L\sigma_b}$$
$$\sigma_s > \frac{N^2}{L}[1 + 2\sqrt{L\sigma_b}/N] \qquad (2)$$

where L is the integrated luminosity, and σ_b is the background cross section within a bin of width $\pm\Delta M_{\mu\bar{\mu}}$ centered at M_ϕ; $N = 2.32$ corresponds to a 99% confidence level and $N = 2.5$ corresponds to a 5σ signal.

To study the observability for the muon pair decay mode, the dominant background from the Drell-Yan (DY) process, $q\bar{q} \to Z, \gamma \to \mu\bar{\mu}$ is considered. We take $\Delta M_{\mu\bar{\mu}}$ to be the larger of the ATLAS muon mass resolution (about 2% of the Higgs bosons mass) [32,29] or the Higgs boson width.[c] The minimal cuts applied are (1) $p_T(\mu) > 20$ GeV and (2) $|\eta(\mu)| < 2.5$ for both the signal and background. For $m_A \gtrsim$ 130 GeV, m_A and m_H are almost degenerate while for $m_A \lesssim$ 100 GeV m_A and m_h are very close to each other [27,28]. Therefore, We sum up the cross sections of the A^0 and the h^0 for $m_A \leq 100$ GeV and those of the A^0 and the H^0 for $m_A > 100$ GeV,

Figure 5 shows the total cross section of $pp \to A^0, H^0 \to \mu^+\mu^- + X$ in fb, as a function of $m_{1/2}$, with $\sqrt{s} = 14$ TeV, for $\mu < 0$, $m_0 = 300$ GeV, and (a) $\tan\beta = 1.8$, (b) $\tan\beta = 10$, (c) $\tan\beta = 35$, and (d) $\tan\beta = 50$, Also shown are m_A versus $m_{1/2}$ and the invariant mass distribution ($d\sigma/dM_{\mu\bar{\mu}}$) of the background from the Drell-Yan process (dash) at $M_{\mu\mu} = m_A$. For $\tan\beta \sim 1.8$, m_A is very large and the cross section of the signal is much smaller than the

[c]The CMS mass resolution will be better than 2% of m_ϕ for $m_\phi \lesssim$ 500 GeV [27,28].

background. For $\tan\beta \gtrsim 35$, and the cross section of the signal is greatly enhanced and can become slight larger than the background. For $\tan\beta \sim 50$ and $m_{1/2} < 200$ GeV, m_A can become close to m_Z and a peak appears in the the invariant mass distribution of the background from the Drell-Yan process.

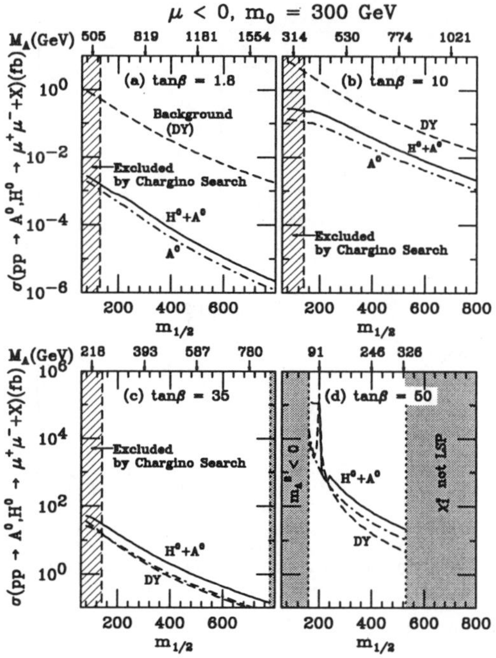

Figure 5: The total cross section of $pp \to A^0, H^0 \to \mu\bar{\mu} + X$ in fb, as a function of $m_{1/2}$.

In Figure 6, we present the LHC discovery contours in the minimal supergravity model, for (a) the $m_{1/2}$ versus $\tan\beta$ plane with $m_0 = 150$ GeV, (b) the $m_{1/2}$ versus $\tan\beta$ plane with $m_0 = 500$ GeV, (c) the $m_{1/2}$ versus m_0 plane with $\tan\beta = 15$, and (d) the $m_{1/2}$ versus m_0 plane with $\tan\beta = 40$. An increase in m_0 raises the mass of Higgs bosons. There are a couple of interesting aspects to note: (i) an increase in $\tan\beta$ leads to a larger m_h but a reduction in m_A and m_H; (ii) increasing m_0 raises m_A, m_H and masses of the other scalars significantly.

The muon pair decay mode is a promising channel to discover the neutral Higgs bosons of minimal supersymmetry and minimal supergravity. This discovery channel might allow precise reconstruction for the Higgs boson masses. The A^0 and H^0 should be observable in a large region of the $(m_{1/2}, m_0)$ parameter space with $\tan\beta \gtrsim 30$. All observable regions are nearly independent of the sign of μ. Adequate subtraction procedures are required to extract the signal for $M_{\mu^+\mu^-}$ close to the M_Z, because it appears near the huge Z peak.

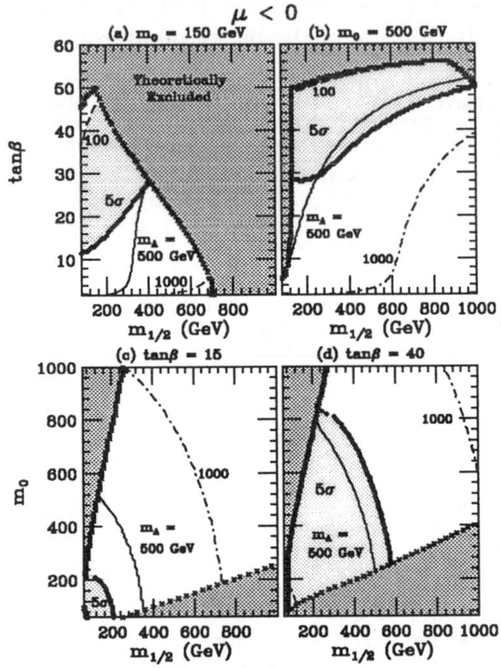

Figure 6: The 5σ contours at the LHC for muon pair discovery channel of neutral Higgs bosons in the minimal supergravity model.

Acknowledgments

This research was supported in part by the U.S. Department of Energy under Grant No. DE-FG02-95ER40896, and in part by the University of Wisconsin Research Committee with funds granted by the Wisconsin Alumni Research Foundation.

References

1. K. Inoue, A. Kakuto, H. Komatsu and H. Takeshita, Prog. Theor. Phys. **68**, 927 (1982) and **71**, 413 (1984).
2. Reviews for the MSSM can be found in: H.P. Nilles, Phys. Rep. **110**, 1 (1984); H. Haber and G. Kane, Phys. Rep. **117**, 75 (1985); X. Tata, in *Campos do Jordao 1997, Particles and fields*, lectures given at 9th Jorge

Andre Swieca Summer School: Particles and Fields, Sao Paulo, Brazil, 16-28 Feb 1997.

3. P. Langacker and M. Luo, Phys. Rev. **D44**, 817 (1991); J. Ellis, S. Kelley and D. Nanopoulos, Phys. Lett. **B260**, 131 (1991); U. Amaldi, W. de Boer and H. Fürstenau, Phys. Lett. **B260**, 447 (1991).
4. J. Gunion, H. Haber, G. Kane and S. Dawson, *The Higgs Hunter's Guide*, (Addison-Wesley Publishing Company, Redwood City, CA, 1990).
5. A. Chamseddine, R. Arnowitt and P. Nath, Phys. Rev. Lett. **49**, 970 (1982); L. Ibañez and G. Ross, Phys. Lett. **B110**, 215 (1982); R. Barbieri, S. Ferrara and C. Savoy, Phys. Lett. **B119**, 343 (1982); L. Hall, J. Lykken and S. Weinberg, Phys. Rev. **D27**, 2359 (1983).
6. V. Barger, M.S. Berger, P. Ohmann, Phys. Rev. D **47**, 1093 (1993); Phys. Rev. D **49**, 4908 (1994); V. Barger, M.S. Berger, P. Ohmann and R.J.N. Phillips, Phys. Lett. B **314**, 351 (1993).
7. G. Ross and R.G. Roberts, Nucl. Phys. **B377**, 571 (1992); R. Arnowitt and P. Nath, Phys. Rev. Lett. **69**, 725 (1992); M. Drees and M.M. Nojiri, Nucl. Phys. **B369**, 54 (1993); S. Kelley *et. al.*, Nucl. Phys. **B398**, 3 (1993); M. Olechowski and S. Pokorski, Nucl. Phys. **B404**, 590 (1993); G. Kane, C. Kolda, L. Roszkowski and J. Wells, Phys. Rev. D **49**, 6173 (1994); D.J. Castaño, E. Piard and P. Ramond, Phys. Rev. D **49**, 4882 (1994); W. de Boer, R. Ehret and D. Kazakov, Z. Phys. C **67**, 647 (1995); H. Baer, M. Drees, C. Kao, M. Nojiri and X. Tata, Phys. Rev. D **50**, 2148 (1994).
8. H. Baer, C.-H. Chen, M. Drees, F. Paige and X. Tata, Phys. Rev. Lett. **79**, 986 (1997).
9. E.W. Kolb and M.S. Turner, *The Early Universe*, (Addison-Wesley, Redwood City, CA, 1989); G. Jungman, M. Kamionkowski and K. Griest, Phys. Rep. **267**, 195 (1996); and reference therein.
10. V. Barger and C. Kao, Phys. Rev. D **57**, 3131 (1998).
11. H. Haber and R. Hempfling, Phys. Rev. Lett. **66**, 1815 (1991); J. Ellis, G. Ridolfi and F. Zwirner, Phys. Lett. B **257**, 83 (1991); Y. Okada, H. Yamaguchi and T. Tanagida, Prog. Theor. Phys. Lett. **85**, 1 (1991); M. Bisset, Ph.D. thesis, University of Hawaii (1994).
12. M. Carena, J.R. Espinosa, M. Quiros, and C.E.M. Wagner, Phys. Lett. **B355**, 209 (1995); M. Carena, M. Quiros, C.E.M. Wagner, Nucl. Phys. **B461**, 407 (1996); H. Haber, R. Hempfling and A. Hoang, Z. Phys. C **75**, 539 (1997).
13. G. Abbiendi et al., the OPAL Collaboration, CERN-EP-98-136, (1998).
14. M.S. Alam et al., the CLEO Collaboration, Phys. Rev. Lett. **74** (1995) 2885.

15. R. Barate et al., the ALEPH Collaboration, Eur. Phys. J. **C2** (1998) 417.
16. H. Baer and M. Brhlik, Phys. Rev. D **55**, 3201 (1997); H. Baer, M. BrhliK, D. Castano and X. Tata, Phys. Rev. D **58**, 015007 (1998); and reference therein.
17. H. Baer, K. Hagiwara and X. Tata, Phys. Rev. D**35**, 1598 (1987); R. Arnowitt and P. Nath, Mod. Phys. Lett. **A2**, 331 (1987); H. Baer and X. Tata, Phys. Rev. D**47**, 2739 (1993); T. Kamon, J. Lopez, P. McIntyre and J. T. White, Phys. Rev. D **50**, 5676 (1994); S. Mrenna, G. Kane, G.D. Kribs and J.D. Wells, Phys. Rev. D **53**, 1168 (1996).
18. H. Baer, C. Kao and X. Tata Phys. Rev. D **48**, 5175 (1993); H. Baer, C-H. Chen, C. Kao and X. Tata Phys. Rev. D **52**, 1565 (1995); H. Baer, C-H. Chen, F. Paige and X. Tata, Phys. Rev. D **54**, 5866 (1996).
19. F. Abe et al., the CDF collaboration, Phys. Rev. Lett. **80**, 5275 (1998); S. Abachi et al., the DØ collaboration, Phys. Rev. Lett. **76**, 4307 (1996).
20. V. Barger, C. Kao and T.-J. Li, Phys. Lett. B **433**, 328 (1998); V. Barger and C. Kao, University of Wisconsin Report, MADPH-98-1085 (1998).
21. H. Baer, C.-H. Chen, M. Drees, F. Paige and X. Tata, Phys. Rev. D **58**, 075008 (1998).
22. J.D. Wells, Mod. Phys. Lett. A **13**, 1923 (1998).
23. H. Baer, F. Paige, S. Protopopescu and X. Tata, *ISAJET 7.37: A Monte Carlo Event Generator for* $pp, \bar{p}p,$ ande^+e^- *Reactions*, Bookhaven National Laboratory Report BNL-HET-98-18 (1998), hep-ph/9804321.
24. H.L. Lai et al., Phys. Rev. D **51**, 4763 (1995).
25. D. Dicus and S. Willenbrock, Phys. Rev. **D39**, (1989) 751.
26. S. Dawson, Nucl. Phys. **B359**, 283 (1991); A. Djouadi, M. Spira and P.M. Zerwas, Phys. Lett. **B264**, 440 (1991); D. Graudenz, M. Spira and P.M. Zerwas, Phys. Rev. Lett. **70**, 1372 (1993); M. Spira, A. Djouadi, D. Graudenz and P.M. Zerwas, Nucl. Phys. **B453**, 17 (1995); S. Dawson, A. Djouadi and M. Spira, Phys. Rev. Lett. **77**, 16 (1996).
27. C. Kao and N. Stepanov, Phys. Rev. D **52** (1995) 5025.
28. CMS Technical Proposal, CERN/LHCC 94-38 (1994).
29. E. Richter-Was, et al., Int. J. Mod. Phys. A **13**, 1371 (1998).
30. V. Barger and R. Phillips, *Collider Physics, updated edition*, (Addison-Wesley Publishing Company, Redwood City, CA, 1997).
31. H. Baer, M. Bisset, C. Kao and X. Tata, Phys. Rev. **D46**, 1067 (1992).
32. Atlas Technical Proposal, CERN/LHCC 94-43 (1994).

Contributed Talks

Penguin contributions in the lifetime difference between B_s and B_d, and a possible New Physics

Yong-Yeon Keum
Asia Pacific Center For Theoretical Physics
207-43 Cheongryangri Dong Dongdaemun-Gu, Seoul 130-012, Korea

Theory Group, KEK, Tsukuba, Ibaraki 305-0801 Japan

We consider penguin contributions to the lifetime splitting between the B_s and the B_d meson. In the Standard Model the penguin effects are found to be opposite in sign, but of similar magnitude as the contributions of the current-current operators, despite of the smallness of the penguin coefficients. We predict

$$\frac{\tau(B_s)}{\tau(B_d)} - 1 = (-1.2 \pm 10.0) \cdot 10^{-3} \cdot \left(\frac{f_{B_s}}{190\,\text{MeV}}\right)^2 ,$$

where the error stems from hadronic uncertainties. Since penguin coefficients are sensitive to new physics and poorly tested experimentally, we analyze the possibility to extract them from a future precision measurement of $\tau(B_s)/\tau(B_d)$. Anticipating progress in the determination of the hadronic parameters ε_1, ε_2 and f_{B_s}/f_{B_d} we find that the coefficient C_4 can be extracted with an uncertainty of order $|\Delta C_4| \simeq 0.1$ from the double ratio $[\tau(B_s) - \tau(B_d)]/[\tau(B^+) - \tau(B_d)]$, if $|\varepsilon_1 - \varepsilon_2|$ is not too small.

1 Introduction

The theoretical achievement of the Heavy Quark Expansion (HQE)[1] has helped a lot to understand the inclusive properties of B-mesons. The comparision of the theoretical predictions with the experimental measurements for the heavy hadron lifetimes and their ratios is an important test of the theory of inclusive decays, and the HQE at the order $(\Lambda_{QCD}/m_b)^3$ which is closely related to the local quark-hadron dualitiy, which is a priori assumption in inclusive non-leptonic decay.

When we neglect higher-order corrections in $1/m_b$, the semileptonic and non-leptonic widths only depend on the CKM-factors and quarks masses. This implies that all the heavy hadrons have the same lifetime and semileptonic width. Lifetime differences aries either from corrections of $\mathcal{O}(1/m_b^2)$ originated by heavy-hadron "wave-function" effects, or from $\mathcal{O}(1/m_b^3)$ corrections induced by non-spectator corrections; i.e., Pauli-interference and W-exchange diagrams.

In the former case, the corrections to lifetime-universality are due to the heavy-quark kninetic energy λ_1 and the chromomagnetic term λ_2 which differ for different hadrons: $\lambda_1(B_s) \neq \lambda_1(B)$ because of $SU(3)_f$ symmetry-breaking effects, the mass of the strange qaurk is much larger than the mass of the u and d quarks; $\lambda_1(B) \neq \lambda_1(\Lambda_b)$ because mesons and baryons have different wave-functions; for the same reason the chromomagnetic term vanishes for the Λ_b but not for the B meson.

The non-spectator contributions, although of $\mathcal{O}(1/m_b^3)$, are enhanced by the factor $16\pi^2$ due to the phase factor for $2 \to 2$ decay and for this reason may give sizeable effects. Note that these corrections may have CKM factors different from those of the leading terms.

By considering lifetime ratios, we can study most conveniently the lifetime difference in B hadrons. In this way, one cancels the dependence on quantities which are poorly known, such as the b-quark mass ($\tau \propto m_b^5$),$|V_{cb}|$, and renormalons[2].

1.1 Experimental Measurement and Theoretical Prediction

The average experimental results for lifetime ratios are[3]:

$$\frac{\tau(B_s)}{\tau(B^0)} = 0.98 \pm 0.07; \qquad \frac{\tau(B^{\pm})}{\tau(B^0)} = 1.07 \pm 0.04; \qquad \frac{\tau(\Lambda_b)}{\tau(B^0)} = 0.78 \pm 0.04. \tag{1}$$

The theoretical predictions is given as follows :

$$\left|\frac{\tau(B_s)}{\tau(B)} - 1\right| < 1\% \tag{2}$$

$$\frac{\tau(B^-)}{\tau(B^0)} = 1 + 16\pi^2 \frac{f_B^2 M_B}{m_b^3}\left[k_1 B_1 + k_2 B_2 + k_3 \epsilon_1 + k_4 \epsilon_2\right] \tag{3}$$

$$\frac{\tau(B^-)}{\tau(B^0)} = 1 - \frac{\lambda_1(\Lambda_b) - \lambda_1(B^0)}{2m_b^2} + c_G \frac{\lambda_2(\Lambda_b) - \lambda_2(B^0)}{m_b^2}$$
$$+16\pi^2 \frac{f_B^2 M_B}{m_b^3}\left[p_1 B_1 + p_2 B_2 + p_3 \epsilon_1 + p_4 \epsilon_2 + (p_5 + p_6 \tilde{B})\right] \tag{4}$$

Using the experimental values of the hadron masses, hadronic parameters are given:

$$\lambda_1(B) - \lambda_1(\Lambda_b) = -(0.001 \pm 0.03) GeV^2, \tag{5}$$

$$\lambda_2(B) \simeq 0.12 GeV^2, \qquad \lambda_2(\Lambda_b) = 0. \tag{6}$$

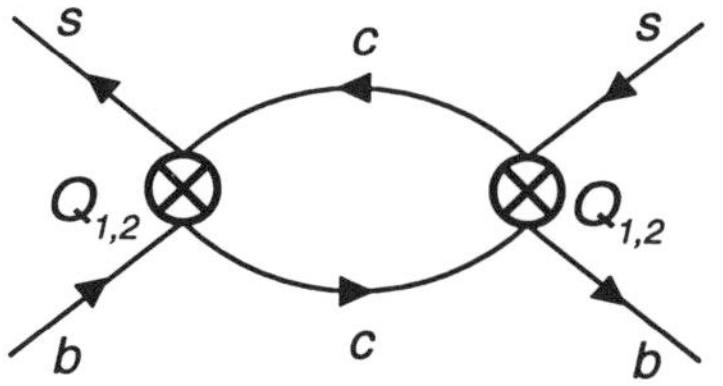

Figure 1: Non-spectator *(weak annihilation)* contribution to the B_s decay rate involving two current-current operators. The corresponding diagram for the B_d decay is obtained by replacing s by d and the upper c by u.

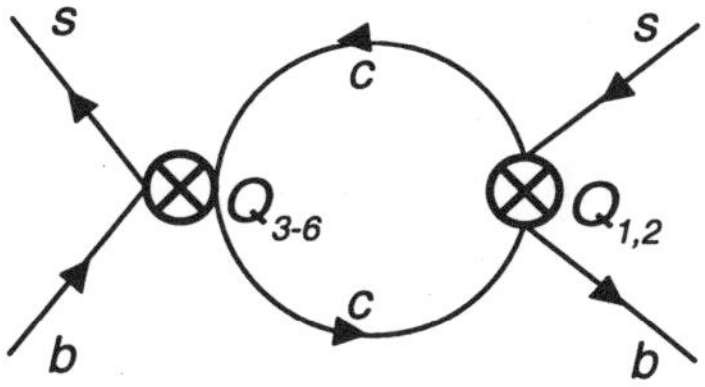

Figure 2: Weak annihilation diagram involving one penguin operator Q_{3-6}. Penguin contributions to the non-spectator rate of the B_d meson are CKM suppressed and therefore negligible.

1.2 Hadronic paremeters

When we define the Operator basis as follows

$$\mathcal{O}_1 = \bar{b}\gamma_\mu(1-\gamma_5)q\bar{q}\gamma^\mu(1-\gamma_5)b; \quad T_1 = \bar{b}\gamma_\mu(1-\gamma_5)T_a q\bar{q}\gamma^\mu(1-\gamma_5)T_a b; \tag{7}$$

$$\mathcal{O}_2 = \bar{b}(1-\gamma_5)q\bar{q}(1+\gamma_5)b; \quad T_2 = \bar{b}(1-\gamma_5)T_a q\bar{q}(1+\gamma_5)T_a b; \tag{8}$$

the matrix elements for B meson are defined in terms of their B-parameters [4]:

$$\langle B|\mathcal{O}_i|B\rangle \equiv f_B^2 M_B^2 B_i(\mu); \qquad \langle B|T_i|B\rangle \equiv f_B^2 M_B^2 \epsilon_i(\mu). \tag{9}$$

For B-baryon, the matrix elements are defined by [4,5]

$$\langle \Lambda_b|\bar{b}_\alpha\gamma^\mu(1-\gamma_5)q_\beta\bar{q}_\beta\gamma_\mu(1-\gamma_5)b_\alpha\rangle \equiv -\tilde{B}\langle\Lambda_b|\mathcal{O}_1|\Lambda_b\rangle; \tag{10}$$

$$\langle\Lambda_b|\mathcal{O}_1|\Lambda_b\rangle = -\frac{f_{B_q}^2 m_{B_q}^2}{6} r; \tag{11}$$

We have to introduce 4 unknown parameters for B meson and 2 unknown parameters for B baryon, which contain non-purtubative property. Now let us consider the penguin contributions to the lifetime splitting between the B_s and the B_d meson.

2 Penguin Contributions

For the non-spectator contributions to the B_s decay rate we need the $|\Delta B| = |\Delta S| = 1$-hamiltonian:

$$H = \frac{G_F}{\sqrt{2}} V_{cb} V_{cs}^* \left[\sum_{j=1}^{6} C_j Q_j + C_8 Q_8 \right] \tag{12}$$

with

$$Q_1 = (\bar{s}^\alpha c^\beta)_{V-A} \cdot (\bar{c^\beta} b^\alpha)_{V-A}, \qquad Q_2 = (\bar{s}c)_{V-A} \cdot (\bar{c}b)_{V-A}$$

$$Q_3 = \sum_{q=u,d,s,c,b} (\bar{s}b)_{V-A} \cdot (\bar{q}q)_{V-A}, \qquad Q_4 = \sum_{q=u,d,s,c,b} (\bar{s}^\alpha b^\beta)_{V-A} \cdot (\bar{q}^\beta q^\alpha)_{V-A}$$

$$Q_5 = \sum_{q=u,d,s,c,b} (\bar{s}b)_{V-A} \cdot (\bar{q}q)_{V+A}, \qquad Q_6 = \sum_{q=u,d,s,c,b} (\bar{s}^\alpha b^\beta)_{V-A} \cdot (\bar{q}^\beta q^\alpha)_{V+A}$$

$$Q_8 = -\frac{g}{8\pi^2} m_b \, \bar{s} \sigma^{\mu\nu} (1 + \gamma_5) T^a b \cdot G^a_{\mu\nu} \, . \tag{13}$$

In (12) we have set $V_{ub} V_{us}^* = O(\lambda^4)$ to zero. The diagram of Fig. 1 has been calculated in [4,6] and yields contributions to the non-spectator part $\Gamma^{\text{non-spec}}$ of the B_s decay rate proportional to $C_2^2, C_1 \cdot C_2$ and C_1^2. $\tau(B_s)/\tau(B_d) - 1$ is proportional to $\Gamma^{\text{non-spec}}(B_d) - \Gamma^{\text{non-spec}}(B_s)$. The main differences between the result of Fig. 1 for these two rates are due to the different mass of u and c and the difference between f_{B_d} and f_{B_s}. Hence the current-current parts of $\tau(B_s)/\tau(B_d) - 1$ proportional to $C_2^2, C_1 \cdot C_2$ or C_1^2 are suppressed by a factor of z or Δ with

$$z = \frac{m_c^2}{m_b^2} = 0.085 \pm 0.023, \qquad \Delta = 1 - \frac{f_{B_d}^2 M_{B_d}}{f_{B_s}^2 M_{B_s}} = 0.23 \pm 0.11. \tag{14}$$

The result for Δ in (14) is the present world average of lattice calculations [7]. There are also $SU(3)_F$ violations in the B-factors, but they are expected to be small from the experience with those appearing in $B^0 - \overline{B}^0$-mixing. In this analysis we use the same B_1, B_2, ϵ_1, and ϵ_2 in $\tau(B_s)$ and $\tau(B_d)$.

Our result for the non-spectator part of the B_c decay rates reads:

$$\Gamma^{\text{non-spec}}(B_s) = -\frac{G_F^2 m_b^2}{12\pi} |V_{cb} V_{cs}|^2 \sqrt{1 - 4z} f_{B_s}^2 M_{B_s} \left[a_1 \varepsilon_1 + a_2 \varepsilon_2 + b_1 B_1 + b_2 B_2 \right] \tag{15}$$

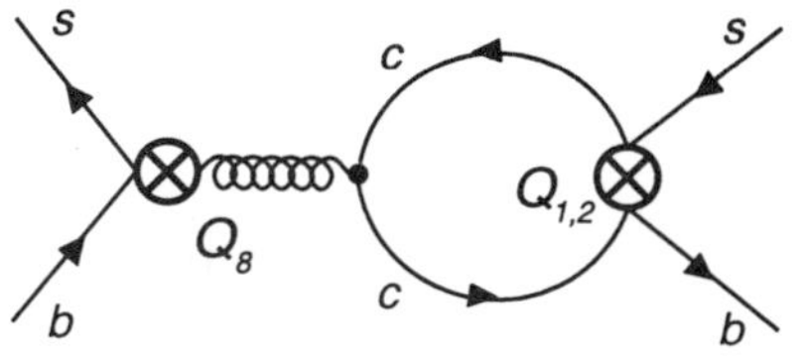

Figure 3: Contribution of Q_8 to $\Gamma^{\text{non-spec}}(B_s)$. In the Standard Model the diagram is of the same order of magnitude as radiative corrections to Fig. 2 and therefore negligible. Yet in models in which quark helicity flips occur in flavour-changing vertices $|C_8|$ can easily be ten times larger than in the Standard Model. The contribution of Q_1 vanishes.

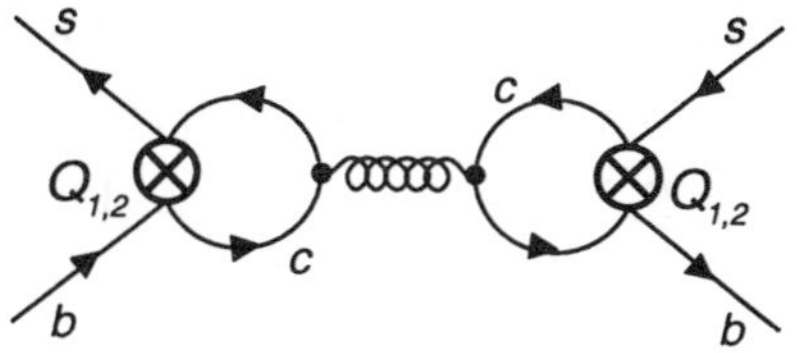

Figure 4: Penguin diagram contribution to $\Gamma^{\text{non-spec}}(B_s)$. The final state corresponds to a cut through either of the $(\overline{c}, c)$-loops. The contributions of Q_1 vanish by colour. This is the only NLO contribution to $\tau(B_s)/\tau(B_d) - 1$ involving $Q_{1,2}$ without suppression factors of Δ or z.

with

$$
\begin{aligned}
a_1 &= \left[2C_2^2 + 4C_2C_4'\right][1-z] + 12zC_2C_6' + [1+2z]\frac{\alpha_s}{\pi}C_2C_8 \\
a_2 &= -[1+2z]\left[2C_2^2 + 4C_2C_4' + \frac{\alpha_s}{\pi}C_2C_8\right] \\
b_1 &= [C_2 + N_cC_1]\left\{(1-z)\left[\frac{C_2}{N_c} + C_1 + 2C_3' + 2\frac{C_4'}{N_c}\right] + 6z\left[C_5' + \frac{C_6'}{N_c}\right]\right\} \\
b_2 &= -[1+2z][C_2 + N_cC_1]\left\{\frac{1}{N_c}[C_2 + N_cC_1] + 2\left[C_3' + \frac{C_4'}{N_c}\right]\right\}
\end{aligned}
\qquad (16)
$$

Here $N_c = 3$ is the number of colours. By setting C_j', $j = 3, \ldots, 6$, and C_8 in (16) to zero one recovers the result of Ref.[4].[c] The result for the non-spectator contributions to the B_d decay rate reads[4]:

$$
\begin{aligned}
\Gamma^{\text{non-spec}}(B_d) = & \frac{G_F^2 m_b^2}{12\pi}|V_{cb}V_{ud}|^2(1-z)^2 f_{B_s}^2 M_{B_s}(\Delta - 1) \\
& \left[a_1^d \varepsilon_1 + a_2^d \varepsilon_2 + b_1^d B_1 + b_2^d B_2\right]
\end{aligned}
\qquad (17)
$$

with[d]

$$
\begin{aligned}
a_1^d &= 2C_2^2\left(1 + \frac{z}{2}\right), & a_2^d &= -2C_2^2(1+2z), \\
b_1^d &= \frac{1}{N_c}(C_2 + N_cC_1)^2\left(1 + \frac{z}{2}\right), & b_2^d &= -\frac{1}{N_c}(C_2 + N_cC_1)^2(1+2z)
\end{aligned}
\qquad (18)
$$

[c] Notice that our notation of C_1 and C_2 is opposite to the one in Ref.[4].

[d] In the large N_c limit one finds $\Gamma^{\text{non-spec}}$ helicity suppressed in analogy to the leptonic decay rate. This shows that one cannot neglect the $O(1/N_c)$ terms.

When we combine (15-18) in order to predict[8] $\tau(B_s)/\tau(B_d) - 1$:

$$\frac{\tau(B_s)}{\tau(B_d)} - 1 = \frac{\Gamma^{\rm non-spec}(B_d) - \Gamma^{\rm non-spec}(B_s)}{\Gamma^{\rm total}} + O(10^{-3})$$
$$= 9.0 \cdot 10^{-4}\epsilon_1 - 1.63 \cdot 10^{-2}\epsilon_2 + 2.0 \cdot 10^{-4} B_1 - 5.0 \cdot 10^{-4} B_2 \quad (19)$$

The experimental world average[3]

$$\frac{\tau(B^+)}{\tau(B_d^0)} = 1.07 \pm 0.04 \quad (20)$$

leads to the following constraint:

$$\varepsilon_1 \simeq (-0.2 \pm 0.1) \left(\frac{0.17\,\text{GeV}}{f_B}\right)^2 \left(\frac{m_b}{4.8\,\text{GeV}}\right)^3 + 0.3\varepsilon_2 + 0.05. \quad (21)$$

Here we consider the range $|\varepsilon_1|$, $|\varepsilon_2| \leq 0.3$, and further obey (20).

From Tab.2 of ref [8] we realize that the penguin contributions are comparable in size, but opposite in sign to the current-current part obtained in Ref.[4]. This makes the experimental detection of any deviation of $\tau(B_s)/\tau(B_d)$ from 1 even more difficult, if the penguin coefficients are really dominated by Standard Model physics. From the results of Tab.2 in Ref.[8], we obtain

$$\frac{\tau(B_s)}{\tau(B_d)} - 1 = (-1.2 \pm 8.0 \pm 2.0) \cdot 10^{-3} \cdot \left(\frac{f_{B_s}}{190\,\text{MeV}}\right)^2 \left(\frac{4.8\,\text{GeV}}{m_b}\right)^3. \quad (22)$$

Here the first error stems from the uncertainty in ε_1 and ε_2 and will be reduced once lattice results for the hadronic parameters are available. The second error summarizes the remaining uncertainties.

3 New Physics Effects

Today we have little experimental information on the sizes of the penguin coefficients. Their smallness in the Standard Model allows for the possibility that they are dominated by new physics. The total charmless inclusive branching fraction $Br(B \to no\ charm)$ is a candidate to detect new physics contributions to C_8 [9], but it is much less sensitive to C_{3-6} [10]. Eq.(17) of Ref.[8] reveals that $\tau(B_s)/\tau(B_d)$ is a complementary observable mainly sensitive to C_4, while C_8 is of minor importance.

New physics contributions $\Delta C_{3-6}(\mu = 200GeV)$ affect $C_4(\mu = 4.8\,\text{GeV})$ by

$$\Delta C_4(\mu = 4.8\,\text{GeV}) = -0.35\,\Delta C_3(200\,\text{GeV}) + 0.99\,\Delta C_4(200\,\text{GeV})$$
$$-0.03\,\Delta C_5(200\,\text{GeV}) - 0.22\,\Delta C_6(200\,\text{GeV}). \quad (23)$$

Observe that $\Delta C_4(200\,\text{GeV}) = -0.05$ already increases $C_4'(m_b)$ by more than a factor of two.

Clearly the usefulness of $\tau(B_s)/\tau(B_d)$ to probe C_{3-6} crucially depends on the size of $|\varepsilon_1 - \varepsilon_2|$ and f_{B_s}. We now investigate the sensitivity of $\tau(B_s)/\tau(B_d)$ to $\Delta C_4(\mu = m_b)$ in a possible future scenario for the hadronic parameters. We assume

$$\begin{aligned} \varepsilon_1 &= -0.10 \pm 0.05, \quad \varepsilon_2 = 0.20 \pm 0.05, \quad B_1, B_2 = 1.0 \pm 0.1, \\ f_{B_s} &= (190 \pm 15)\ \text{MeV}, \quad \Delta = 0.23 \pm 0.05, \quad \text{m}_\text{b} = (4.8 \pm 0.1)\ \text{GeV}. \end{aligned} \tag{24}$$

A cleaner observable which can see a new physics effects is the double ratio:

$$\frac{\tau(B_s) - \tau(B_d)}{\tau(B^+) - \tau(B_d)} = \frac{B_{SL}(B_s) - B_{SL}(B_d)}{B_{SL}(B^+) - B_{SL}(B_d)}, \tag{25}$$

which depends on $\varepsilon_1, \varepsilon_2$ and Δ, while the dependence on f_B and m_b cancels. The corresponding plot for the parameter set of (24) can be found in Fig. 5 If $\Delta C_4 < -0.075$ or $\Delta C_4 > 0.140$, we find the allowed range for $[\tau(B_s) - \tau(B_d)]/[\tau(B^+) - \tau(B_d)]$ incompatible with the Standard Model. The experimental detection of a sizeable negative lifetime difference $\tau(B_s) - \tau(B_d)$ may reveal non-standard contributions to C_4' of similar size as its Standard Model value. Fig. 5 shows that e.g. the bound $\tau(B_s) - \tau(B_d) < -0.20[\tau(B^+) - \tau(B_d)]$ would indicate $\Delta C_4 > 0.051$. We conclude that the detection of new physics contributions to C_4 of order 0.1 is possible with precision measurements of $\tau(B_s)/\tau(B_d)$.

4 Conclusions

We have calculated the contributions of the penguin operators Q_{3-6}, of the chromomagnetic operator Q_8 and of penguin diagrams with insertions of Q_2 to the lifetime splitting between the B_s and B_d meson. In the Standard Model the penguin effects are found to be roughly half as big as the contributions from the current-current operators Q_1 and Q_2, despite of the smallness of the penguin coefficients. Yet they are opposite in sign, so that any deviation of $\tau(B_s) - \tau(B_d)$ from zero is even harder to detect experimentally. Assuming a reasonable progress in the determination of the hadronic parameters a precision measurement of $\tau(B_s)/\tau(B_d)$ can be used to probe the coefficient C_4 with an accuracy of $|\Delta C_4| = 0.1$. Hence new physics can only be detected, if C_4 is dominated by non-standard contributions. The sensitivity to C_4 depends crucially on the difference of the hadronic parameters ε_1 and ε_2. For the extraction of C_4 the double ratio $[\tau(B_s) - \tau(B_d)]/[\tau(B^+) - \tau(B_d)]$ turns out to be very useful to see a new physics effects.

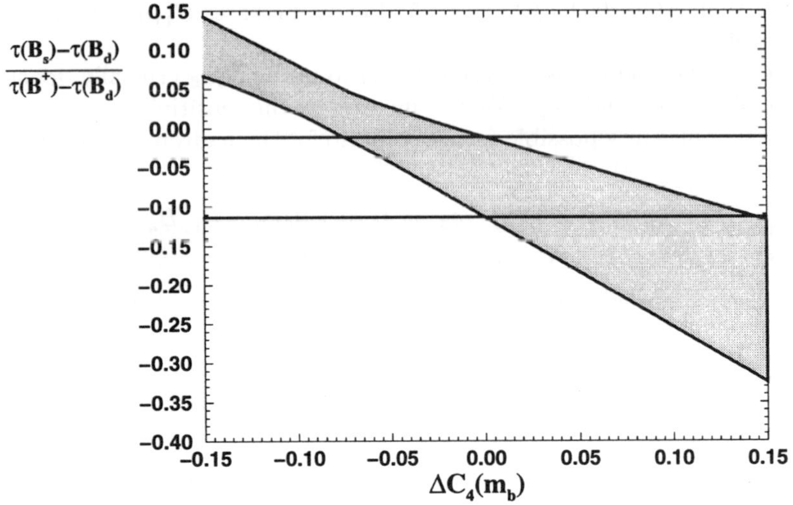

Figure 5: Dependence of $(\tau(B_s) - \tau(B_d))/(\tau(B^+) - \tau(B_d))$ on ΔC_4 for the parameter set in (24). This double ratio depends on f_{B_s} and f_{B_d} only through Δ, and the factor of m_b^{-3} cancels.

Acknowledgements

The author would like to thank the organizers of the fourth international workshop on particle physics phenomenology at Kaoshiung. Y.-Y.Keum thanks Chris Sachrajda and Hai-Yang Cheng for helpful discussions and U. Nierste for enjoyable collaboration on this subject. He is grateful to Prof. M. Kobayashi and Prof. W. Buchmüller for their hospitality and encouragement. This work is supported in part by the Basic Science Research Institute Program, Ministry of Education, Project No. BSRI-97-2414, and in part by the Grant-in Aid for Scientific from the Ministry of Education, Science and Culture, Japan.

References

1. I.I. Bigi, N. Uraltsev, and A. Vainshtein, Phys. Lett. B 293, 430 (1992); Erratum *ibid.* 297, 477 (1993).

B. Blok and M. Shifman, Nucl. Phys. B399 (1993) 441; *ibid.* 399 (1993) 459. A. Manohar and M. Wise, Phys. Rev. D 49, (1994) 1310.
B. Blok, L. Koyrakh, M. Shifman and A.I. Vainshtein, Phys. Rev. D49 (1994), 3356; Erratum *ibid.* D50 (1994) 3572.
T. Mannel, Nucl. Phys. B413 (1994) 396.
I.I. Bigi, M.A. Shifman, N.G. Uraltsev, A.I. Vainshtein, Int. J. Mod. Phys. A9 (1994) 2467.

2. M. Neubert and C. Sachrajda, Nucl. Phys. B438 (1995) 235.
3. M. Feindt, talk at the *International Europhysics Conference on High Energy Physics*, Jerusalem, August 1997.
4. M. Neubert and C. Sachrajda, Nucl. Phys. B483 (1997) 339.
5. H.Y. Cheng, Phys. Rev. D56 (1997) 2783.
6. M. Beneke, G. Buchalla and I. Dunietz, Phys. Rev. D54 (1996) 4419.
7. H. Wittig, talk at the DESY seminar, Hamburg 1997, and Int. J. Mod. Phys. A12 (1997) 4477.
   C. Sachrajda, talk at the *XVIII International Symposium on Lepton Photon Interactions*, Hamburg, July 1997.
   C. Bernard et. al, Proceedings in Chicago 1997, Twenty Beautiful years of Bottom Physics.
8. Y.-Y. Keum and U. Nierste, Phys. Rev. D57 (1998) 4282.
9. S. Bertolini, F. Borzumati and A. Masiero, Nucl. Phys. B294 (1987) 321.
   A.L. Kagan, Phys. Rev. D51 (1995) 6196.
   M. Ciuchini, E. Gabrielli and G.F. Giudice, Phys. Lett. B388 (1996) 353.
   A.L. Kagan and J. Rathsman, (unpublished) hep-ph/9701300 .
10. A. Lenz, G. Ostermaier and U. Nierste, Phys. Rev. D56 (1997) 7228.

Exclusive charmless B_s hadronic decays into η' and η

B. Tseng

Institute of Physics, Academia Sinica, Taipei, Taiwan 115, R.O.C.
E-mail: btseng@phys.sinica.edu.tw

Using the next-to-leading order QCD-corrected effective Hamiltonian, charmless exclusive nonleptonic decays of the B_s meson into η or η' are calculated within the generalized factorization approach. Nonfactorizable contributions are included with two different treatments. Some subtleties involved are discussed.

1 Introduction

Stimulated by the recent observations of the large inclusive and exclusive rare B decays by the CLEO Collaboration [1], there are considerable interests in the charmless B meson decays [2]. To explain the abnormally large branching ratio of the semi-inclusive process $B \to \eta' + X$, several mechanisms have been advocated [3,4,5,6] and some tests of these mechanisms have been proposed [7]. It is now generally believed that the QCD anomaly [3,4,5] plays a vital role. The understanding of the exclusive $B \to \eta' K$, however, relies on several subtle points. First, the QCD anomaly does occur through the equation of motion [8,9] when calculating the $(S-P)(S+P)$ penguin operator and its effect is found to reduce the branching ratio. Second, the mechanism of $c\bar{c} \to \eta'$, although proposed to be large and positive originally [10,11], is now preferred to be negative and smaller than before as implied by a recent theoretical recalculation [12] and several phenomenological analyses [9,13]. Third, the running strange quark mass which appears in the calculation of the matrix elements of the $(S-P)(S+P)$ penguin operator, the $SU(3)$ breaking effect in the involved η' decay constants and the normalization of the $B \to \eta^{(\prime)}$ matrix element involved raise the branching ratio substantially. Finally, nonfactorizable contributions, which are parametrized by the N_c^{eff}, gives the final answer for the largeness of exclusive $B \to \eta' K$ [14,15]. It is very interesting to see the impacts of these subtleties mentioned above on the the exclusive charmless B_s decays to an η' or η [16]. That is the main purpose of this talk [17].

2 *Theoretical Framework*

We begin with a brief description of the theoretical framework. The relevant effective $\Delta B = 1$ weak Hamiltonian is

$$\mathcal{H}_{\rm eff} = \frac{G_F}{\sqrt{2}}\Big[V_{ub}V_{uq}^*(c_1O_1^u+c_2O_2^u)+V_{cb}V_{cq}^*(c_1O_1^c+c_2O_2^c)-V_{tb}V_{tq}^*\sum_{i=3}^{10}c_iO_i\Big], \quad (1)$$

where $q = d, s$, and O_{3-6} are QCD penguin operators and O_{7-10} are electroweak penguin operators. $C_i(\mu)$ are the Wilson coefficients, which have been evaluated to the next-to-leading order (NLO) [18,19]. One important feature of the NLO calculation is the renormalization-scheme and -scale dependence of the Wilson coefficients (for a review, see [20]). In order to ensure the μ and renormalization scheme independence for the physical amplitude, the matrix elements, which are evaluated under the factorization hypothesis, have to be computed in the same renormalization scheme and renormalized at the same scale as $c_i(\mu)$. However, as emphasized in [14], the matrix element $\langle O\rangle_{\rm fact}$ is scale independent under the factorization approach and hence it cannot be identified with $\langle O(\mu)\rangle$. Incorporating QCD and electroweak corrections to the four-quark operators, we can redefine $c_i(\mu)\langle O_i(\mu)\rangle = c_i^{\rm eff}\langle O_i\rangle_{\rm tree}$, so that $c_i^{\rm eff}$ are renormaliztion scheme and scale independent. Then the factorization approximation is applied to the hadronic matrix elements of the operator O at the tree level. The numerical values for $c_i^{\rm eff}$ are shown in the last column of Table I, where $\mu = m_b(m_b)$, $\Lambda^{(5)}_{\overline{\rm MS}} = 225$ MeV, $m_t = 170$ GeV and $k^2 = m_b^2/2$ are used [14].

In general, there are contributions from the nonfactorizable amplitudes. Because there is only one single form factor (or Lorentz scalar) involved in the decay amplitude of $B\,(D) \to PP$, PV decays (P: pseudoscalar meson, V: vector meson), the effects of nonfactorization can be lumped into the effective parameters $a_i^{\rm eff}$ [21]:

$$a_{2i}^{\rm eff} = c_{2i}^{\rm eff} + c_{2i-1}^{\rm eff}\left(\frac{1}{N_c} + \chi_{2i}\right), \qquad a_{2i-1}^{\rm eff} = c_{2i-1}^{\rm eff} + c_{2i}^{\rm eff}\left(\frac{1}{N_c} + \chi_{2i-1}\right), \quad (2)$$

where $c_{2i,2i-1}^{\rm eff}$ are the Wilson coefficients of the 4-quark operators, and nonfactorizable contributions are characterized by the parameters χ_{2i} and χ_{2i-1}. We can parametrize the nonfactorizable contributions by defining an effective number of colors $N_c^{\rm eff}$, called $1/\xi$ in [22], as $1/N_c^{\rm eff} \equiv (1/N_c) + \chi$. Different factorization approach used in the literature can be classified by the effective number of colors $N_c^{\rm eff}$. The so-called "naive" factorization discards all the nonfactorizable contributions and takes $1/N_c^{\rm eff} = 1/N_c = 1/3$, whereas the "large-$N_c$ improved" factorization [23] drops out all the subleading $1/N_c$ terms and takes $1/N_c^{\rm eff} = 0$. In principle, $N_c^{\rm eff}$ can vary from channel to channel, as in the case of charm decay. However, in the energetic two-body B decays, $N_c^{\rm eff}$ is expected to be process insensitive as supported by data [24]. If $N_c^{\rm eff}$ is process independent, then we have a generalized factorization. In this paper, we will treat the nonfactorizable contributions with two different phenomenological ways : (i) the one with "homogenous" structure, which assumes that $(N_c^{\rm eff})_1 \approx (N_c^{\rm eff})_2 \approx \cdots \approx (N_c^{\rm eff})_{10}$, and (ii) the "heterogeneous" one, which considers the possibility of $N_c^{\rm eff}(V+A) \neq N_c^{\rm eff}(V-A)$. The consideration of the "homogenous" nonfactorizable contributions, which is commonly used in the literature, has its advantage of simplicity. However, as argued in [14], due to the different Dirac structure of the Fierz transformation, nonfactorizable effects in the matrix elements of $(V-A)(V+A)$ operators are *a priori* different from that of $(V-A)(V-A)$ operators, i.e. $\chi(V+A) \neq \chi(V-A)$. Since $1/N_c^{\rm eff} = 1/N_c + \chi$, theoretically it is expected that

$$N_c^{\rm eff}(V-A) \equiv \left(N_c^{\rm eff}\right)_1 \approx \left(N_c^{\rm eff}\right)_2 \approx \left(N_c^{\rm eff}\right)_3 \approx \left(N_c^{\rm eff}\right)_4 \approx \left(N_c^{\rm eff}\right)_9 \approx \left(N_c^{\rm eff}\right)_{10},$$
$$N_c^{\rm eff}(V+A) \equiv \left(N_c^{\rm eff}\right)_5 \approx \left(N_c^{\rm eff}\right)_6 \approx \left(N_c^{\rm eff}\right)_7 \approx \left(N_c^{\rm eff}\right)_8, \qquad (3)$$

To illustrate the effect of the nonfactorizable contribution, we extrapolate $N_c(V-A) \approx 2$ from $B \to D\pi(\rho)$ [25] to charmless decays.

Table 1: Numerical values of effective coefficients a_i at $N_c^{\rm eff} = 2, 3, 5, \infty$, where $N_c^{\rm eff} = \infty$ corresponds to $a_i^{\rm eff} = c_i^{\rm eff}$. The entries for $a_3,\ldots,a_{10}$ have to be multiplied with 10^{-4}.

	$N_c^{\rm eff}=2$	$N_c^{\rm eff}=3$	$N_c^{\rm eff}=5$	$N_c^{\rm eff}=\infty$
a_1	0.986	1.04	1.08	1.15
a_2	0.25	0.058	-0.095	-0.325
a_3	$-13.9-22.6i$	61	$120+18i$	$211+45.3i$
a_4	$-344-113i$	$-380-120i$	$-410-127i$	$-450-136i$
a_5	$-146-22.6i$	-52.7	$22+18i$	$134+45.3i$
a_6	$-493-113i$	$-515-121i$	$-530-127i$	$-560-136i$
a_7	$0.04-2.73i$	$-0.7-2.73i$	$-1.24-2.73i$	$-2.04-2.73i$
a_8	$2.98-1.37i$	$3.32-0.9i$	$3.59-0.55i$	4
a_9	$-87.9-2.73i$	$-91.1-2.73i$	$-93.7-2.73i$	$-97.6-2.73i$
a_{10}	$-29.3-1.37i$	$-13-0.91i$	$-0.04-0.55i$	19.48

The $N_c^{\rm eff}$-dependence of the effective parameters a_i's are shown in Table I, from which we see that a_1, a_4, a_6 and a_9 are $N_c^{\rm eff}$-stable, and the remaining ones are $N_c^{\rm eff}$-sensitive. We would like to remark that while a_3 and a_5 are both $N_c^{\rm eff}$-sensitive, the combination of $(a_3 - a_5)$ is rather stable under the

Updated Analysis of a_2/a_1 in Exclusive B Decays [a]

Hai-Yang Cheng and Kwei-Chou Yang

Institute of Physics, Academia Sinica, Taipei, Taiwan 115, R.O.C.

Using recent experimental data and various theoretical calculations on form factors, we reanalyze the effective parameters a_1 and a_2 and their ratio.

Nonleptonic two-body decays of mesons were conventionally studied in the generalized factorization approach, in which the factorized decay amplitude is associated with the effective coefficients

$$a_{1,2}^{\text{eff}} = c_{1,2}^{\text{eff}} + c_{2,1}^{\text{eff}}[(1/N_c) + \chi_{1,2}], \tag{1}$$

where $c_{1,2}^{\text{eff}}$ are the effective Wilson coefficients and the nonfactorized terms are expressed by $\chi_{1,2}$. For simplicity, in what follows we will drop the superscript "eff" of $a_{1,2}^{\text{eff}}$. Based on the generalized factorization assumption, one can catalog the decay processes into three different classes. For class-I decays, the decay amplitudes, dominated by color-allowed external W-emission, are proportional to the a_1 parameter. For class-II decays, the decay amplitudes, governed by color-suppressed internal W-emission, are described by a_2. The decay amplitudes of the class-III involve a linear combination of a_1 and a_2. From the data analysis, contrary to the case of charmed meson decays in which a_2 is negative and $\chi \sim -1/N_c$, a_2 is positive in two-body decays of the B meson. $\chi_{2(1)}$ coming from the contribution of the matrix element of the color octet-octet and singlet-singlet currents, since it is made of nonfactorizable soft gluon effects in itself, will become more important if the final particles move slower [1,2]. Therefore, we naïvely expect that $|\chi_i(D \to V\ P)| \gtrsim |\chi_i(D \to P\ P)| > |\chi_i(B \to V\ P)|$, where P is the pseudoscalar meson and V the vector meson. The data of $B \to J/\Psi\ K^{(*)}$ which are class-II decay modes can be used to study the absolute value of a_2. However, the experimental results are quite difficult to account for the observed longitudinal polarization fraction Γ_L/Γ and the ratio $R \equiv \Gamma(B \to J/\Psi\ K^*)/\Gamma(B \to J/\Psi\ K)$ simultaneously using the existing form factor calculations. To accommodate the data, it has been advocated [2] a modification to the generalized factorization hypothesis so that a_2 is process dependent. In this talk, we focus on the extraction of the value a_1, $|a_2|$, and a_2/a_1 from the current experimental data [3] and various form factor calculations [4], the BSWI model, the BSWII model [5], the LF model [6],

[a] presented by K.-C. Yang.
Email address: `kcyang@phys.sinica.edu.tw`

Table 1: The effective parameter a_1 extracted from $B \to D^{(*)}\ D_s^{(*)}$ using different form-factor models.

	BSWI	BSWII	LF	SR+HQET	NS
$\bar{B}^0 \to D^+ D_s^-$	1.05±0.16	1.05±0.16	1.06±0.16	1.17±0.18	1.32±0.20
$B^- \to D^0 D_s^-$	0.80±0.15	0.80±0.15	0.81±0.15	0.89±0.17	1.01±0.19
$\bar{B}^0 \to D^+ D_s^{*-}$	1.24±0.31	1.10±0.27	1.11±0.28	1.26±0.31	1.43±0.36
$B^- \to D^0 D_s^{*-}$	1.14±0.25	1.10±0.22	1.02±0.23	1.16±0.26	1.32±0.29
$\bar{B}^0 \to D^{*+} D_s^-$	1.34±0.24	1.21±0.21	1.11±0.28	1.09±0.19	1.03±0.18
$B^- \to D^{*0} D_s^-$	1.46±0.30	1.31±0.28	1.02±0.23	1.18±0.25	1.12±0.23
$\bar{B}^0 \to D^{*+} D_s^{*-}$	0.88±0.15	0.90±0.16	0.83±0.15	0.86±0.15	0.85±0.15
$B^- \to D^{*0} D_s^{*-}$	1.05±0.20	1.07±0.20	0.99±0.18	1.03±0.19	1.02±0.19
Average	1.03±0.07	1.01±0.07	0.97±0.07	1.03±0.07	1.06±0.07

the NS model [7], the QCD sum rule results (Yang) [8], and the light-cone sum rule calculations (LC) [9]. Since the accuracy of these values still suffers from reliability of the form factor calculations, we desire that an objective estimation can be obtained.

Here will consider the decay amplitudes of processes which have been measured experimentally:

- Class I: $B^- \to D^{(*)0}\ D_s^{(*)-}$, $\bar{B}_d^0 \to D^{(*)+}\ D_s^{(*)-}$, and $\bar{B}_d^0 \to D^{(*)+}\ \pi^-(\rho^-)$,
- Class II: $B^- \to J/\Psi\ K^{(*)-}$ and $B^0 \to J/\Psi K^{(*)0}$.
- Class III: $B^- \to D^{(*)0}\ \pi^-(\rho^-)$,

The value of $a_{(1,2)}$ can be extracted from the above data of the class-(I,II) type transitions, while their ratio can be obtained by the the following ratios:

$$R_1 \equiv \frac{\mathcal{B}(B^- \to D^0\pi^-)}{\mathcal{B}(\bar{B}^0 \to D^+\pi^-)}, \quad R_2 \equiv \frac{\mathcal{B}(B^- \to D^0\rho^-)}{\mathcal{B}(\bar{B}^0 \to D^+\rho^-)},$$
$$R_3 \equiv \frac{\mathcal{B}(B^- \to D^{*0}\pi^-)}{\mathcal{B}(\bar{B}^0 \to D^{*+}\pi^-)}, \quad R_4 \equiv \frac{\mathcal{B}(B^- \to D^{*0}\rho^-)}{\mathcal{B}(\bar{B}^0 \to D^{*+}\rho^-)}. \tag{2}$$

The hadronic matrix element of effective operators $O_{1,2}$ can be redefined in the scheme- and scale(μ)-dependent ways $\langle O(\mu)\rangle = g(\mu)\langle O\rangle_{\rm tree}$, where $\langle O\rangle_{\rm tree}$ is μ-independent. We thus can rewrite the matrix element of the effective Hamiltonian $\langle H_{\rm eff}\rangle$ as the product of $c^{\rm eff}$ and $\langle O\rangle_{\rm tree}$, both of which are scheme- and μ-independent. The only relevant scale, which is implicit and of order m_b, separating $c^{\rm eff}$ and $\langle O\rangle_{\rm tree}$ is the energy release of B decays. Here we use $c_1^{\rm eff} = 1.149$ and $c_2^{\rm eff} = -0.325$, to the next-to-leading order [10]. Using the

Table 2: The effective parameter a_1 extracted from $\bar{B}_d^0 \to D^{(*)+}\ \pi^-(\rho^-)$ using different form-factor models.

	BSWI	BWSII	LF	SR+HQET	NS
$\bar{B}^0 \to D^+\pi^-$	0.89±0.06	0.89±0.06	0.88±0.06	1.15±0.08	1.31±0.09
$\bar{B}^0 \to D^+\rho^-$	0.91±0.08	0.90±0.08	0.89±0.08	1.13±0.10	1.29±0.11
$\bar{B}^0 \to D^{*+}\pi^-$	0.98±0.04	0.98±0.04	0.84±0.03	0.93±0.04	0.89±0.03
$\bar{B}^0 \to D^{*+}\rho^-$	0.86±0.21	0.87±0.21	0.75±0.18	0.83±0.20	0.80±0.20
Average	0.95±0.03	0.94±0.03	0.85±0.03	0.98±0.03	0.96±0.03

Table 3: The effective parameter $|a_2|$ extracted from $B \to J/\Psi\ K^{(*)}$ using different form-factor models.

	BSWI	BSWII	LF	NS	Yang	LC
$B^+ \to J/\Psi K^+$	0.33±0.02	0.22±0.01	0.28±0.01	0.36±0.02	0.37±0.02	0.30±0.02
$B^0 \to J/\Psi K^0$	0.32±0.02	0.21±0.01	0.27±0.02	0.35±0.02	0.36±0.02	0.29±0.02
Average	0.32±0.01	0.22±0.01	0.28±0.01	0.36±0.01	0.36±0.01	0.30±0.01
$B^+ \to J/\Psi K^{*+}$	0.20±0.02	0.21±0.02	0.25±0.02	0.24±0.02	0.39±0.04	0.20±0.02
$B^0 \to J/\Psi K^{*0}$	0.19±0.01	0.21±0.01	0.25±0.02	0.25±0.02	0.39±0.03	0.19±0.01
Average	0.20±0.01	0.22±0.01	0.26±0.01	0.25±0.01	0.40±0.02	0.20±0.01

theoretical results of form factors and the experimental data from the Particle Data Group (PDG) [3], one can easily extract the values of a_1, $|a_2|$, and a_2/a_1. The results are listed in Tables 1, 2, 3, and 5. In Tables 1 and 2, the updated data are used for the parameters of NS, while the results of SR+HQET [b] are gotten by the NS formula but the parameters are obtained from the sum rule results and the predictions of the heavy quark effective theory [11]. All the results [c] in Tables 1 and 2 indicate that the value of a_1, which is very close to 1, seems to follow the pattern $a_1(B \to D^{(*)}D_s^{(*)}) \gtrsim a_1(B \to D^{(*)}\pi(\rho))$.

For $B \to J/\Psi\ K^{(*)}$, since they are color suppressed and sensitive to the nonfactorizable contributions, therefore, they are very good examples to explore the generalized factorization hypothesis to see if a_2 is universal. On the other hand, since the energy release in $B \to J/\Psi\ K$ is close to that in $B \to J/\Psi\ K^*$, we expect that the extracted values of $|a_2|$ from these two kinds of processes can be slightly different (the so-called minimal modified factorization hypothesis). However, from Table 3, we know that only BSWII, LF, and

[b]We will use the $B \to D^{(*)}$ form-factor results of SR+HQET as subsidiaries of the Yang's [8] and LC [9] calculations, since Yang's sum rules and LC only show the form factors of the B meson into a light meson.

[c]The class-I transitions determine the absolute value of a_1. However, since this kind of processes is color allowed, we expect the sign of a_1 is the same as that of $c_1^{\rm eff}$ and is therefore positive.

Table 4: R, Γ_L/Γ, and $|P|^2$ from various form factor calculations based on the generalized factorization hypothesis together with the CLEO new data.

	BSWI	BSWII	LF	NS	Yang	LC	CLEO
R	4.15	1.58	1.79	3.17	1.30	3.48	1.45 ± 0.26
Γ_L/Γ	0.57	0.36	0.53	0.49	0.42	0.47	0.52 ± 0.08
$\|P\|^2$	0.09	0.24	0.09	0.12	0.19	0.23	0.16 ± 0.09

Table 5: a_2/a_1 extracted from the PDG data.

	BSWI	BSWII	LF	NS	Yang	LC
R_1	0.29±0.10	0.29±0.10	0.38±0.13	0.28±0.10	0.29±0.10	0.27±0.10
R_2	0.59±0.30	0.52±0.27	0.56±0.28	0.43±0.22	1.21±0.62	0.34±0.17
R_3	0.23±0.06	0.20±0.06	0.31±0.09	0.30±0.09	0.28±0.08	0.25±0.07
R_4	0.55±0.40	0.67±0.49	0.85±0.62	0.79±0.58	1.50±1.09	0.62±0.45
Average	0.26±0.05	0.23±0.05	0.35±0.07	0.31±0.06	0.30±0.06	0.27±0.05

Yang's sum rules can satisfy our expectation.

The other ways to examine the generalized factorization hypothesis and/or the quality of various form factor results are to study the production ratio $R = \mathcal{B}(B \to J/\Psi\ K^*)/\mathcal{B}(B \to J/\Psi\ K)$, the fraction of longitudinal polarization Γ_L/Γ , and the $P-$wave transverse polarization $|P|^2$ measured in the transversity basis in $B \to J/\Psi\ K^*$ decays. The predictions from various form factor calculations based on the generalized factorization hypothesis together with the CLEO new data [12] are shown in Table 4. We find again that only BSWII, LF, and Yang's sum rules can accommodate the data, expect that the value of Γ_L/Γ in BSWII is a little smaller than the CLEO data. Furthermore, if allowing the nonfactorizable contribution χ_2 to be slightly different in $B \to J/\Psi\ K^{(*)}$ and in $B \to D^{(*)}\ \pi(\rho)$, we find that the values of $a_2(B \to J/\Psi\ K^{(*)})$ of BSWII/LF are consistent with that obtained from the analyses of R_{1-4} (see Tables 5 and 6) if assuming $a_1 \approx 1$. However, note that the Yang's sum rule results do not behave like the BSWII or LF model. The value of a_2 in the sum rule analysis of Yang is larger. There are two possibilities to explain it. One is $\chi_2 \approx 0.3$, a lager value compared to that in BSWII or LF ($\chi_2 \approx$0.14 in BWSII, 0.18 in LF if we have adopted $a_2 > 0$). The other possibility is both the nonfactorizable values χ_2 and a_2 become negative in $B \to J/\Psi\ K^{(*)}$, as the D meson decays. The QCD sum rule calculation on the nonfactorizable contribution in $B \to J/\Psi\ K$ was reported to be negative [13], contrary to the generalized nonfactorization hypothesis. However, it is difficult to understand why a_2 is negative in $B \to J/\Psi\ K^{(*)}$ while it becomes positive in $B^- \to D^{(*)}\ \pi^-(\rho^-)$. One way to solve directly this problem is to evaluate the class-III or class-II decay channels of $B \to D\ \pi(\rho)$, within framework of

Table 6: a_2/a_1, if using the CLEO II data.

	BSWI	BSWII	LF	NS	Yang	LC
R_1	0.33±0.13	0.33±0.13	0.43±0.16	0.32±0.12	0.33±0.13	0.31±0.12
R_2	0.10±0.24	0.09±0.21	0.09±0.23	0.07±0.18	0.20±0.49	0.06±0.14
R_3	0.16±0.08	0.14±0.07	0.22±0.10	0.22±0.10	0.20±0.09	0.18±0.08
R_4	0.33±0.14	0.40±0.18	0.51±0.22	0.47±0.21	0.90±0.39	0.37±0.16
Average	0.22±0.06	0.20±0.05	0.29±0.08	0.25±0.07	0.27±0.07	0.21±0.06

the QCD sum rules to see if χ_2 becomes positive.

From the analyses of R_{1-4} defined in Eq. (2) one can obtain a_2/a_1. Since we have adopted that a_1 is positive, we can thus determine the sign of a_2. The results are listed in Table 5 if using PDG data. All of the results show that a_2 is positive. Two remarks are in order. First, a_2/a_1 of $R_{1,3}$ for all of form factor results are quantitatively stable, while that of $R_{2,4}$ have large central values and errors. Second, the prediction of a_2/a_1 in LF is slightly large is because a_1 in LF is a smaller value (see Tables 1 and 2).

CLEO II [14] has also reported the data on various class-III and class-I branching ratios of $B \to D^{(*)}\ \pi(\rho)$. We show that the analysis results of a_2/a_1 in Table 6. But now the values of a_2/a_1 from $R_{2,4}$ is much smaller the corresponding values in Table 5. Eventually, if we account of the fact that a_1 is little small in LF, then a_2 is lying in $0.2 - 0.3$, which is a reasonable result for all of form factor calculations.

To conclude, we have used the current experimental data and various theoretical results of form factors to analyze the effective coefficients a_1 and a_2. Our results have shown that if allowing a minimal modification to the generalized factorization hypothesis, *i.e.*, the nonfactorizable contribution $\chi_{1,2}$ can be slightly different in different decay processes, then we have the following conclusions: (1) $a_1 \approx 1$ and $a_1(B \to D^{(*)} D_s^{(*)}) \gtrsim a_1(B \to D^{(*)}\pi(\rho))$. (2) Only the BSW II model, the light-front model, and Yang's sum rules can satisfy this factorization assumption in $B \to J/\Psi\ K^{(*)}$ analyses. The results are shown in Tables 3 and 4. (3) From the results of a_2/a_1 listed in Tables 5 and 6, we obtain that a_2 is lying in $0.2 - 0.3$, which is a reasonable result for all of form factor calculations in the $B \to D^{(*)}\ \pi(\rho)$ decay processes.

This work was supported in part by the National Science Council of R.O.C. under Grant No. NSC88-2112-M-001-006.

1. H.Y. Cheng and B. Tseng, *Phys. Rev.* D **51**, 6529 (1995); H.Y. Cheng, *Z. Phys.* C **69**, 647 (1996).
2. H.Y. Cheng, *Phys. Lett.* B **395**, 345 (1997).
3. C. Cao *et al.* (Particle Data Group), Eur. Phys. J. **C3**, 1 (1998).

4. For a recent similar work, see F.M. Al-Shamali and A.N. Kamal, hep-ph/9806270.
5. M. Bauer, B. Stech, and M. Wirbel, *Z. Phys.* C **34**, 103 (1987).
6. H.Y. Cheng, C.Y. Cheung, and C.W. Hwang, *Phys. Rev.* D **55**, 1559 (1997).
7. M. Neubert, B. Stech, CERN-TH-97-099, in second edition of Heavy Flavours, ed. by A.J. Buras and M. Lindner, World Scientific, Singapore.
8. The detailed results will be published in a separate work. The basic formulas can be found in K.C. Yang, *Phys. Rev.* D **57**, 2983 (1998).
9. P. Ball and V.M. Braun, CERN-TH/98-162 [hep-ph/9805422].
10. H.Y. Cheng and B. Tseng, *Phys. Rev.* D **58**, 094005 (1998).
11. M. Neubert, CERN-TH-97-024 [hep-ph/9702375]; Phys. Rep. **245**, 259 (1994).
12. CLEO Collab, C.P. Jessop *et al.*, *Phys. Rev. Lett.* **79**, 4533 (1997).
13. A. Khodjamirian and R. Rückl, WUE-ITP-97-049 [hep-ph/9801443].
14. J.L. Rodriguez, hep-ex/9801028.

THE Λ_b POLARIZATION PROBLEM AND ITS SOLUTION

Tsung-Wen Yeh
Institute of Physics, Academia Sinica,
Taipei, Taiwan, R.O.C.
twyeh@tcts1.seed.net.tw
twy@phys.sinica.edu.tw

The problem of Λ_b polarization in the inclusive semileptonic decays of a polarized Λ_b baryon is investigated using the modified perturbative QCD formalism. The contradiction between the theoretical calculation and the experimental data for the Λ_b polarization is reconciled. Specifically, we have calculated the Λ_b polarization using four different models which are the naive quark model (QM), the modified quark model (MQM), the naive parton model (PM), and the modified parton model (MPM). The corresponding Λ_b polarization, denoted as P's, are found to be -0.23 (QM), -0.94 (MQM), -0.37 (PM), and -0.68 (MPM), respectively. We note that P_{MQM} (equal to -0.94) is very close to the b quark polarization asymmetry, $A_{RL} = -0.94$, calculated at the Z vertex in $Z \to b\bar{b}$ process, and that P_{MPM} (equal to -0.68) is also very close to the Λ_b baryon polarization (equal to -0.68) which was estimated from the fragmentation processes under the heavy quark limit.

A recent observation by the ALEPH Collaboration [1] indicates that the Λ_b polarization is -0.23. The Λ_b polarization was measured through the semileptonic decays of a polarized Λ_b baryon. This result shows that the Λ_b baryon only carries one fourth of the b-quark polarization, which is in contradiction to the Standard Model expectation that the Λ_b polarization should be -0.94 [2]. Such a discrepancy implies that nonperturbative effects might play an important role in the polarization retention of the b quark as it fragments into the Λ_b baryon. A careful analysis is thus needed.

Here, we would like to re-interpret the measurement by the ALEPH Collaboration from the perturbative QCD point of view. The measured Λ_b polarization, denoted as P, could be related to the experimentally measured quantity, $R = 1.12 \pm 0.10$, defined by [5]

$$R = \frac{< E_l >}{< E_\nu >} \ , \tag{1}$$

where $< E_l >$ and $< E_\nu >$ are respectively the averaged energies of the electron and neutrino in the $\Lambda_b \to X_c e \nu$ decay. Since R is invariant under the Lorentz boost from the Lab. frame to the Λ_b rest frame where P is best defined, we can recast R into

$$R = \frac{< E_\nu^* > (< E_l^* > + < P_l^*(\mathrm{P}) >)}{< E_l^* > (< E_\nu^* > + < P_\nu^*(\mathrm{P}) >)} \ , \tag{2}$$

where all the averaged quantities are defined in the Λ_b rest frame. As specified, the quantities $< P_l^* >$ and $< P_\nu^* >$ are related to the Λ_b polarization, and are theoretically calculable. Thus, for example, one could calculate $< E_l^* >$, $< E_\nu^* >$, $< P_l^* >$ and $< P_\nu^* >$ in the Λ_b rest frame using the naive quark model, and then extract P from the experimentally measured value of R. The resulting P is equal to -0.23 [1]. We stress that the value of P, as determined from the above method, is model dependent. This is because that the averaged quantities which arise in Eq. (2) are model dependent. To illustrate the model dependence of P, we employ four different models to calculate P. They are the naive quark model (QM), the modified quark model (MQM), the naive parton model (PM), and the modified parton model (MPM). The modified quark model and modified parton model take into account the effects of Sudakov suppression.

We adopt the following factorization formula for $\Lambda_b(P) \to X_c(P_q)e(p_l)\nu(p_\nu)$ [4]:

$$\frac{1}{\Gamma^{(0)}} \frac{d^3\Gamma}{dxdydy_0 d\cos\theta_l} = \frac{M^2}{2} \int_x^1 dz \int \frac{d^2\mathbf{b}}{(2\pi)^2} \\ \times \tilde{S}(z, \mathbf{b}, \mu) \tilde{J}(z, P_q^-, \mathbf{b}, \mu) H(z, P_q^-, \mu) \ . \tag{3}$$

where $\Gamma^{(0)} = \frac{G_F^2}{16\pi^3}|V_{cb}|^2 M^5$ with M the mass of the Λ_b baryon, and μ is the renormalization and factorization scale. The variables x, y and y_0 are defined as $x = 2E_l/M$, $y = q^2/M^2$, and $y_0 = 2q_0/M$ with E_l the electron energy, and $q^2 = (p_l + p_\nu)^2$; P_q^- is the minus component of the charm quark momentum; θ_l is the angle between the third component of p_l and that of the b quark polarization vector, $S_b = (S_b^+, S_b^-, \mathbf{S}_{b\perp})$. Furthermore, $\tilde{S}$, $\tilde{J}$, and H are soft, jet and hard functions which are used to describe the dynamics for soft, collinear and hard partons at the scale μ. Finally, we note that the $\mathbf{b}$ space is introduced to facilitate the factorization of parton transverse momenta.

The transverse degrees of freedom of partons inside the Λ_b baryon are crucial to show the model dependence of P. The reasons are as following: The transverse momenta not only regularize the divergences which occur as the outgoing light quarks approach to the end point kinematics[6], but also enhance the contributions from the longitudinal component of the Λ_b baryon spin vector. These effects indicate that the intrinsic b dependence of the parton distribution function is important. The form for the b dependence of the parton distribution function could be determined by exploiting the infrared (IR) renormalon contributions[7,8] , such that one can parameterize the distribution function $f(z,b)$ as $\exp[-tM^2b^2]f(z)$. The parameter t is determined from the charged lepton spectra.

Let us now list the charged-lepton spectra given by the above four different models. Without giving the details, we arrive at[4]

Naive quark model :

$$\frac{1}{\Gamma^{(0)}}\frac{d^2\Gamma^{\rm T}_{\rm QM}}{dx d\cos\theta_\ell} = \frac{1}{\Gamma^{(0)}}\frac{d^2\Gamma^{\rm U}_{\rm QM}}{dx d\cos\theta_\ell} + {\rm P_{QM}}\cos\theta_\ell \frac{1}{\Gamma^{(0)}}\frac{d^2\Gamma^{\rm S}_{\rm QM}}{dx d\cos\theta_\ell}\,, \quad (4)$$

with

$$\frac{1}{\Gamma^{(0)}}\frac{d^2\Gamma^{\rm U}_{\rm QM}}{dx d\cos\theta_\ell} = \frac{x^2}{6}(3-2x)\,, \quad (5)$$

and

$$\frac{1}{\Gamma^{(0)}}\frac{d^2\Gamma^{\rm S}_{\rm QM}}{dx d\cos\theta_\ell} = \frac{x^2}{6}(1-2x)\,. \quad (6)$$

The superscripts U and S denote the unpolarized and polarized spectra respectively; $\rm P_{QM}$ is the Λ polarization associated with the current model.

Modified quark model :

$$\frac{1}{\Gamma^{(0)}}\frac{d^2\Gamma^{\rm T}_{\rm MQM}}{dx d\cos\theta_\ell} = \frac{1}{\Gamma^{(0)}}\frac{d^2\Gamma^{\rm U}_{\rm MQM}}{dx d\cos\theta_\ell} + {\rm P_{MQM}}\cos\theta_\ell \frac{1}{\Gamma^{(0)}}\frac{d^2\Gamma^{\rm S}_{\rm MQM}}{dx d\cos\theta_\ell}\,, \quad (7)$$

$$\begin{aligned}\frac{1}{\Gamma^{(0)}}\frac{d^2\Gamma^{\rm U}_{\rm MQM}}{dx d\cos\theta_\ell} = {} & M\int_0^x dy \int_0^{1/\Lambda} db e^{[-t^{\rm U}M^2b^2]} e^{-S(P_q^-,b)}(x-y)\eta \\ & \times\Big[(1+y-x)J_1(\eta M b) - \frac{2}{Mb}\eta J_2(\eta M b) \\ & + \eta^2 J_3(\eta M b)\Big]\,, \end{aligned} \quad (8)$$

$$\begin{aligned}\frac{1}{\Gamma^{(0)}}\frac{d^2\Gamma^{\rm S}_{\rm MQM}}{dx d\cos\theta_\ell} = {} & M\int_0^x dy \int_0^{1/\Lambda} db e^{[-t^{\rm S}M^2b^2]} e^{-S(P_q^-,b)}(x-y)\eta \\ & \times\Big[(1+y-x-2\frac{y}{x})J_1(\eta M b) - \frac{2}{Mb}\eta J_2(\eta M b) \\ & + \eta^2 J_3(\eta M b)\Big]\,, \end{aligned} \quad (9)$$

where $P_q^- = (1 - y/x)M/\sqrt{2}$, $\eta = \sqrt{(x-y)(1/x-1)}$ and J_1,J_2,J_3 are the Bessel functions of the order 1, 2 and 3, respectively. The parameters $t^{\rm U}$ and $t^{\rm S}$ are chosen as $t^{\rm U} = 0.005$ and $t^{\rm S} = 0.0423$.

Naive parton model :

$$\frac{1}{\Gamma^{(0)}}\frac{d^2\Gamma^{\rm T}_{\rm PM}}{dx d\cos\theta_\ell} = \frac{1}{\Gamma^{(0)}}\frac{d^2\Gamma^{\rm U}_{\rm PM}}{dx d\cos\theta_\ell} + {\rm P}_{\rm PM}\cos\theta_\ell \frac{1}{\Gamma^{(0)}}\frac{d^2\Gamma^{\rm S}_{\rm PM}}{dx d\cos\theta_\ell} \tag{10}$$

with

$$\frac{1}{\Gamma^{(0)}}\frac{d^2\Gamma^{\rm U}_{\rm PM}}{dx d\cos\theta_\ell} = \int_0^x dy \int_x^1 dz f(z)(x-y)(y+z-x) \tag{11}$$

and

$$\frac{1}{\Gamma^{(0)}}\frac{d^2\Gamma^{\rm S}_{\rm PM}}{dx d\cos\theta_\ell} = \int_0^x dy \int_x^1 dz f(z)(x-y)(y+z-x-2z\frac{y}{x}) \,. \tag{12}$$

Note that $f(z)$ is a part of the parton distribution defined before.

Modified parton model :

$$\frac{1}{\Gamma^{(0)}}\frac{d^2\Gamma^{\rm T}_{\rm MPM}}{dx d\cos\theta_\ell} = \frac{1}{\Gamma^{(0)}}\frac{d^2\Gamma^{\rm U}_{\rm MPM}}{dx d\cos\theta_\ell} + {\rm P}_{\rm MPM}\cos\theta_\ell \frac{1}{\Gamma^{(0)}}\frac{d^2\Gamma^{\rm S}_{\rm MPM}}{dx d\cos\theta_\ell} \,, \tag{13}$$

$$\begin{aligned}\frac{1}{\Gamma^{(0)}}\frac{d^2\Gamma^{\rm U}_{\rm MPM}}{dx d\cos\theta_\ell} &= M\int_0^x dy \int_0^{1/\Lambda} db e^{-S(P_q^-,b)} e^{-t^{\rm U}M^2b^2} f(z)(x-y)\eta \\ &\times \Big[(1+y-x)J_1(\eta M b) - \frac{2}{Mb}\eta J_2(\eta M b) \\ &+ \eta^2 J_3(\eta M b)\Big] \,,\end{aligned} \tag{14}$$

$$\begin{aligned}\frac{1}{\Gamma^{(0)}}\frac{d^2\Gamma^{\rm S}_{\rm MPM}}{dx d\cos\theta_\ell} &= M\int_0^x dy \int_0^{1/\Lambda} db e^{-S(P_q^-,b)} e^{-t^{\rm S}M^2b^2} f(z)(x-y)\eta \\ &\times \Big[(z+y-x-2z\frac{y}{x})J_1(\eta M b) - \frac{2}{Mb}\eta J_2(\eta M b) \\ &+ \eta^2 J_3(\eta M b)\Big] \,,\end{aligned} \tag{15}$$

with $\eta = \sqrt{(x-y)(z/x-1)}$. The parameters $t^{\rm U,S}$ are set to be $t^{\rm U} = 0.003$ and $t^{\rm S} = 0.024$.

Table 1: The values of the Λ_b polarization are determined from the quark model, the modified quark model, the parton model and the modified parton model.

P_{QM}	P_{MQM}	P_{PM}	P_{MPM}
-0.23	-0.94	-0.37	-0.68

The corresponding Λ_b polarizations denoted as P_{QM}, P_{MQM}, P_{PM} and P_{MPM} are list in table 1. Our result is consistent with those estimates from the Standard Model calculation of the b quark polarization asymmetry[2] and the polarization retention of the Λ_b baryon produced from the b quark fragmentation processes[3]. Finally, we emphasize the correctness of the measurement by the ALEPH Collaboration[1]. Of course, a more consistent measurement of Λ_b polarization should also include a measurement of the lepton spectrum from the semileptonic Λ_b decay.

1. ALEPH Collaboration, *Phys. Lett.* **B 365**, 437 (1996).
2. F.E. Close, J.G. Körner, R.J.N. Philips and D.J. Summers, *J. Phys.* **G18**, 1719 (1992).
3. A. Falk and M. Peskin, *Phys. Rev.* **D 49**, 3320 (1994).
4. Tsung-Wen Yeh, hep-ph/9806452.
5. G. Bonvicini and L. Randall, *Phys. Rev. Lett.* **73**, 392 (1994).
6. H.-n. Li and H.-Y. Yu, *Phys. Rev.* **D 53**, 4970 (1996).
7. H.-n. Li and H.-Y. Yu, *Phys. Rev. Lett.* **74**, 4388 (1995).
8. G. P. Korchemsky and G. Sterman, *Phys. Lett.* **B 340**, 96 (1994).

The $\Lambda_b \to pl\bar{\nu}$ decay in perturbative QCD

Hsien-Hung Shih and Shih-Chang Lee
Institute of Physics, Academia Sinica, Taipei, Taiwan

Hsiang-nan Li
Department of Physics, National Cheng-Kung University, Tainan, Taiwan

We develop perturbative QCD (PQCD) factorization theorem for the exclusive semileptonic baryon decay $\Lambda_b \to pl\bar{\nu}$, in which the relevant hadronic form factor is expressed as the convolution of a hard subamplitude with the Λ_b baryon wave function and the proton wave function. The Sudakov resummation for the wave functions, including that for a heavy-light system, is incorporated to improve the applicability of PQCD around the energy scale of the Λ_b baryon mass. The specific evolution scale for the proton wave function, determined from the best fit to the data of the proton form factor, is adopted.

1 INTRODUCTION

We apply the perturbative QCD (PQCD) to heavy baryon decays, taking the simplest mode $\Lambda_b \to pl\bar{\nu}$ as a starting point. It has been shown that the decay spectra or decay rates can be factorized into a hard subamplitude, which can be computed in perturbation theory at the parton level, and hadron wave functions, which absorb nonperturbative dynamics of processes. The Sudakov resummation plays an important role in the analyses. It has been shown that PQCD is applicable to exclusive processes for energy scale higher than few GeV [1]. The extension of PQCD from much higher energies down to the few GeV is due to the inclusion of the parton transverse momenta k_T into factorization formulas. Resumming $\ln k_T$ from radiative corrections into an exponential factor, the resultant Sudakov suppression ensures that main contributions to the decay rates come from the short-distance region. The first attempt to apply PQCD to the $\Lambda_b \to pl\bar{\nu}$ decay is made by Akhoury *et al.* [2]. However, they did not consider the Sudakov suppression associated with the Λ_b wave function, and their evaluation of the hard subamplitudes is not complete. In our analysis we shall improve these points by including the resummation for a heavy-light system, and a complete evaluation of the hard subamplitude. Furthermore, we shall adopt the King-Sachrajda (KS) mode [3], along with the specific choice of the evolution scale, for the proton wave function, which were determined from the best fit to the data of the proton form factor [4,5].

2 FACTORIZATION THEOREMS

The semileptonic decay amplitude is written as

$$\mathcal{M} = \frac{G_F}{\sqrt{2}} V_{ub} \mathcal{M}_\mu [\bar{l}\gamma^\mu (1-\gamma_5)\nu_l], \tag{1}$$

All QCD dynamics is contained in the hadronic matrix element,

$$\begin{aligned} \mathcal{M}_\mu &= \langle P(p')|\bar{u}\gamma_\mu(1-\gamma_5)b|\Lambda_b(p)\rangle & (2)\\ &= L_1(q^2)\bar{P}(p')\gamma_\mu(1-\gamma_5)\Lambda_b(p) & (3)\end{aligned}$$

where p (p') is the momentum of the Λ_b baryon (proton). We only consider the form factor L_1 which dominates.

There are three leading momentum regions for radiative corrections, from which important contributions proportional to $\ln M_{\Lambda_b}$ arise. These regions are collinear, when the loop momentum is parallel to an energetic light quark, soft, when the loop momentum is much smaller than M_{Λ_b}, and hard, when the loop momentum is of order M_{Λ_b}. Here M_{Λ_b} is assumed to be large and serves as an ultraviolet cutoff of loop integrals. We associate small transverse momenta k_T with the valence quarks, which are taken as an infrared cutoff. The collinear and soft contributions may overlap to give double logarithms $\ln^2(M_{\Lambda_b}/k_T^2)$, while the hard contribution gives only single logarithms $\ln(M_{\Lambda_b}/\mu)$, μ being the renormalization scale.

Note that all the transverse momenta are much smaller than M_{Λ_b}, since the valence quarks are near or on the mass shell. Otherwise, the quark lines should be absorbed into the hard part.

3 SUDAKOV RESUMMATION

The Sudakov resummation of large logarithms contained in a heavy-light system, such as the B meson, has been developed in [6]. The resummation for the Λ_b baryon wave function is

$$\begin{aligned}\Phi(x_i, b_i, p, \mu) = \exp[-\sum_{l=2}^{3} s(cw, x_l M_{\Lambda_b}) - 3\int_{cw}^{\mu}\frac{d\bar{\mu}}{\bar{\mu}}\gamma_q(\alpha_s(\bar{\mu}))] \\ \times\phi(x_i, cw)\ . \end{aligned} \tag{4}$$

Note the lack of the exponent s for the b quark. Since the b quark is heavy and radiative corrections associated with it do not produce collinear divergences, double logarithms are not present.

Employing a series of permutations of the valence quark kinematic variables, the summation of the contributions reduces to two terms. We can obtain the factorization formula,

$$L_1 = \frac{2\pi}{27}\int_0^1 [dx'][dx]\int_0^{1/\Lambda} b_1 db_1 b_2 db_2 \int_0^{2\pi} d\theta f_P(cw) f_{\Lambda_b} \\ \times \sum_{j=1}^{2} H_j(x_i', x_i, b_i, M_{\Lambda_b}, t_{jl}) \Psi_j(x_i', x_i, cw) \\ \times \exp[-S(x_i', x_i, cw, M_{\Lambda_b}, t_{jl})] \,. \tag{5}$$

The Sudakov exponent S is written as

$$S(x_i, x_i', cw, M_{\Lambda_b}, t_{jl}) = \sum_{l=2}^{3} s(cw, x_l M_{\Lambda_b}) + 3\int_{cw}^{t_{j1}} \frac{d\bar{\mu}}{\bar{\mu}} \gamma_q(\alpha_s(\bar{\mu})) \\ + \sum_{l=1}^{3} s(cw, x_l' \rho M_{\Lambda_b}) + 3\int_{cw}^{t_{j2}} \frac{d\bar{\mu}}{\bar{\mu}} \gamma_q(\alpha_s(\bar{\mu})) \,, \tag{6}$$

where t_{jl} are the hard scales.

The infrared cutoff cw' is chosen to be the inverse of a typical transverse distance among the three valence quarks. The recent PQCD analysis of the proton form factor shows that a reasonable choice is

$$w' = \min\left(\frac{1}{b_1'}, \frac{1}{b_2'}, \frac{1}{b_3'}\right) \,, \tag{7}$$

and that the best fit to the experimental data of the proton form factor determines the parameter $c = 1.14$ [5].

4 CONCLUSION

We compute L_1 in Eq. (5) numerically. Setting $\Lambda = 0.2$ GeV and $m_q = 0.8$ GeV, results of L_1 for $\beta = 1.0$, with b_1 and b_2 integrated up to the same cutoff $b_{1c} = b_{2c} = b_c$. It is found that at $\rho = 0.8$ approximately 45.7% of the contribution to L_1 comes from the region with $\alpha_s(1/b_c)/\pi < 0.5$, or equivalently, $b_c < 0.6\Lambda$. At $\rho = 0.9$, 48.4% of the contribution is accumulated in this perturbative region. As $\rho = 1.0$, the perturbative contribution has reached 49.9%. It implies that the PQCD analysis of the decay $\Lambda_b \to p l \bar{\nu}$ in the range of $\rho > 1.0$ is relatively self-consistent.

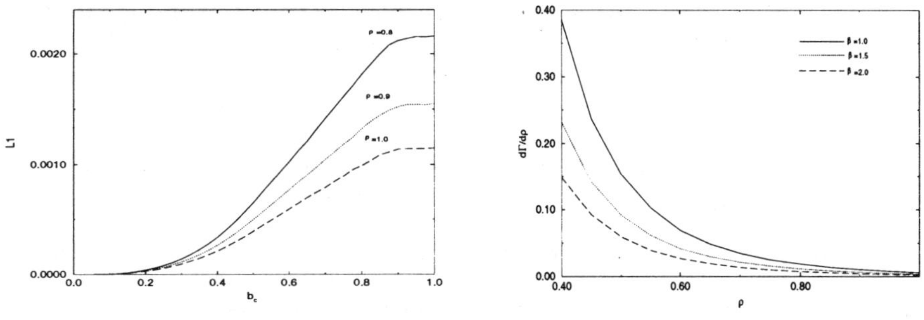

(a) (b)

Figure 1: (a) Dependence of L_1 on ρ for $m_q = 0.8$, $\beta = 2.0$. (b) Behavior of $d\Gamma/d\rho$ in units of 10^{-18} GeV^{-1}.

We then investigate the variation of L_1 with the parameter $\beta = 2.0$, and results are displayed in Fig. 1 (a). For $\beta = 2.0$, the perturbative analysis becomes reliable for $\rho > 0.8$. This is because ϕ_{Λ_b} becomes broader, which does not enhances the soft contribution.

With results of L_1, the differential decay rate of $\Lambda_b \to pl\bar{\nu}$ for massless leptons is derived

$$\frac{d\Gamma}{d\rho} = |V_{ub}|^2 \frac{G_F^2 M_{\Lambda_b}^5}{96\pi^3} \rho^2 (3 - 2\rho) |L_1|^2 \,, \tag{8}$$

and shown in Fig. 1 (b).

Our approach can be applied to the similar semileptonic decay $\Lambda_b \to \Lambda_c l\bar{\nu}$, for which experimental data are available. We shall determine the parameter β from the above decay, and then apply the PQCD formalism to nonleptonic Λ_b baryon decays.

References

1. H.-n. Li and G. Sterman, *Nucl. Phys.* B **381**, 129 (1992).
2. W. Loinaz and R. Akhoury, *Phys. Rev.* D **53**, 1416 (1996).
3. I.D. King and C.T. Sachrajda, *Nucl. Phys.* B **279**, 785 (1987).
4. H.-n. Li, *Phys. Rev.* D **48**, 4243 (1993).
5. B. Kundu, H-n. Li, J. Samuelsson, and P. Jain, hep-ph/9806419.
6. H-n. Li and H.L. Yu, *Phys. Rev.* D **53**, 2480 (1996).

A MULTI-PURPOSE CsI(Tl) DETECTOR FOR THE PILOT EXPERIMENT OF NEUTRINO PHYSICS IN TAIWAN

W. P. Lai
Institute of Physics, Academia Sinica, Nan-Kang Taipei, Taiwan
E-mail: wpl@phys.sinica.edu.tw

A mulit-purpose CsI(Tl) crystal calorimeter of 500 kg is being built for conducting a pilot experiment of neutrino physics in Taiwan. The goal of the pilot experiment is to address some interesting issues of neutrino and astro-particle physics , and to demonstrate the feasibilities of building a large scale detector and organizing an international collaboration as a pioneering effort by Taiwan. The overview of the experiment and its physics menu, as well as the highlights of experimental details, are presented. Future directions are also discussed.

1 The Pilot Experiemnt

A Collaboration "TEXONO" [a] has been formed since 1996 in Taiwan to initial and pursue an experimental program in neutrino and astro-particle physics in Taiwan. Such a "pilot" experiment[1] can demonstrate the feasibilities of building a large scale detector and organizing an international collaboration in Taiwan. Currently, the Collaboration comprises scientists from Taiwan, mainland China, and the United States.

Well matured operation of 3 nuclear power plants and excellent mountainous landscape dotted with mines and tunnels make it conceivable to perform experimental studies on reactor neutrino physics in Taiwan. An underground site at Nuclear Power Plant II, with about 30 m water-equivalent overburden, was chosen as the laboratory for conducting the pilot experiment.

A CsI(Tl) crystal calorimeter will be placed about 30 m from one of the two reactor cores of Nuclear Power Plant II. Intensive preparation for the pilot experiment is under way. Conceptual design and studies for future projects have also been started and will be pursued, such as long baseline experiments, solar neutrino studies or dark matter searches.

1.1 Motivations

Gamma-ray spectroscopy has been a standard technique in nuclear sciences for the MeV physics detection, but a large scale CsI(Tl) gamma detector has never been built or attempted in other reactor neutrino experiments,[2] most of which were liquid-scintillator technology based to study the ($\bar{\nu}_e$ p) interaction.

[a] Taiwan **EX**periment **O**n Reactor **N**eutrin**O**.

Table 1: Characteristic properties of the common crystal scintillators and their comparison with typical liquid, plastic and glass scintillators.

Properties	CsI(Tl)	NaI(Tl)	BGO	Liquid	Plastic	Glass
Density	4.51	3.67	7.13	0.9	1.0	∼3.5
Relative Light Yield	0.45	1.00	0.15	0.4	0.35	0.15
Radiation Length (cm)	1.85	2.59	1.12	∼45	∼45	4
Emission Peak (nm)	565	410	480	425	425	395
Decay Time (ns)	1000	230	300	2	2	100
Refractive index	1.80	1.85	2.15	1.5	1.6	1.55
Hygroscopic	slightly	yes	no	no	no	no

The difficulties are due to the large target mass required and the high cost. In the past few years, large electro-magnetic calorimeters using CsI(Tl) crystals have been built for high energy physics experiments, especially up to 40 tons of crystals for the forthcoming B-factory detector. This makes it conceivable to build a CsI(Tl) crystal calorimeter, with photo-diodes readout, of the order of half ton in target mass for the pilot experiment.

A comparison between CsI(Tl) and other commonly used scintillators are listed in Table 1. The CsI(Tl) crystal provides certain advantages: relatively high light yield and photon absorption, mechanically stable and ease to machine, and only weakly hygroscopic. Its emission spectra well matches the response of silicon photo-diodes. This makes possible a compact detector and shield design.

1.2 Physics Menu

Reactor neutrinos can serve as a nice probe for particle and nuclear physics at the MeV range, in addition to the neutrino oscillation search. While high energy (GeV) neutrino beams from accelerators has been very used to successfully conduct electro-weak precision measurements, QCD effects and nuclear structure physics, rooms for intertesting physics at low energy (MeV) are still widely open using the CsI crystal calorimeter. The following physics channels of neutrino interactions are good candidates for the pilot experiment.

- *Neutrino-Electron Scattering* is suitable for probing Electroweak Physics (g_V, g_A, and $\sin^2\theta_W$) at MeV and search for neutrino magnetic moment[3]. The sensitivity of measuring this quantity is explained in Figure 1(a).
- *Neutrino-Deuteron Scattering* provides measurements of Charged-Current (CC) and Neutral-Current (NC) cross-sections and demonstrates the CC/NC technique in neutrino oscillation searches. Both are the physics bases of the forthcoming SNO solar neutrino experiment.[4]

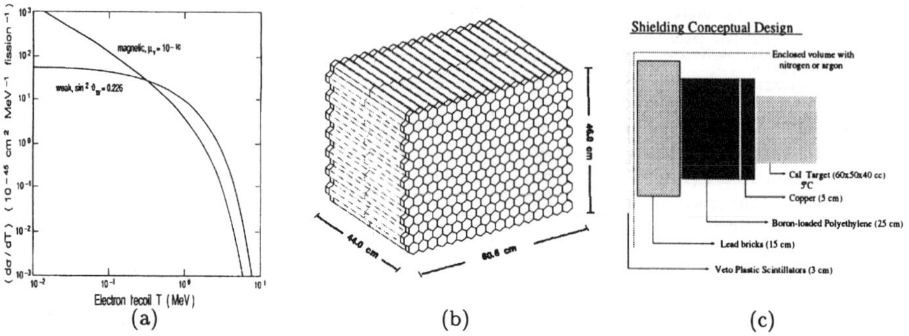

Figure 1: (a) Differential cross section showing the electron recoil energy spectrum in $\bar{\nu_e}$-e scatterings with reactor neutrinos, for Standard Model predictions and for the case with a neutrino magnetic moment of 10^{-10} Bohr magneton, the present experimental limit. (b) Schematic layout of the CsI(Tl) target/detector (active type). (c) Schematic layout of the inner CsI(Tl) targer/detector and the outer shielding. The coverage is 4π, but only one face is depicted.

- *Neutral Current Excitation on* ^{10}B *and* ^{11}B was theoretically predicted to be sensitive to the axial isoscalar component of NC interactions and the strange quark content of the nucleon.[5]

More details of these processes are discussed elsewhere.[6]

1.3 Highlights of Experimental Details

The Collaboration will first conduct the neutrino-electron scattering experiment. This requires an "active" configuration with about 500 CsI(Tl) crystal modules, as both the target and the detector, stacked into a cube, as depicted schematically in Figure 1(b). Each crystal module is 1 kg in mass and hexagonal in shape with 2 cm sides and 20 cm long. Two modules are optically coupled together and read out at both ends by two photo-diodes, followed by pre-amplifiers, main amplifiers and pulse shapers. The electron recoil energy can be measured from the sum of pulse heights from the two ends, and their difference derives the longitudinal position. Pulses after shaping are digitized by FADC's to be read out by a data acquisition system. The achieved energy resolution is about 16% FWHM at 660 keV, which is electronic noise-limited and hence improves linearly with energy. Pulse shape discrimination between γ/e and α events can be achieved to better than the 99% level.

The shielding consists of lead, boron-loaded polyethylene and copper, as shown schematically in Figure 1(c). The outmost is convered by layers of plastic scintillators to veto cosmic rays. The outer modules of the CsI(Tl) target can be used as active veto if necessary. The entire inner target will be immersed in a dry nitrogen environment to eliminate the radon gas, and will be operated at 5°C to reduce electronic noise. The background studies on a 3 kg sample crystal in a underground site indicates the absence of α-peaks above 3 MeV.[7] Contaminations from ^{238}U and ^{232}Th are less than 10^{-12} g/g, and less than ppb for ^{40}K.

2 Outlook

The CsI(Tl) detector offers a potential new tool to study various reactor neutrino physics at MeV. A successful performance of the pilot experiment can stimulate longer term experimental projects in other neutrino or astro-particle physics. Conceptual design and discussions on a long baseline experiment, solar neutrino physics, and dark matter searches are under way.

References

1. C.Y. Chang, S.C. Lee and H.T. Wong, *Procs. of XVI Int. Workshop on Weak Interactions and Neutrinos 1997*, to be published in *Nucl. Phys. (Procs. Suppl.)* B (1998).
2. P. Vogel, *Neutrino '94 Procs., Nucl. Phys. (Procs. Suppl.)* B **38**, 204 (1995).
3. F. Reines, H.S. Gurr and H.W. Sobel, *Phys. Rev. Lett.* **37**, 315 (1976); P.Vogel and J.Engel, *Phys. Rev.* D **39**, 3378 (1989); A.I. Derbin et al., *JETP Lett.* **57**, 769 (1993); A.G. Beda et al., Preprint *hep-ex/9706004* (1997); B. Kayser et al., *Phys. Rev.* D **20**, 87 (1997);
4. T.L. Jenkins, F.E. Kinard, and F. Reines, *Phys. Rev.* **185**, 1599 (1969); E. Pasierb et al., *Phys. Rev. Lett.* **43**, 96 (1979); G.T. Ewan et al. SNO Proposal (1987); G.S. Vidyakin et al., *JETP Lett.* **49**, 151 (1988); G.S. Vidyakin et al., *JETP Lett.* **51**, 279 (1990).
5. H.C. Lee, *Nucl. Phys.* A **294**, 473 (1978); T.W. Donnelly and R.D. Reccei, *Phys. Rep.* **50**, 1 (1979); R.S. Raghavan and S. Pakvasa, *Phys. Rev.* D **37**, 849 (1988). J. Bernabéu et al., *Nucl. Phys.* B **378**, 131 (1992); K. Kubodera and S. Nozawa, *Int. J. Mod. Phys.* E **3**, 101 (1994).
6. H.T. Wong and J. Li, Preprint *hep-ex/9805024*.
7. U. Kilgus, R. Kotthaus, and E. Lange, *Nucl. Instrum. Methods* A **297**, 425, (1990); R. Kotthaus, *Nucl. Instrum. Methods* A **329**, 433 (1993).

Vector Quark Model and B Meson Radiative Decay

Chia-Hung V. Chang and Darwin Chang
Physics Department, National Tsing-Hua University,
Hsinchu 30043, Taiwan, R.O.C.

Wai-Yee Keung
Physics Department, University of Illinois at Chicago,
Chicago IL 60607-7059, USA

We study the B meson radiative decay $B \to X_s\gamma$ in the vector quark model. Deviation from the Standard Model arises from the non-unitarity of the charged current KM matrix and related new FCNC interactions. The complete calculation including leading logarithmic QCD correction is carefully carried out. Using the most updated data and NLO theoretical calculation, the branching fraction of the observed B meson radiative decay places a limit on the mixing angles as stringent as that of the process $B \to X\mu\bar{\mu}$.

1 Introduction

A simple extension of Standard Model (SM) is to enlarge the particle content by adding vector quarks, whose right-handed and left-handed components transform in the same way under the weak $SU(2) \times U(1)$ gauge group. This extension is acceptable because the anomalies generated by the vector quarks cancel automatically and vector quarks can be heavy naturally. Vector quarks also arise in some Grand Unification Theory (GUT).

In this article, we discuss the B meson radiatve decay in the context of vector quark model and show that the experimental data can be used to constrain the mixing angles.

2 Vector Quark Model

We consider the model in which the gauge structure of SM remains while one charge $-\frac{1}{3}$ and one charge $\frac{2}{3}$ isosinglet vector quarks are introduced. Denote the charge $-\frac{1}{3}$ vector quark as D and the charge $\frac{2}{3}$ vector quark as U. Large Dirac masses of vector quarks, invariant under $SU(2)_L$, naturally arise. Gauge invariant Yukawa couplings between vector quarks and ordinary quarks are possible, which give rise to mixing between quarks of the same charge. In general, U will mix with the up-type quarks and D with down-type quarks.

3 B Meson Radiative Decay

The $B \to X_s\gamma$ decay, which already exists via one-loop W-exchange diagram in SM, is known to be extremely sensitive to the structure of fundamental interactions at the electroweak scale and serve as a good probe of new physics beyond SM because new interaction contribution can well be at the one-loop level also.

In SM, $b \to s\gamma$ arises at the one loop level from the various W meson mediated penguin diagrams. It is convenient to discuss weak decays using the Effective Hamiltonian formalism [1]. To avoid the uncertainty in m_b, it is customary to calculate the ratio R between the radiative decay and the dominant semileptonic decay. The ratio R is given, to LLA, by

$$R \equiv \frac{\Gamma(b \to s\gamma)}{\Gamma(b \to ce\bar{\nu}_e)} = \frac{1}{|V_{cb}|^2} \frac{6\alpha}{\pi g(z)} \left|V_{ts}^* V_{tb} C_7(m_b)\right|^2 .$$

In the vector quark model, deviations from SM result come from various sources: (1) charged current KM matrix non-unitarity, (2) Flavor Changing Neutral Current (FCNC) effects in neutral meson mediated penguin diagrams. We concentrate on these two contributions, which have been discussed in Refs.[4,5,6], but none of them presented the complete result without errors.

KM matrix is not unitary in the presence of a isosinglet vector quark. The mass-independent contributions from the various up-type quarks no longer cancel and could give rise to a significant deviation from SM prediction. The extra contribution to the Wilson coefficient C_7 is given by

$$\frac{(V^\dagger V)_{23}}{V_{ts}^* V_{tb}} \frac{23}{36} = \frac{\delta}{V_{ts}^* V_{tb}} \frac{23}{36} . \tag{1}$$

with $\delta = (V^\dagger V)_{23} = z_{sb}$.

The $b \to s\gamma$ transitions also arise from FCNC Z meson and Higgs boson mediated penguin diagrams. The contribution from internal $i = b, s$ quark, $f_{s,b}^Z$, is given by ($y_i \equiv m_i^2/M_Z^2$):

$$f_b^Z + f_s^Z \approx -\frac{1}{9} - \frac{1}{27} \sin^2\theta_W \simeq -0.12 \tag{2}$$

The Z-mediated penguin diagram with internal D quark can also be calculated.

$$f_D^Z = -\frac{5}{72 y_D} + O(\frac{1}{y_D^2}) . \tag{3}$$

It approaches zero when $y_D \to \infty$ and thus f_D^Z can be ignored, as the leading order approximation. For a gauge invariant result, the unphysical neutral

Higgs χ mediated penguin need to be considered together with the Z meson penguin.

$$f_i^{\chi} = -\frac{Q_d}{8} y_i \left[(2 - y_i) \ln y_i - \frac{5}{6} y_i^3 + 4 y_i^2 - \frac{13}{2} y_i + \frac{10}{3} \right] \frac{1}{(1 - y_i)^4} \qquad (4)$$

It is obvious that the light quark contributions are suppressed by the light quak masses and thus negligible. The situation is quite different for the heavy D quark. For $y_D \to \infty$, $f_D^{\chi} \to -\frac{5}{144} \sim -0.035$. We expand f_D^{χ} in powers of $1/y_D$ and keep also the next leading term.

$$f_D^{\chi} \approx -\frac{5}{144} + \frac{1}{36} \frac{1}{y_D} + O(\frac{1}{y_D^2}) \ . \qquad (5)$$

The Higgs boson H mediated penguin is similar to that of unphysical Higgs χ and can be ignored since $m_s, m_b \ll m_H$. For f_D^H, we again expand it in powers of $1/w_D$ and keep up to the next leading term:

$$f_D^H \approx +\frac{7}{144} - \frac{1}{18} \frac{1}{w_D} + O(\frac{1}{w_D^2}) \qquad (6)$$

The leading term is +0.048, again comparable to the Z penguin.

Put together, the Wilson coefficient $C_7(M_W)$ in the vector quark model is given by

$$C_7^{\rm SM}(M_W) + \frac{z_{bs}}{V_{tb} V_{ts}^*} \left(\frac{23}{36} - \frac{1}{9} - \frac{1}{27} \sin^2\theta_W + \frac{5}{72 y_D} + \frac{5}{144} - \frac{1}{36} \frac{1}{y_D} - \frac{7}{144} \right.$$
$$\left. + \frac{1}{18} \frac{1}{w_D} \right) \to -0.193 + \frac{z_{bs}}{V_{tb} V_{ts}^*} \times 0.506 \quad . \qquad (7)$$

Here we have used the unitariy relations $z_{4b} z_{4s}^* = -|U_{44}|^2 z_{sb} \approx -z_{sb}$ to leading order in FCNC due to the unitariry of U_L^d and $\delta = z_{sb}$ from Eq. (3).

Similarly the Wilson coefficient of the gluonic magnetic-penguin operator O_8 is modified by the vector quark.

$$C_8^{\rm SM}(M_W) \to -0.096 + \frac{z_{bs}}{V_{tb} V_{ts}^*} \times 0.942 \quad . \qquad (8)$$

The above deviation from SM does not include QCD evolution. Actually it is trivial to incorporate LL QCD corrections to these deviations in the framework of effective Hamiltonian. The correction to the ratio R in the vector quark model including its LL QCD contributions is given by

$$\Delta R = \frac{6\alpha}{\pi g(z)} \times 0.163 \times {\rm Re} \left[\frac{V_{cs}^*}{V_{cb}} z_{sb} \right]$$

to leading order in δ.

4 Constraints

The inclusive $B \to X_s\gamma$ branching ratio has been measured by CLEO with the branching ratio[7]

$$\mathcal{B}(b \to s\gamma)_{\text{EXP}} = (2.50 \pm 0.61) \times 10^{-4} \tag{9}$$

This branching ratio could be used to put a constraint on the mixing in the vector quark model. SM theoretical prediction up to Next-to-Leading Order (NLO) is first calculated by Misiak el al[2] and later by Buras el al[3]

$$\mathcal{B}(b \to s\gamma)_{\text{NLO}} = (3.28 \pm 0.33) \times 10^{-4} \tag{10}$$

$$\mathcal{B}(b \to s\gamma)_{\text{NLO}} = (3.60 \pm 0.33) \times 10^{-4} \tag{11}$$

It gives a range of possible vector quark model deviation and hence on z_{sb}:

$$-0.0130 < z_{sb} < 0.0012 \tag{12}$$

$$-0.0154 < z_{sb} < -0.0012 \tag{13}$$

The previously strongest bound on z_{sb} is from Z-mediated FCNC effect in the mode $B \to X\mu^+\mu^-$:

$$-0.0012 < z_{sb} < 0.0012 \tag{14}$$

Our higher bound using [2] is as strong as that from FCNC. It shows that even though the quark model deviation contributes to the radiative decay rate through one loop, as SM, the data could still put strong bound. Our bound using[3] is stranger. It dictates that z_{sb} should be negative and smaller than -0.0012, which is almost outside the bound from FCNC (14). Maybe it indicates we need a model with KM matrix non-unitarity but very small FCNC, which is not the case in vector quark model.

1. G. Buchalla, A. J. Buras and M. Lautenbacher, Rev. Mod. Phys. **68** 1125 (1996).
2. K. Chetyrkin, M. Misiak and M. Munz, Phys. Lett. B **400** 206 (1997).
3. A. J. Buras, A. Kwiatkowski and N. Pott, Phys. Lett. B **414** 157 (1997).
4. G. Bhattacharyya, G. C. Branco and D. Choudhury, Phys. Lett. B **336** 487 (1994).
5. L. T. Handoko and T. Morozumi, Mod. Phys. Lett. A **10** 309 (1995).
6. V. Barger, M.S. Berger and R. J. N. Phillips, Phys. Rev. D **52** 1663 (1995).
7. S. Glenn (representing the CLEO Collaboration), talk presented at the meeting of the American Physics Society, Columbus, Ohio, 18-21 March 1998.

Light Cone analysis of B to Light meson decay

Bum-Hoon Lee
Department of Physics, Sogang University
Seoul 121-742, Korea
E-mail: bhl@ccs.sogang.ac.kr

We calculate the form factors of $B \to \pi$ and $B \to \rho$ transition matrix elements by using the light cone formalism. In the limit of $m_\pi/M_B = 0$, $m_\rho/M_B = 0$, $M_b/M_B = 1$ and $(1-x) << 1$, we show the pole types of each form factors and show also that there exists relations among form factors with 2 unknown functions.

1 Introduction

For the decays involving $b \to c$ transition, we can apply the heavy quark symmetry and it is possible to understand reliably through the heavy quark effective theory (HQET) [1]. However, for those involving $b \to u$ it is less likely that the heavy quark symmetry applies, and the determination of the matrix element has heavily relied on the models for the form factors. The theoretical calculation of the form factors involving $b \to u$ transition is a difficult task, since it is concerned with the nonperturbative realm of QCD. Recently there have been active investigations of the form factors of $B \to \pi$ and $B \to \rho$ by using quark model, QCD sum rule and lattice calculations [2].

In this work[3], we will calculate the form factors of $B \to \pi$ and $B \to \rho$ transitions by using the method [4] based on the meson theory of Brodsky and Lepage [5]. We obtain the q^2 dependences hence the pole types of the form factors and also show the relations among the form factors in the limit of $m_\pi/M_B = 0$, $m_\rho/M_B = 0$, $M_b/M_B = 1$.

2 Form Factors and their relations

From Lorentz invariance one finds the decomposition of the hadronic matrix element for $B \to \pi$ transition in terms of hadronic form factors[6]:

$$
\begin{aligned}
&< \pi^-(p_\pi)|V^\mu|B^0(p_B) > \\
&= \left(r^\mu - \frac{m_B^2 - m_\pi^2}{q^2} q^\mu\right) F_1^{B\pi}(q^2) + \frac{m_B^2 - m_\pi^2}{q^2} q^\mu F_0^{B\pi}(q^2), \qquad (1)
\end{aligned}
$$

where $V^\mu = \bar{u}\gamma^\mu b$, $q^\mu = (p_B - p_\pi)^\mu$, $r^\mu = (p_B + p_\pi)^\mu$. In the similar way, the hadronic matrix element for $B \to \rho$ transition is written [6] in terms of hadronic form factors V, A_0, A_1, and A_2.

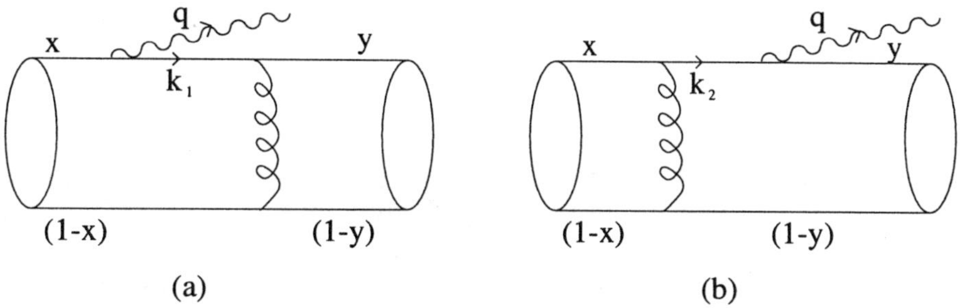

Figure 1: Feynman diagrams to the first order in α_s.

We calculate the heavy to light transition matrix element by using the PQCD factorization of exclusive amplitudes. To the first order in $\alpha_s = \alpha_s(Q^2)$, two Feynman diagrams in Fig. 1 give the following amplitude:

$$\begin{aligned}
< \pi^-(p_\pi)|V^\mu|B^0(p_B) >= & \frac{8\pi\alpha_s}{3}\int_0^1 dx \int_0^{1-\epsilon} dy\, \phi_B(x) \\
\times [& \frac{\mathrm{Tr}\{(\not{p}_\pi + m_\pi)\gamma_5\gamma^\nu \not{k}_1 \gamma^\mu(\not{p}_B + g(x)M_B)\gamma_5\gamma_\nu\}}{k_1^2 Q^2} \\
+ & \frac{\mathrm{Tr}\{(\not{p}_\pi + m_\pi)\gamma_5\gamma^\mu(\not{k}_2 + M_b)\gamma^\nu(\not{p}_B + g(x)M_B)\gamma_5\gamma_\nu\}}{(k_2^2 - M_b^2)Q^2}]\, \phi_\pi(y), \qquad (2)
\end{aligned}$$

and the similar expression for the decay into ρ. Here, $Q^\mu = (1-x)p_B^\mu - (1-y)p_\pi^\mu$, $k_1^\mu = -(1-x)p_B^\mu + p_\pi^\mu$, $k_2^\mu = p_B^\mu - (1-y)p_\pi^\mu$. The variables x and y represent the momentum fraction carried by the correspondig quarks. In (2) $g(x)$ is a phenomenological parameter for B meson wave function, and $\phi_\pi(y)$ and $\phi_B(x)$ are the distribution amplitudes for π and B mesons. In (2) we took the upper limit of the integration over momentum fraction y of a quark in the light meson as $1-\epsilon$. The integration in the interval $1-\epsilon \le y \le 1$ corresponds to the Drell-Yan-West [7] end-point region and is also expected to be suppressed by a Sudakov form factor [4].

We now evaluate the form factors and show the results in the limit of $m_\pi/M_B = 0$, $m_\rho/M_B = 0$, $M_b/M_B = 1$ and $(1-x) << 1$. The approximations $m_\pi/M_B = 0$, $m_\rho/M_B = 0$ and $M_b/M_B = 1$ are reasonable. For the B meson, it is expected that the momentum fraction of the light quark, $1-x$, is roughly given by m_{light}/M_B and small. For this dominant range of x, the approximation $(1-x) << 1$ can be applied in the integrand safely if the range of q^2 not too big.

The form factors can be organized in the following form:

$$F_i(q^2) = \frac{32\pi\alpha_s}{3M_B^2}\int_0^1 dx \int_0^{1-\epsilon} dy\, \phi_B(x)\ \phi_i(y)\frac{1}{(1-x)(1-y)^2}f_i, \qquad (3)$$

where F_i=F_0,F_1,V,A_0,A_1,A_2, and $\phi_i(y) = \phi_\pi(y)$ for F_0 and F_1, and $\phi_i(y) = \phi_\rho(y)$ for V, A_0, A_1 and A_2. In (3) all f_i are given by

$$f_i = c1_i\frac{1}{z} + c2_i\frac{1}{z^2} \qquad (4)$$

where ci_i and $c2_i$ are function of y only, and $z \equiv 1 - \frac{q^2}{M_B^2}$. The integration does not change the q^2 dependence and can be summarized as

$$F_0(q^2) = (a_\pi + b_\pi)\frac{1}{z}, \quad F_1(q^2) = 2b_\pi\frac{1}{z} + (a_\pi - b_\pi)\frac{1}{z^2}, \qquad (5)$$
$$A_1(q^2) = a_\rho\frac{1}{z}, \quad A_2(q^2) = -V(q^2) = a_\rho\frac{1}{z^2}, \quad -A_0(q^2) = (a_\rho + b_\rho)\frac{1}{z^2},$$

where

$$a_i = \frac{32\pi\alpha_s}{3M_B^2}\int_0^1 dx \int_0^{1-\epsilon} dy\, \phi_B(x)\ \phi_i(y)\frac{2g-1}{(1-x)(1-y)^2}, \qquad (6)$$
$$b_i = \frac{32\pi\alpha_s}{3M_B^2}\int_0^1 dx \int_0^{1-\epsilon} dy\, \phi_B(x)\ \phi_i(y)\frac{1}{(1-x)(1-y)}$$

with $i = \pi$ or ρ.

We emphasize that the q^2 dependences of the form factors given in (5) are independent of the shapes of the distribution amplitudes $\phi_B(x)$, $\phi_\pi(y)$, $\phi_\rho(y)$ and the value of the parameter ϵ. Their dependences appear only in the values of the constants a_i and b_i in (6), which affect the normalizations of the form factors. From (5) we find that $F_0(q^2)$ and $A_1(q^2)$ have the simple pole q^2 dependence, and $A_2(q^2)$, $V(q^2)$ and $A_0(q^2)$ have the dipole q^2 dependence. $F_1(q^2)$ has the mixture of the simple pole and dipole q^2 dependences, but the dipole q^2 dependence is dominant.

From (5) we find the relations among the form factors:

$$F_1(q^2) = F_0(q^2)(2 - \frac{1}{z}) + 2\frac{a_\pi}{a_\rho}A_1(q^2)(-1 + \frac{1}{z}) \qquad (7)$$

$$F_0(q^2)\frac{1}{z} = -\frac{a_\pi + b_\pi}{a_\rho + b_\rho}A_0(q^2) \qquad (8)$$

$$A_1(q^2)\frac{1}{z} = A_2(q^2) = -V(q^2). \qquad (9)$$

For $q^2 = 0$, this can be checked by measuring the differential branching ratios at $q^2 = 0$.

3 Conclusion

We calculated the form factors of $B \to \pi$ and $B \to \rho$ heavy to light transition matrix elements by using the factorization formalism of perturbative QCD. We obtained the q^2 dependences of the form factors and the relations among them in the limit of $m_\pi/M_B = 0$, $m_\rho/M_B = 0$, $M_b/M_B = 1$ and $(1-x) << 1$. $F_0(q^2)$ and $A_1(q^2)$ have the simple pole q^2 dependence, and $A_2(q^2)$, $V(q^2)$ and $A_0(q^2)$ have the dipole q^2 dependence. $F_1(q^2)$ has the mixture of the simple pole and dipole q^2 dependences. The pole types and the relations among the form factors are independent of the shapes of the distribution amplitudes $\phi_B(x)$, $\phi_\pi(y)$, $\phi_\rho(y)$ and the value of the parameter ϵ.

Acknowledgments

This work was supported in part by the Basic Science Research Institute Program, Ministry of Education, Project No. BSRI-97-2414, in part by Korea Science and Engineering Foundation through the SRC Program of SNU-CTP.

References

1. N. Isgur and M.B. Wise, Phys. Lett. B232 (1989) 113; B237 (1990) 527
2. H. Wittig, OUTP-97-59-P, hep-lat/9710088 (1997)
3. Dae Sung Hwang and Bum-Hoon Lee, hep-ph/9801421, To appear in Eur. Phys. J. C.
4. A. Szczepaniak, E.M. Henley and S.J. Brodsky, Phys. Lett. B243 (1990) 287; S.J. Brodsky, in *QCD 20 Years Later*, P.M. Zerwas and H.A. Kastrup, Ed. (World Scientific, 1993)
5. G.P. Lepage and S.J. Brodsky, *Phys. Lett.* B **87**, 359 (1979) ; Phys. Rev. D22 (1980) 2157; S.J. Brodsky and G.P. Lepage, Phys. Rev. D24 (1981) 1808; in *Perturbative Quantum Chromodynamics*, A.H. Mueller, Ed. (World Scientific, 1989)
6. M. Wirbel, B. Stech and M. Bauer, Z. Phys. C29 (1985) 637; C34 (1987) 103; M. Bauer and M. Wirbel, Z. Phys. C42 (1989) 671
7. S.D. Drell and T.M. Yan, Phys. Rev. Lett. 24 (1970) 181; G. West, Phys. Rev. Lett. 24 (1970) 1206
8. V.L. Chernyak and I.R. Zhitnitsky, Phys. Rept. 112 (1984) 173; V.L. Chernyak and A.R. Zhitnitsky, Nucl. Phys. B345 (1990) 137

A NEW UNIFIED EVOLUTION EQUATION

Jyh-Liong Lim [a]
Department of Electrophysics, National Chiao-Tung University,
Hsinchu 300, Taiwan
E-mail: u8221805@cc.nctu.edu.tw

We propose a new unified evolution equation for parton distribution functions appropriate for both large and small Bjorken variable x. Compared with the Ciafaloni-Catani-Fiorani-Marchesini equation, the cancellation of sofe poles between virtual and real gluon emissions is made explicit without introducing infrared cutoffs, and next-to-leading contributions to the Sudakov resummation are included systematically.

1 Introductions

As we know, the conventional evolution equations appropriate for different kinematic regions are different, such as the Dokshitzer-Gribov-Lipatov-Altarelli-Parisi (DGLAP) equation that organizes large logarithmic corrections at high momentum transfer and intermediate Bjorken variable x, and the Balitsky-Fadin-Kuraev-Lipatov (BFKL) equation that organizes large logarithmic corrections at small x. The Ciafaloni-Catani-Fiorani-Marchesini (CCFM) equation as a unification of the above two equations is appropriate for both large and small x. These evolution equations are derived by summing ladder diagrams to all orders, in which various leading logarithms are produced by specific orderings of the momentum flows through rung gluons [1,2,3].

Recently, we have developed an alternative approach to the derivation of the evolution equations, which is based on the Collins-Soper resummation technique [4]. This technique has been used to resum the double logarithmic corrections to a parton distribution function in hard QCD scattering processes [5]. By extending the resummation technique to the single-logarithm cases, we have easily reproduced all the above equations. We have also proposed a modified BFKL equation, in which the extra dependence on the momentum transfer Q is introduced [6]. This dependence is essential for the explanation of the HERA data of the structure function $F_2(x, Q)$.

In this talk we discuss a new unified evolution equation derived from the viewpoint of the resummation. Compared to the CCFM equation, the cancellation of sofe poles between virtual and real gluon emissions is made explicit

[a] This work is a collaboration with Hsiang-nan Li of Department of Physics, National Cheng-Kung University, Tainan 701, Taiwan

without introducing infrared cutoffs, and next-to-leading contributions to the Sudakov resummation are included systematically.

2 New Unified Evolution Equation

2.1 Resummation Formalism

We study the unintegrated gluon distribution function in the axial gauge $n\cdot A = 0$,

$$F(x, k_T, p^+) = \frac{1}{p^+}\int \frac{dy^-}{2\pi}\int \frac{d^2y_T}{4\pi} e^{-i(xp^+y^- - \mathbf{k}_T\cdot\mathbf{y}_T)} \times \frac{1}{2}\sum_\sigma \langle p,\sigma|F_\mu^+(y^-,y_T)F^{\mu+}(0)|p,\sigma\rangle\ , \tag{1}$$

where $|p,\sigma\rangle$ denotes the incoming hadron with light-like momentum $p^\mu = p^+\delta^{\mu+}$ and spin σ, and F_μ^+ is the field tensor. The key step is to obtain the derivative p^+dF/dp^+. Because of the scale invariance of F in the gauge vector n, F must depend on p^+ via the ratio $(p\cdot n)^2/n^2$. Hence, we have the chain rule relating p^+d/dp^+ to d/dn :

$$p^+\frac{d}{dp^+}F = -\frac{n^2}{v\cdot n}v_\alpha\frac{d}{dn_\alpha}F\ , \tag{2}$$

with $v_\alpha = \delta_{\alpha+}$ a dimensionless vector along p. The operator d/dn_α applies to a gluon propagator, giving a special vertex [4],

$$\hat{v}_\alpha = \frac{n^2 v_\alpha}{v\cdot n n\cdot l}\ , \tag{3}$$

to which the differentiated gluon attaches.

Employing the Ward identities, Eq. (2) reduces to [4,5]

$$p^+\frac{d}{dp^+}F(x,k_T,p^+) = 2\bar{F}(x,k_T,p^+)\ , \tag{4}$$

as shown in Fig. 1(a).

2.2 New Unified Evolution Equation

We Fourier transform Eq. (4) into the b space, b being the conjugate variable of k_T. The leading regions of the loop momentum flowing through the special vertex are soft and hard. Therefore, we factorize the subdiagrams containing

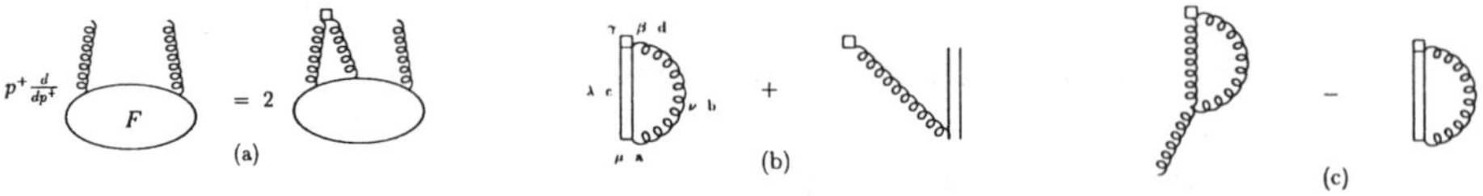

Figure 1: (a) The derivative p^+dF/dp^+ in the axial gauge. (b) The lowest-order subdiagrams for K. (c) The lowest-order subdiagrams for G.

the special vertex order by order in these leading regions according to Figs. 1(b) and 1(c). For the contribution from Fig. 1(b), we express the function F as

$$\begin{aligned} F(x+l^+/p^+,b,k^+) &= \theta((1-x)p^+ - l^+)F(x,b,k^+) \\ &+[F(x+l^+/p^+,b,k^+) - \theta((1-x)p^+ - l^+)F(x,b,k^+)] \,. \end{aligned} \tag{5}$$

The contribution from the first term of Eq. (5) is combined with virtual gluon correction, leading to the soft function

$$K(x,b\mu,\alpha_s(\mu)) = -\bar{\alpha}_s(\mu)\left[K_0(2(1-x)p^+\nu b) + \ln\frac{b\mu}{2} + \gamma_E\right] , \tag{6}$$

which is infrared finite. The second term in Eq. (5), being also infrared finite, gives

$$\bar{F}_s' \approx \bar{\alpha}_s(k^+)\int_x^1 \frac{d\xi}{\xi}\frac{F(\xi,b,k^+) - F(x,b,k^+)}{1-x/\xi} \,. \tag{7}$$

Fig. 1(c) gives the hard function

$$G(k^+/\mu,\alpha_s(\mu)) = -\bar{\alpha}_s(\mu)\left[\ln\frac{2k^+}{\mu} + \ln\nu\right] \,. \tag{8}$$

It can been shown that the ultraviolet divergences of K and G cancel each other, and that their renormalization-group evolution is written as

$$\begin{aligned} &K(x,b\mu,\alpha_s(\mu)) + G(k^+/\mu,\alpha_s(\mu)) = \\ &\qquad K(x,1,\alpha_s(k^+)) + G(1,\alpha_s(k^+)) - s(x,b,k^+) \,, \end{aligned} \tag{9}$$

with

$$K(x,1,\alpha_s(k^+)) = \bar{\alpha}_s(k^+)\left[\ln(1-x) + \ln(p^+b) + \ln 2\nu\right] , \tag{10}$$

$$G(1,\alpha_s(k^+)) = -\bar{\alpha}_s(k^+)\ln 2\nu , \tag{11}$$

$$s(x,b,k^+) = \int_{1/b}^{k^+} \frac{d\bar{\mu}}{\bar{\mu}}\left[\gamma_K(\alpha_s(\bar{\mu})) + \beta\frac{\partial}{\partial g}K(x,1,\alpha_s(\bar{\mu}))\right] , \tag{12}$$

and the gauge factor $\nu = \sqrt{(v\cdot n)^2/|n^2|}$.

Absorbing the term $\bar{\alpha}_s \ln(1-x)$ in Eq. (10) into Eq. (7), we arrive at the splitting function $1/(1-x/\xi)_+$. The new evolution equation then becomes

$$p^+\frac{d}{dp^+}F(x,b,k^+) = -2\left[s(b,k^+) - \bar{\alpha}_s(k^+)\ln(p^+b)\right]F(x,b,k^+) + 2\bar{\alpha}_s(k^+)\int_x^1 \frac{dz}{z}\frac{F(x/z,b,k^+)}{(1-z)_+} . \tag{13}$$

The first term of the above expression will lead to the Sudakov form factor with the accuracy up to next-to-leading logarithms.

3 Conclusion

We have proposed a new unified evolution equation appropriate for both large and small x in the resummation framework. The infrared finiteness is explicit in both the Sudakov resummation and the splitting function. The Sudakov form factor includes the next-to-leading contribution systematically.

Acknowledgements

This work was supported by National Science Council of R.O.C. under the Grant No. NSC87-2112-M-006-018.

References

1. V.N. Gribov and L.N. Lipatov, Sov. J. Nucl. Phys. **15**, 438 (1972); G. Altarelli and G. Parisi, Nucl. Phys. **B126**, 298 (1977); Yu.L. Dokshitzer, Sov. Phys. JETP **46**, 641 (1977).
2. E.A. Kuraev, L.N. Lipatov and V.S. Fadin, Sov. Phys. JETP **45**, 199 (1977); Ya.Ya. Balitskii and L.N. Lipatov, Sov. J. Nucl. Phys. **28**, 822 (1978); L.N. Lipatov, Sov. Phys. JETP **63**, 904 (1986).
3. M. Ciafaloni, Nucl. Phys. **B296**, 49 (1988); S. Catani, F. Fiorani, and G. Marchesini, Phys. Lett. B **234**, 339 (1990); Nucl. Phys. **B336**, 18 (1990); G. Marchesini, Nucl. Phys. **B445**, 49 (1995).

4. J.C. Collins and D.E. Soper, Nucl. Phys. **B193**, 381 (1981).
5. H-n. Li, Phys. Rev. D **55**, 105 (1997).
6. H-n. Li, Phys. Lett. B **416**, 192 (1998); J.L. Lim and H-n. Li, hep-ph/9712408.

Symmetry Breaking Effects in Heavy Mesons

Chi-Yee Cheung, Chien-Wen Hwang, and Wei-Min Zhang
Institute of Physics, Academia Sinica, Taipei, Taiwan 115, R.O.C.

Using a recently proposes covariant effective theory of heavy mesons, we study various symmetry breaking effects in heavy meson properties. Heavy quark symmetry violating effects are calculated in accordance with heavy quark effective theory to order $(1/m_Q)$. In addition, we also discussed $SU(2)$ symmetry violations due to the non-equality of light-quark masses and electromagnetic interactions.

1 A Covariant Model for Heavy Mesons

Although the development of heavy quark effective theory (HQET) from QCD has simplified the analysis of heavy hadron physics, many properties of hadrons, for example, their decay constants, axial coupling constants, magnetic moments, and transition form factors *etc.*, are still not calculable directly from QCD. To study these quantities, one unavoidably has to use phenomenological models to describe the structures of hadrons. To overcome some drawbacks mentioned in [1], a covariant field theoretical model has recently been proposed for the heavy meson bound state problem [2,3]. This model is fully covariant and satisfies HQS, at the same time, it retains the simplicity of the quark model picture. This theory allows us to formulate theoretical calculations in terms of the standard Feynman diagrams, therefore the lack of Z-diagrams [4] in the ordinary LFQM is no longer a problem. Combining this model with HQET, we can systematically study various $1/m_Q$ corrections to heavy meson properties.

The dominant character of the covariant model is the additional Feynman rule which illurstated in Figure 1. The relevant matrix elements can be written from all the Feynman rules. The wave function we used here is $F(v \cdot p) = \varphi(v \cdot p)(\bar{\Lambda} - v \cdot p)$ where $\varphi(v \cdot p) = \frac{1}{(v \cdot p + \omega - i\epsilon)^n}$ $(n = \text{integer})$. ω is a parameter.

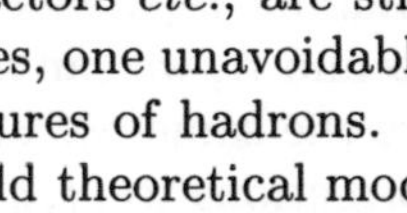

Figure 1: Feynman rules of the heavy quark effective theory.

2 $1/m_Q$ Symmetry Breaking

To fix the parameters (quark masses m_q and m_Q, ω, α_s) of the covariant model, we follow the strategy described below. First of all, we note that we can use the experimental ratios of radiative to pionic decay rates [5,6]

$$r_1 = \frac{\Gamma(D^{*0} \to D^0\gamma)}{\Gamma(D^{*0} \to D^0\pi^0)} = 0.616 \pm 0.076, r_2 = \frac{\Gamma(D^{*+} \to D^+\gamma)}{\Gamma(D^{*+} \to D^+\pi^0)} = 0.055 \pm 0.017, \tag{1}$$

to estimate the light quark mass m_q. Then we get $m_q = (0.230 \pm 0.140)$ GeV. This solution agrees very well with typical light-quark masses $m_q = 0.200 \sim 0.250$ GeV used in a relativistic formulism [7]. Of course, $\mathcal{O}(1/m_Q)$ corrections will change this estimate and some fine tunings would be necessary to fit the experimental data. We eventually settle on $m_q = 0.241$ GeV. After fixing m_q and given a m_Q, we can use $f_M \simeq f_B \simeq 180$ MeV [8] to deterimine the wave function parameter ω. Subsequently we can calculate the hyperfine mass parameters. From these parameters and the hyperfine mass splittings [9] $\delta m_{HF}(B^*B) = 0.046$ GeV, $\delta m_{HF}(D^*D) = 0.142$ GeV, we can deterimine the ratio α_s/m_Q. Now, given a value for m_Q, we can get a corresponding g_0. On the other hand, knowing m_q, ω and $\bar{\Lambda}$, we can also calculate g_0. A consistency is reached if both methods give the same g_0. Final

Next, we list the $1/m_Q$ corrections to (f, g) and (d, d'') for $n = 10$,

system	f_0	δf_1	δf_2	f	g_0	δg_1	δg_2	g	R_g
D^*D	−1.37	0.171	−0.141	−1.34	−0.683	0.0854	0.0705	−0.528	−21 %
B^*B	−1.37	0.0666	−0.0456	−1.35	−0.683	0.0333	0.0228	−0.627	−6.8 %

Table 1: $1/m_Q$ corrections to f and g.

system	d_0	δd_1	δd_2	d	d''_0	$\delta d''_1$	$\delta d''_2$	d''	R_d
D^*D	1.42	−0.120	0.146	1.44	2.84	−0.240	−0.292	3.07	−20 %
B^*B	1.42	−0.0477	0.0472	1.42	2.84	−0.0954	−0.0944	2.65	−6.7 %

Table 2: $1/m_Q$ corrections to d and d''

where $R_g \equiv \frac{2g-f}{f}$, $R_d \equiv \frac{d''-2d}{2d}$. In calculating the $1/m_Q$ corrections due to $\mathcal{O}_1$, infrared divergences are encountered, so that our model cannot give precise predictions for $\delta f_1^{(g)}$, $\delta g_1^{(g)}$, $\delta d_1^{(g)}$, and $\delta d_1''^{(g)}$. In order to estimate the effects, we have introduced an effective gluon mass of $m_g \simeq \Lambda_{QCD} \simeq 300$ MeV. The results are not very sensitive to the choice of m_g; for instance, doubling m_g to 600 MeV reduces $\delta f_1^{(g)}$ by a factor of $\frac{2}{3}$. We see that for B mesons,

$\left(\frac{\delta g_1}{g}\right) = 4.8\ \%, \left(\frac{\delta g_2}{g}\right) = 3.3\ \%$ are consistent with the rough estimate of $\alpha_s \frac{\Lambda_{QCD}}{m_b} \sim (2 \sim 3)\ \%$ for $\alpha_s = 0.4$ and $\Lambda_{QCD} \simeq 300\ MeV$. For D meson, due to the smaller m_c, $1/m_Q$ corrections are about a factor 3 larger.

In the D^*Dsystem, we find $R_g = \frac{2g-f}{f} = -21\ \%$, which much bigger than $R_g = 2.7\ \%$ found in [11]; morever, the signs disagree. QCD sum rules give $R_g = 0.6\ GeV/m_c \simeq 38\ \%$ [10] which again has the opposite sign to ours. In Chapter 3, we will recalculate R_g in the MIT bag model, and obtain $R_g = -4.1\ \%$, which is much smaller than what we have obtained here, but the signs agree. Therefore, a consistent picture has not emerged from these works. As mentioned in Chapter 1, in order to explain the experimental mass difference $\delta m_{_{HF}}(D_s^* D_s) - \delta m_{_{HF}}(D^* D) = 2.42 \pm 0.4\ MeV$ in chiral perturbation theory, one would need a large and positive R_g, namely $R_g \simeq 40 \sim 50\ \%$. If what we have found here were not too far from reality, it would mean either (1) chiral perturbation series is not converging fast enough, or (2) it is not enough to retain only the non-analytic contributions from meson loops, and effects from higher order chiral Lagrangian.

3 $SU(2)$ Symmetry Breaking

$SU(2)$ or isospin symmetry originates from treating the up and down quarks as an isospin doublet. This symmetry is broken by the up-down quark mass difference, and also by electromagnetic interactions which distinguish the different charges carried by the up and down quarks.

Taking into account the electromagnetic Coulomb energy $e_Q e_q \delta m_{coul}$, we can write the mass of a $(Q\bar{q})$ meson as $M(Q\bar{q}) = m_Q + \bar{\Lambda}_q + e_Q e_q \delta m_{coul}$, where we have ignored $SU(2)$ breaking of the $1/m_Q$ contributions since they are higher order effects. Thus, $M(Q\bar{d}) - M(Q\bar{u}) = \delta\bar{\Lambda}_{du} - e_Q \delta m_{coul}$ where $\delta\bar{\Lambda}_{du} = \bar{\Lambda}_d - \bar{\Lambda}_u$. From the experimental values of $M(D^+) - M(D^0) = 4.78 \pm 0.1\ MeV$ and $M(B^0) - M(B^-) = 0.35 \pm 0.29\ MeV$, we get $\delta m_{coul} = -4.43 \pm 0.31\ MeV$, $\delta\bar{\Lambda}_{du} = 1.83 \pm 0.31\ MeV$. We see that although $m_d - m_u \sim 5\ MeV$ [9], $\delta\bar{\Lambda}_{du} = \bar{\Lambda}_d - \bar{\Lambda}_u$ is only $1.83\ MeV$. The reason is similar to the case of $SU(3)$ symmetry breaking difference $\bar{\Lambda}_s - \bar{\Lambda}_{u,d}$ discussed in the previous subsection. Namely, since the d-quark is heavier, it is also more tightly bound, so that part of the mass difference $m_d - m_u$ is cancelled by the larger binding energy of the d-quark. The Coulomb energy δm_{coul} can be easily calculated in the covariant model. The predicted δm_{coul} for difference choices of the wave function $\varphi_n(v{\cdot}p)$ are listed in Table 3.

The excellent agreement with the experimental value is a triumph of the covariant model.

$m_q(GeV)$	n	$\omega(GeV)$	$\delta m_{coul}(MeV)$
0.241	10	1.96	−4.42
0.241	12	2.54	−4.48
0.241	14	3.12	−4.51

Table 3: Coulomb energy of heavy meson.

4 Summary

In this article, we have used a fully covariant model to calculate symmetry breaking effects coming from $1/m_Q$ corrections, as well as $SU(2)$ symmetry breakings. First of all, for $1/m_Q$ corrections, the sizes of the symmetry breaking effects are much larger than those found in a nonrelativistic quark model [11] and with opposite signs. Secondly, for $SU(2)$ symmetry breaking, we studied the QED Coulomb energy of the heavy mesons, and the $SU(2)$ violating transition $D_s^* \to D_s + \pi^0$. The predicted values are in excellent agreements with experiments. This provides a vote of confidence for the validity of the covariant model.

References

1. C.W. Hwang Ph.D. Thesis.
2. C.Y. Cheung and W.M. Zhang, hep-ph/9712258.
3. H.Y. Cheng, C.Y. Cheung and W.M. Zhang, hep-ph/9804324.
4. H.Y. Cheng, C.Y. Cheung and Chien-Wen Hwang, Phys. Rev. D **55**,1559 (1997).
5. CLEO Collab., J. Bartelt *et al.,* Phys. Rev. Lett. **80**, 3919 (1998).
6. I.W. Stewart, hep-ph/9803227.
7. P.C. Tiemeijer and J.A.Tjon, Nucl-th/9211003.
8. C.R. Allton *et al.,* Phys. Let. B**405**, 133 (1997).
9. Particle Data Group, Phys. Rev. D**54**, 1 (1996).
10. N.Di Bartolomeo, R.Gatto, F.Feuiglio, and G. Nardulli. Phys. Lett.B**347**, 405(1995).
11. Guey-Lin Lin, Chin. J. Phys. **34**, 1056 (1996).

Gauge Mediated Supersymmetry Breaking and Spontaneous CP Violation as a Solution to the Strong CP Problem

Otto C.W. Kong
Department of Physics and Astronomy,
University of Rochester, Rochester NY 14627-0171.

The origin of CP violation is a major mystery, especially in relation to the strong CP problem. CP being a spontaneously broken symmetry could provide an elegant solution. However, such models have difficulty making themselves compatible with low-energy supersymmetry, which is popularly accepted as the solution to the hierarchy problem. We demonstrate that a certain class of low scale supersymmetric "Nelson-Barr" type models can solve the strong and supersymmetric CP problems while at the same time generating sufficient weak CP violation in the $K^0 - \bar{K}^0$ system. Gauge-mediated supersymmetry breaking is used to provide the needed squark mass degeneracies and A-term proportionalities; though that proves to be still insufficient for a generic Nelson-Barr model. The workable model we consider here, essentially a supersymmetric version of the aspon model, has the Nelson-Barr mass texture enforced by a U(1) gauge symmetry, broken at the TeV scale. The resulting model is predictive with rich phenomenology soon to be available. Feasibility of the model considered is established by a detailed renormalization group studies.

The sources of CP violation is still a major puzzle. While the only experimentally observed CP violating effect, in the $K^0 - \bar{K}^0$ system, is compatible with the Kabayashi-Maskawa (KM) prescription within the standard model, there are evidents for a more elaborate theory of CP violation. Among them, the strong CP problem is the most pressing. A potential strong interaction contribution to the neutron electric dipole moment and leads to the very stringent experimental contraint

$$\bar{\theta} < 10^{-9} \; . \tag{1}$$

Here, $\bar{\theta}$ is the physical combination of the strong CP parameter θ and the CP phase(s) of the color fermion mass matrices. In the case of the supersymmetric standard model, this is to be given by

$$\bar{\theta} = \theta - \arg\det M_q - 3\arg M_g \; , \tag{2}$$

where M_q is the quark mass matrix and M_g the gluino mass. The incredible smallness of this *a priori* arbitary parameter demands an explanation. With supersymmetry (SUSY), there are also a large number of admissible CP violating phases among the soft SUSY breaking parameters (SSBP), which are also constrained to be quite small ($\sim 10^{-3}$).

Using spontaneous CP violation to solve the strong CP problem was pioneered by Nelson and Barr[1]. Extending the approach to a SUSY model could take care of the SUSY CP phases easily. Attempts to realize the Nelson-Barr mechanism in SUSY models[2] have, however, run up against a formidable difficulty: There generically exist potentially large 1-loop contributions to $\bar{\theta}$ in these models. As discussed at length in Ref. [3], one requires an exceptionally high degree of proportionality of the soft SUSY breaking trilinear scalar couplings to their associated Yukawa couplings as well as degeneracy among the soft squark mass terms for each charge and color sector, if these contributions are to be sufficiently suppressed. The degree of proportionality and degeneracy required among the SSBP is very difficult to maintain due to the effects of renormalization.

A particularly interesting modified version of the Nelson-Barr type model is given by the aspon model[4]. Here the extra symmetry needed to enforce the Nelson-Barr mass texture is promoted to a new $U(1)$ gauge symmetry. This $U(1)_A$ gauge boson (aspon) then mediates a tree level CP violating $K^0 - \bar{K}^0$ mixing. This allows the effective KM phase to be small. A preliminary study suggested a SUSY version of the model has a much better chance to provide a feasible solution[5]. The detail model is constructed in the framework of gauge mediated SUSY breaking (GMSB), and demonstrated to work, through a careful renormalization group (RG) study[6].

The model have CP spontaneously broken at low energies (of order a TeV). It has an extra right-handed down quark superfield $\bar{D}$ together with its mirror D coupling to the ordinary down quarks via the superpotential

$$W_d = Y_d^{ij} Q_j \bar{d}_i H_d + \mu_D D\bar{D} + \gamma^{ia} D\chi_a \bar{d}_i \ , \tag{3}$$

where the VEVs of the scalar components of the two χ_a's contain a relative phase, thus breaking CP. D and χ_a have opposite $U(1)_A$ charges. The down-sector quark mass matrix can be written as

$$m_q = \begin{pmatrix} m_d & x\mu_D \boldsymbol{a} \\ 0 & \mu_D \end{pmatrix} , \tag{4}$$

and squark mass matrices as

$$\begin{aligned} \widetilde{M}^2_{RL} = & \begin{pmatrix} \bar{A}_d m_q (1+\mu\tan\beta) & \bar{A}_\gamma x\mu_D \boldsymbol{a} \\ 0 & \bar{A}_\gamma \mu_D \end{pmatrix} \\ & + \begin{pmatrix} \langle H_d\rangle \,\delta A_d & x\mu_D \delta A_\gamma \boldsymbol{c} - \gamma^{ia}\langle F_{\chi_a}\rangle \\ 0 & \mu_D(B_D - \bar{A}_\gamma) \end{pmatrix} , \end{aligned} \tag{5}$$

$$\widetilde{M}^2_{RR} = \tilde{m}^2_R I + \begin{pmatrix} \delta\tilde{m}^2_R & 0 \\ 0 & \delta\tilde{M}^2_R \end{pmatrix} + m_q m_q^\dagger \ , \tag{6}$$

$$\widetilde{M}^2_{LL} = \tilde{m}^2_L I + \begin{pmatrix} \delta\tilde{m}^2_L & 0 \\ 0 & \delta\tilde{M}^2_L \end{pmatrix} + m^\dagger_q m_q \,, \tag{7}$$

where $a^i = \frac{1}{x\mu_D}\gamma^{ia}\langle\chi_a\rangle$, such that $\boldsymbol{a}^\dagger\boldsymbol{a}= 1$, and

$$\delta A_\gamma c^i = \frac{1}{x\mu_D} h^{ia}_\gamma \langle\chi_a\rangle - \bar{A}_\gamma a^i \,, \tag{8}$$

with $\boldsymbol{c}^\dagger\boldsymbol{c}= 1$. Hence, the complex phases are all contained in three-vectors $\boldsymbol{a}$ and $\boldsymbol{c}$, and the most dangerous $\langle F_{\chi_a}\rangle$'s.

The spontaneous CP violation (SCPV) is obtained from a superpotential of χ_a's together with two oppositely charged $\bar{\chi}_a$'s and a neutral $\aleph$ superfield,

$$W_\chi = \bar{\chi}_a \mu^{ab}_\chi \chi_b + \aleph\bar{\chi}_a\lambda^{ab}\chi_b + \lambda_\aleph \aleph^3 + \mu_\aleph \aleph^2 \,, \tag{9}$$

which generally gives a SUSY perserving vacuum with complex VEV's breaking CP. The sector is minimal for anomaly cancellation and SCPV. Without nonzero SSBP for superfields of the sector, the $\langle F_{\chi_a}\rangle$'s are then vanishing. A minimal GMSB scenario[7] blind to the $U(1)_A$ is used to suppress proportionality violations δA_d, δA_γ, and (B_D-$\bar{A}_\gamma$), degeneracy violations, as well as the SSBP for that lifts the zero $\langle F_{\chi_a}\rangle$'s. The latter is then estimates both analytically and diagrammatically, based on RG analysis of all SSBP. The 1-loop $\bar{\theta}$ contributions from each terms is calculated through the mass insertion method. The model predicts a measureable neutron EDM, which could be close to the present experimental bound for $x > 0.01$, with some region of the admissible x-values (of the non-SUSY version) plausibly rule out by having the $\langle F_{\chi_a}\rangle$'s too large, and also a very rich spectrum of new particles around the TeV scale.

References

1. A. Nelson, *Phys. Lett.* **136B**, 387 (1984); S. M. Barr, *ibid.* **53**, 329 (1984); *Phys. Rev.* **D 30**, 1805 (1984); A. Nelson, *Phys. Lett.* **143B**, 165 (1984).
2. S.M. Barr and A. Masiero, *Phys. Rev.* **D 38**, 366 (1988); S.M. Barr and G. Segrè, *ibid.* **D 48**, 302 (1993).
3. M. Dine, R. Leigh, and A. Kagan, *Phys. Rev.* **D 48**, 2214 (1993).
4. P.H. Frampton and T.W. Kephart, *Phys. Rev. Lett.* **66**, 1666 (1991); P.H. Frampton and D. Ng, *Phys. Rev.* **D 43**, 3034 (1991).
5. P.H. Frampton and O.C.W. Kong, *Phys. Lett.* **B402**, 297 (1997).
6. O. C. W. Kong, B. D. Wright, *Phys. Rev.* **D 58**, *015002* (1998).
7. See, for example, R. Rattazzi and U. Sarid, *Nucl. Phys.* **N501**, 297 (1997); F.M. Borzumati, WIS-96/50/Dec.-PH, hep-ph/9702307.

List of Participants

M. Bisset	Rochester, USA
Y. Q. Chen	Ohio State, USA
K. Cheung	Davis, USA
E. J. Chun	KIAS, Korea
S. Donati	Pisa, Italy
R. Fleischer	CERN, Switzerland
X. G. He	U. of Melbourne, Australia
D. S. Hwang	Sejong , Korea
S. K. Kang	KIAS, Korea
C. Kao	Madison, USA
Y. Y. Keum	APCTP, Korea
P. Ko	KAIST, Korea
O. Kong	Rochester, USA
B. H. Lee	Sogang , Korea
T. Morozumi	Hiroshima, Japan
Y. Okada	KEK, Japan
C. Sachrajda	Southampton, UK
J. Soffer	CNRS, France
A. Soni	BNL, USA
S. Stone	Syracuse, USA
H. Yamamoto	Hawaii, USA
C. P. Yuan	Michigan State U., USA
P. Chang	Academia Sinica
P. F. Chen	Academia Sinica
H. Y. Cheng	Academia Sinica
C. Y. Cheung	Academia Sinica
M. H. Huang	Academia Sinica
W. P. Lai	Academia Sinica
S. C. Lee	Academia Sinica
K. W. Ng	Academia Sinica
H. H. Shih	Academia Sinica
P. K. Teng	Academia Sinica
T. K. Wong	Academia Sinica
K. C. Yang	Academia Sinica
T. W. Yeh	Academia Sinica

H. L. Yu	Academia Sinica
W. M. Zhang	Academia Sinica
C. T. Chan	National Taiwan U.
C. K. Chua	National Taiwan U.
J. A. Gu	National Taiwan U.
G. W. S. Hou	National Taiwan U.
C. W. Hwang	National Taiwan U.
R. S. Lu	National Taiwan U.
K. Ueno	National Taiwan U.
C. C. Wang	National Taiwan U.
M. Z. Wang	National Taiwan U.
W. K. Sze	National Taiwan Normal U.
C. C. Lih	National Central U.
H. H. Liu	National Central U.
C. W. Chung	National Chiao-Tung U.
T. K. Chyi	National Chiao-Tung U.
J. L. Lim	National Chiao-Tung U.
G. L. Lin	National Chiao-Tung U.
S. Y. Lin	National Chiao-Tung U.
J. J. Tseng	National Chiao-Tung U.
D. Chang	National Tsing Hua U.
W. F. Chang	National Tsing Hua U.
C. H. Chou	National Tsing Hua U.
C. Q. Geng	National Tsing Hua U.
I. J. Hsu	National Tsing Hua U.
H. L. Lai	National Tsing Hua U.
H. S. Liao	National Tsing Hua U.
T. H. Wu	National Tsing Hua U.
W. C. Su	National Chung Cheng U.
Y. H. Chen	National Cheng Kung U.
R. C. Hsieh	National Cheng Kung U.
H. n. Li	National Cheng Kung U.
S. L. Nyeo	National Cheng Kung U.
B. Tseng	National Cheng Kung U.
C. H. V. Chang	National Center for Theoretical Sciences
H. F. Kao	National Center for Theoretical Sciences
C. Soo	National Center for Theoretical Sciences
W. C. Kuo	National Sun Yat-Sen U.

R. S. Guo	National Kaohsiung Normal U.
C. H. Wang	National Lien-Ho College
C. Lin	Tai-chung Science Museum